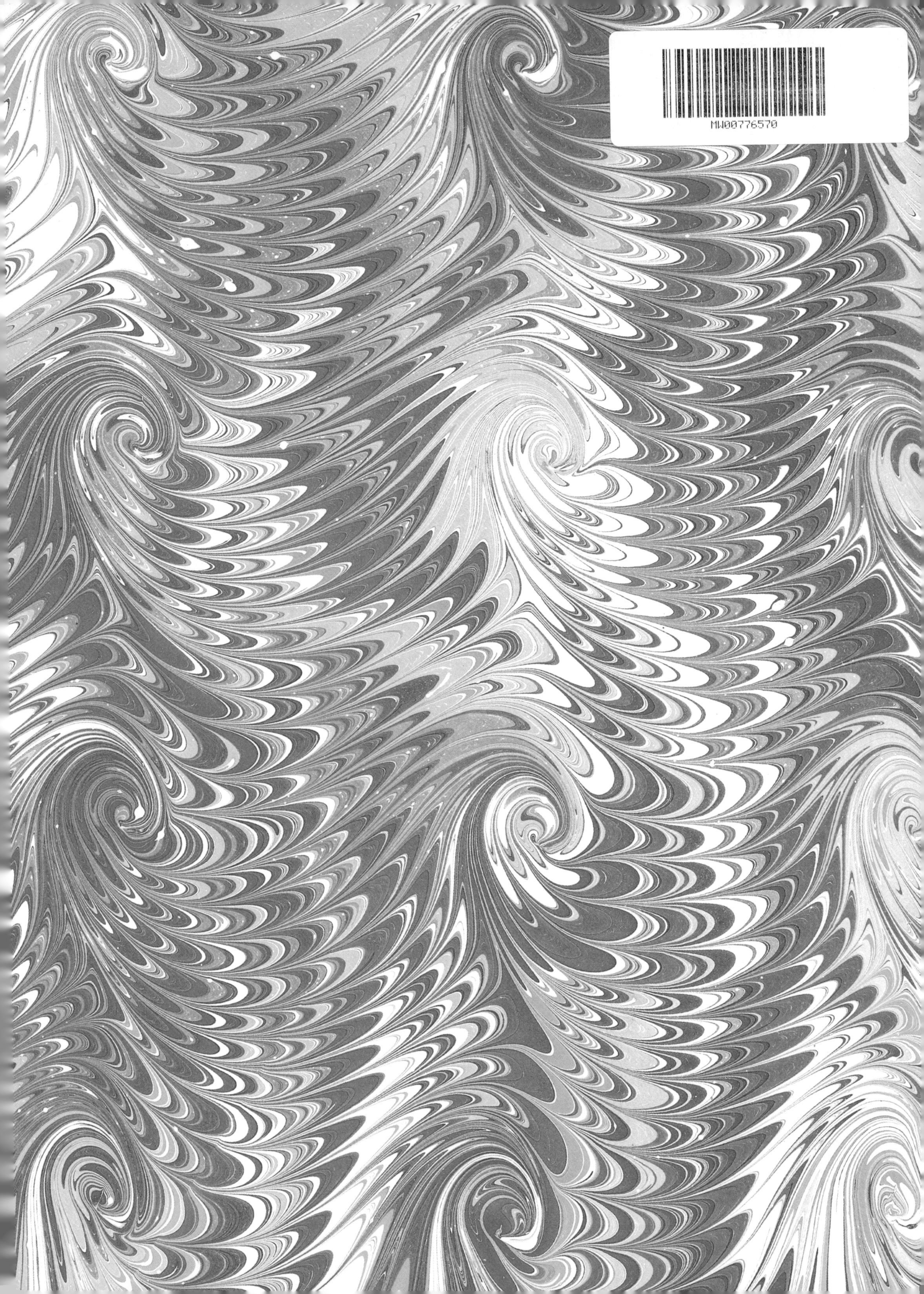
MW00776570

Victorian Pencils

TOOLS TO JEWELS

Deborah Crosby

4880 Lower Valley Rd. Atglen, PA 19310 USA

To my mother:

All that I am that's good is because of her,
And to my daughters,
who I hope may someday say the same.

Library of Congress Cataloging-in-Publication Data

Crosby, Deborah,
Victorian pencils : tools to jewels / Deborah Crosby.
p. cm.
Includes bibliographical references and index.
ISBN 0-7643-0413-5 (hardcover : alk. paper)
1. Pencils--Collectors and collecting--Catalogs. 2. Victoriana--Collectors and collecting. I. Title.
TS1268.C76 1998
681'.6--dc21 98-18727
CIP

All photographs, except those noted, are by the author.
The author wishes to thank Chris Odgers for the photographs he contributed.
Pen and ink illustrations by Deborah Crosby.

The author is always interested in information regarding writing implements. She may be contacted through the publisher.

Designed by Bonnie M. Hensley
Layout by Randy L. Hensley
Typeset in Shelley Allegro BT/Bodini Bk BT

ISBN: 0-7643-0413-5
Printed in China
1 2 3 4

Published by Schiffer Publishing Ltd.
4880 Lower Valley Road Atglen, PA 19310
Phone: (610) 593-1777; Fax: (610) 593-2002
e-mail: schifferbk@aol.com

In Europe, Schiffer books are distributed by Bushwood Books
6 Marksbury Avenue Kew Gardens Surrey TW9 4JF England
Phone: 44 (0)181 392-8585; Fax: 44 (0)181 392-9876
e-mail: bushwd@aol.com

Please write for a free catalog. This book may be purchased from the publisher.
Please include $3.95 for shipping. Please try your bookstore first.
We are interested in hearing from authors with book ideas on related subjects.

Contents

The faithful pencil has designed
Some bright ideas of the master's mind,
Where a new world leaps out at his command,
And ready nature waits upon his hand.

—Alexander Pope

Foreword

BY ED FINGERMAN

The funny thing about antiques and collectibles is that when they are very scarce, too few people can be exposed to them and, accordingly, very few collectors of such objects are ever created who care enough to seek additional examples. To put it another way, there must first be a sufficient number of varieties of a particular item to stir demand, thereby developing a collector base and then satiating the demand then created. This holds true with both vintage and modern collectibles—from old mainstay collectibles such as stamps and coins, to vintage wrist watches, cigarette lighters, Coca Cola memorabilia and the like, to today's collector plates, Swatch watches and Beanie Babies. In each of these examples (and hundreds of others), a significant collector base developed by virtue of a large supply of merchandise first being available to arouse and then satisfy collector demand.

The field of Victorian writing instruments is one of those many areas that, until the last few years, have been largely overlooked, at least within the United States, for no apparent reason other than a proven scarcity of the more interesting examples of these items. Although as an avid collector of vintage fountain pens, commencing around 1975, I invariably would find a handful of collectible fountain pens at most large flea markets, I would rarely encounter any Victorian pencils. In 1978, I happened upon four or five at a dealer's stall and was immediately taken by their beauty and mechanical know-how. The dealer's quote of $40 for the group seemed reasonable enough for these "neat" little objects. My quest over the next several years for additional examples was, unfortunately, stymied by their general scarcity. It was only after making some contacts with British dealers that my collection began to grow, though not at any significant pace. It appeared that either large collections had already been assembled which had removed most pencils from circulation or they simply were as scarce as "hen's teeth." Recently, however, more pencils are being brought to market by astute dealers who have begun to "turn over every rock" to locate them. Now that more of these wonderful objects are being made available, new collectors are being born who are pursuing them.

In viewing the array of interesting and exquisite pencils appearing throughout this volume, one would be hard pressed to come away without a new appreciation for the craftsmanship, artistic talents, and technological sophistication of our ancestors. Even the occasional mundane example still reflects a degree of industriousness that is basically unmatched by modern artisans. It took the energies of many collectors to amass a sufficient number of these little treasures and the inquisitiveness and dedication of Deb Crosby to consolidate these various efforts into this skillfully presented text. All lovers of fine writing instruments are indebted to Deb for this brilliant and entertaining history of the pencil.

As *Victorian Pencils: Tools to Jewels* is distributed, this "sleeping giant" of a hobby will no longer be the special secret of those who quietly sequestered every interesting example they encountered. Likely, prices will begin to spiral upwards as new collectors realize how affordable many of these gems are in relation to their beauty and rarity. I envy the joy awaiting the newcomer who is being exposed to the field of Victorian pencils for the first time in this text. A note of warning, however...it can be addictive!

Edward Fingerman
March 1998

Acknowledgments

Sincere thanks go to the following individuals and institutions, for providing access to original documents and information that allowed me to complete my research:

Shannon Whitt-Lawrence at the A. T. Cross Archives; Chris Bailey, Horologist, and Snowden Taylor, President, with the American Clock and Watch Museum; Brent M. Sverdloff, Reference Archivist at the Baker Business Library, Harvard University (Dun Collection); Kathryn Atkin of Filofax, who provided information about Yard-O-Led; Mark N. Brown, Curator of Manuscripts at the John Hay Library, Brown University (Gorham Collection); the New York Public Library; Ann Clifford, Assistant Archivist at the Society for the Preservation of New England Antiquities; Louisa Bann, Archives Coordinator, and Annamarie Sandecki, Archivist, at the Tiffany Archives; the United States Patent Office; the Warshaw Collection of Business Americana, Archives Center, National Museum of American History, Smithsonian Institution; The Victorian House, Tarrytown, New York; Westchester County Libraries; and Neville Brown, Librarian, at Winterthur.

There are also friends and family members who have helped me in more ways than I can express. Friendships with Lynn Brant, Chuck Cohn, Rachel Marks, Larry Reich, and Charles Starke have sustained me. D. Albert Soeffing unstintingly shared his knowledge, gave me information about the many resources available for the kind of research I was doing, and inspired me with his deep love of research, particularly in the field of American silver. Thanks to Don Erenberg, for his encouragement, Peter Schiffer, for his enthusiasm, and my editor, Dawn Stoltzfus, for being so supportive. My family: brothers, brother in-law, sister, sisters in-law, nieces, and nephews, and Bill Berglas, cheered me on. My lovely daughters, Maris and Rachel, helped me with research, tolerated the amount of time I spent working on this book, and offered their enthusiasm when I needed it most. My sister, Lisa Metzger, shared her talents as an editor. More importantly, though, she always supported and encouraged me. My mother, Bert Crosby, taught me the importance of education, and, through her example, courage.

The following people have generously loaned pencils for photography or have helped me in other ways: R. George Adams, Anita Allen, Antique Source of New York City, Judie Barnett of Fountain Pen Hospital (Kingwood, Texas), Neville Bedford of Quill, Geoffrey Berliner of Berliner Pens, George Brown, Lynn and Nancy Brant, Martin Burke, Classic Consignments of Chappaqua, Chuck Cohn, Cornelia Cotton, Neil Davis, Paul Erano, Ed Fingerman, Megan Gentilesco, Brenda Glass, Fred and Eviva Gorstein, James Heusinger, Alan Hirsch, Richard Jenkins, Dick Johnson, Arline Kimerling, Barry Klampert, Carl Linnemeyer, John Loring, Peter Markman, Rachel Marks, Jane and Jim Marshall, John McKenzie, Len Nessen, David Nishamura, Bob Novak, Chris Odgers, Stephen Overbury of Pens Unlimited, Edith Peroff, Jim Rainer, Mark Reichbach of Mark's Time, Mt. Kisco, Julia and Boris Rice, Lidia Rubinstein, Stuart Schneider, Abe Schwartz, Guido Staltari, Donna Swartz, Osman Sümer, Lou Vion, and John Woo.

The author is also grateful for the gracious hospitality and generosity extended by Gerald Sattin and Neil Davis.

Introduction

In the realm of writing implements, there is nothing quite like the pencil. The pencil is used for thinking, figuring, dreaming. It's used to record ideas as they evolve and images as they appear. It's a tool used for learning and experimenting because the line it makes can be rethought and easily changed. It's an implement that's been essential since it was invented. No other writing instrument serves the variety of purposes it does as well and as matter-of-factly as it does.

Today we take the wooden pencil for granted, usually having one immediately at hand and only wondering at all about them when we can't find one. We rarely think twice about the availability of mechanical pencils, knowing that if we wanted one it would only be as far away as the closest market. Most of the pencils we use are utilitarian and few are made of anything but wood or plastic, with perhaps a couple of inexpensive metal parts. They are generally considered disposable. But as recently as two hundred years ago (while the wooden pencil was still relatively new), the mechanical or propelling pencil was just beginning to emerge as a viable tool. For many reasons: including strong cultural influences, the cost of graphite, the Industrial Revolution, and the nature of inventors intent on improving recent inventions (their own or someone else's), not only was the practical purpose of the mechanical pencil explored, but its aesthetic or decorative functions, as well. From the earliest patents for "ever-pointed" pencils at the turn of the nineteenth century to the proliferation of "Eversharp" pencils a century later, both the inner workings and outer casings have been the focus of inventors, designers, and manufacturers. Clearly the market supported this research and development, as is evident in the numbers of patents awarded for mechanisms and pencil-cases in the nineteenth century (the term "pencil-case" refers to the outer body of the mechanical pencil, not to a box made to hold pencils). Although the quantity of handsome pencils (mostly from the Victorian era) that appear at antique shows and in antique shops is diminishing, there are still wonderful finds to be made by the educated and determined collector.

Nineteenth century trade card, with a woman's hand holding a mechanical pencil.

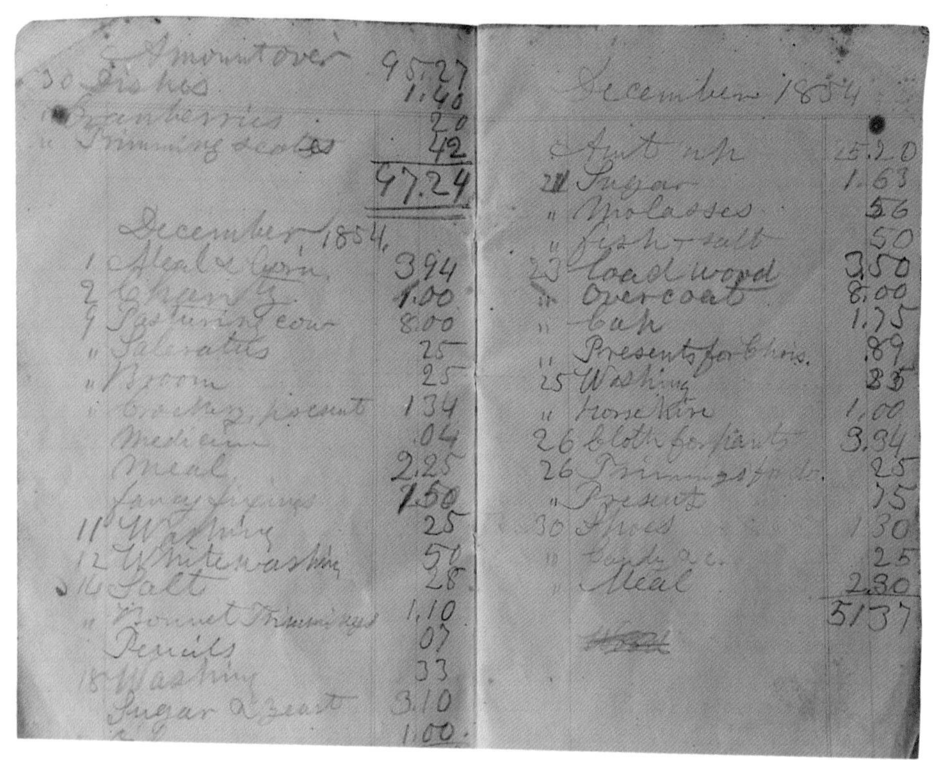

Amount over 95.27
30 Dishes 1.40
Cranberries .20
Trimming scales .42
47.24

December 1854

1	Meal & Corn	3.94
2		1.00
4	Pasturing cow	8.00
"	Saleratus	.25
"	Broom	.25
"	Crockery, discount	1.34
	Medicine	.04
	Meal	2.25
	fancy fixings	1.50
11	Washing	.25
12	Whitewashing	.50
14	Salt	.28
"	Bonnet Trimmings	1.10
	Pencils	.07
18	Washing	.33
	Sugar & yeast	3.10
		1.00

December 1854

	Amt up	25.20
21	Sugar	1.63
"	Molasses	.56
"	fish & salt	.50
23	load wood	3.50
"	Overcoat	8.00
"	Cap	1.75
"	Presents for Chris.	.89
25	Washing	.33
"		1.00
26	Cloth for pants	3.34
26	Trimmings for	.25
"	Present	.75
30	Shoes	1.30
"		.25
"	Meal	2.30
		51.37

Leather-covered journal from Cape Cod, written in pencil, documenting expenses of December 1854. Pencils are listed as costing $.07, along with "fancy fixings" at $1.50 and Bonnet Trimmings costing $1.10.

In the nineteenth century, mechanical pencils were made of gold, rolled gold, silver, ivory, and other materials. They were mounted with precious or semi-precious stones, engine-turned, hand-engraved, or enameled. Some are pen and pencil combinations, some are combinations of pencils, knives, rulers, toothpicks, buttonhooks, watch keys, postal scales, and other implements that would have been considered useful. There are figural pencils shaped like pharaohs, obelisks, owls, nails, screws, tennis rackets, pigs, dogs, cats, butter knives, matches, crosses, pipes, acorns, and caricatures of people. Pencils are ornamented with flowers, snakes, a variety of animals and various symbols (a shamrock for good luck, a sheet of music for song, etc.). They were prized by their owners and were often worn as jewelry.

The humble wood cased pencil ostensibly serves the same purpose as its gold and enamel, diamond mounted, or figural relative. All are used to write or draw with and use the same substance to mark with. But while the line or tone one type of pencil makes may be indistinguishable from the line or tone made by another, the impression made by the different pencils is markedly different. One item is merely a tool. It acts as a vehicle to express ideas, concepts, visions, and perceptions in a generally erasable way. Its purpose is primary; it is designed to serve, to act as an intermediary between its user's private and public thought. That this makes it a powerful tool can't be denied, but a wooden pencil is made to be used and to be used up. When its tiny stub is too small to write with, it's replaced with a clone of itself, which in turn is used up and discarded.

But the other type, while still functioning as a pencil, can be, in itself, a work of art. It can be an expression of who uses it, a declaration of how that person wants to be seen, a reflection of the era in which it was created. It

School boy's wooden pencil box, nineteenth century, names written in pencil on the inside of the lid, along with "old trappers of the mountains." Includes the original key, magnifier, sealing wax, pen-wipe, and wooden pencil with an extender.

Stunning slide pen/pencil combination, circa mid-19th, approx. 5-1/2". Unmarked. Gold, with dark blue enamel inlay and diamonds set in and around the finial. $3,800-$4,200.

can contain a complex mechanism that disguises and protects the graphite point or it can consist of a wooden pencil in a simple gold sheath. It can be worn on a watch chain or a sautoir. It is replenishable. It is jewelry that functions, ornamentation that can be put to use.

Although most of the pencils illustrated and described in this book contain some form of mechanism, the designation "mechanical pencil" often connotes something purely utilitarian. These pencils, however, are beautiful, elegant, and sometimes whimsical. The perception of a mechanical instrument as being devoid of grace will be rethought when one sees the stunning pencils represented in this book.

Some of the mechanisms are relatively simple, consisting of a slider that allows the pencil to be pushed in or out of a metal case. Others are far more intricate, involving many parts that cause the pencil end to emerge when the other end is pulled. The phrase "magic pencil" is well suited to describe these delightful pieces, which most people pass by without ever realizing what they're missing. One of the most intriguing aspects of researching this book was discovering and then being able to identify the vast array of configurations of these pencils, particularly the figural pencils. Another interesting element was discovering the variety of mechanisms that were employed to allow the pencil point to be withdrawn into the pencil case and propelled out when the pencil was to be used. The creativity of the nineteenth century inventors and designers, along with the fine craftsmanship of the people who made these pencils, led to the creation of some of the most remarkable pieces of functional jewelry one will ever encounter. Whether called ever-pointed, telescoping, propelling, screw, automatic, magic or mechanical, these pencils are fascinating. Also interesting are the precious metal sheath, scabbard, or vest-pocket pencils and pencil protectors or extenders. The history of these pieces, the evolution of "tools to jewels" (primarily in the United States, although it would be impossible to relate the history of the mechanical pencil without discussing particular British and continental makers), will be illustrated and described in this book.

These writing implements were clearly treasured by their nineteenth century owners. Many are worn from use. Some of the mechanisms have been unable to withstand the repeated push-pull, and no longer function as originally intended. Some of the telescoping pencils seem forever frozen shut. Many of the pencils, however, still work after more than one hundred and fifty years. A working magic pencil (so called because it appears to contain no pencil at all, until its end is pulled) is a noteworthy find, especially if it's a figural one. It's easy to understand why people enjoyed these charming pencils; a sterling silver sculpture of a dog, one inch long, becomes a pencil when its tail is pulled, or a miniature silver broom reveals a pencil when a pin is pushed. Inventions and improvements in mechanical pencils eventually led to a product that is so common today we hardly acknowledge its presence. Yet, a Victorian pencil was too interesting to be ignored when it was new. It still is today.

Wood and celluloid pencil in the shape of a cane, which holds the tablet closed when it's looped through the holders on the open side. 9" x 3", circa 1880. $50-$75.

Staedtler pencil point protector, circa early 20th century, 2-1/2". Notice the contrast between this worn and chewed pencil, and the elegant piece in the preceding photograph.

Silica or slate books were marked with slate pencils, circa 1870. $25-$50 for books with original slate pencils.

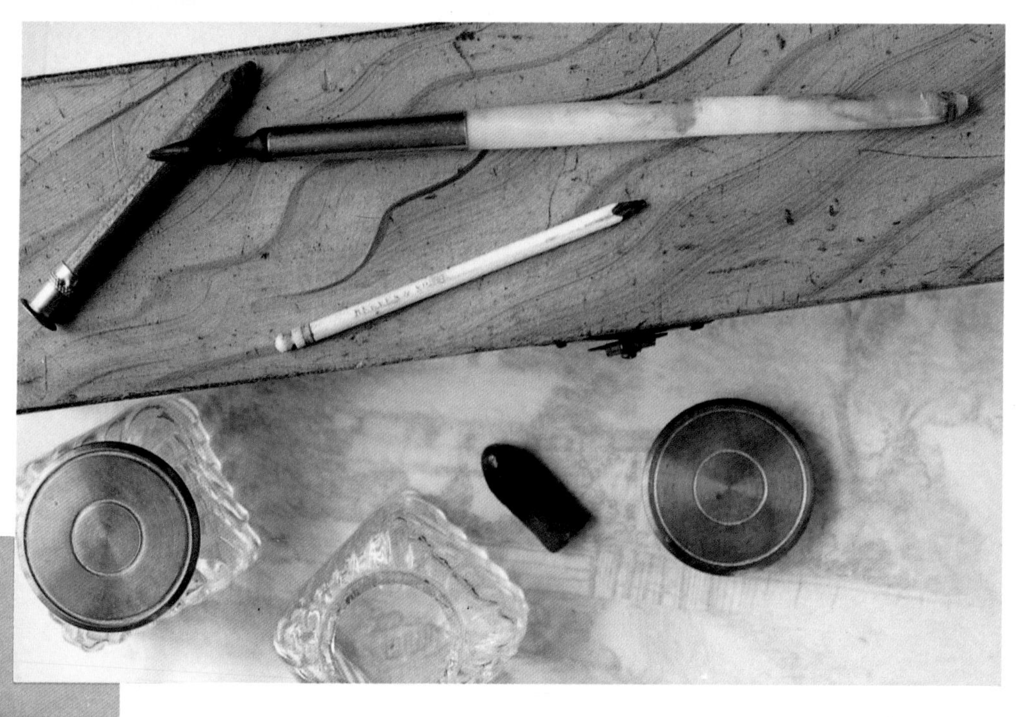

Trompe l'eoil pencil box with ivory ended pencil made by Reeves, a wooden pencil made in England, early metal nib with a quill staff, inkwells, and sealing wax, circa 1860. $175-$225.

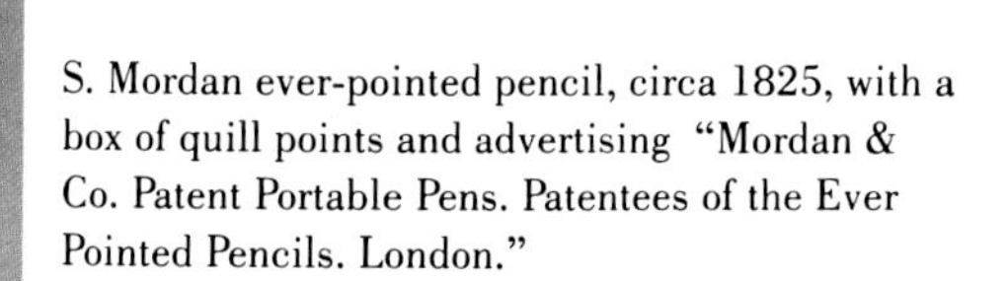

S. Mordan ever-pointed pencil, circa 1825, with a box of quill points and advertising "Mordan & Co. Patent Portable Pens. Patentees of the Ever Pointed Pencils. London."

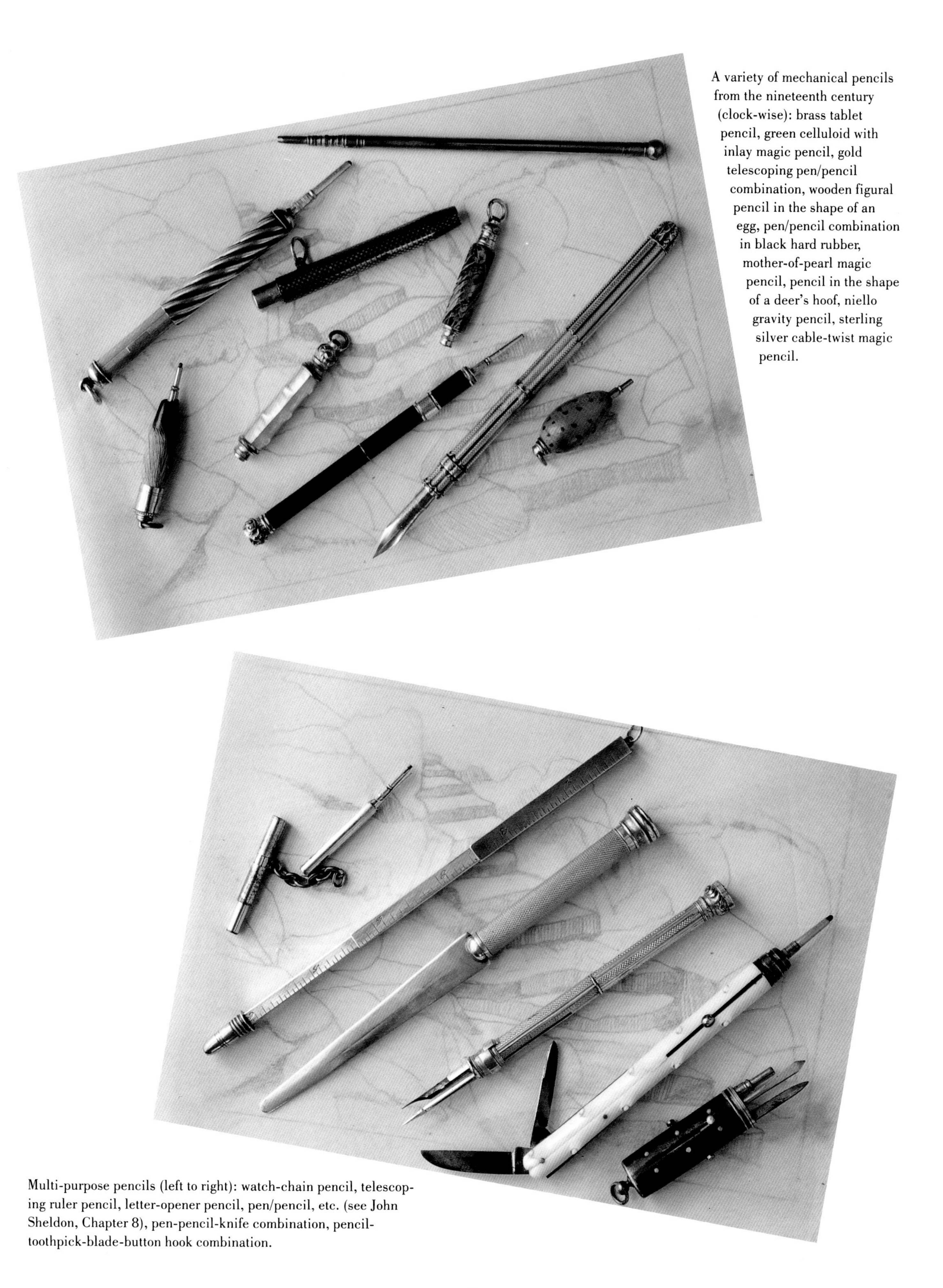

A variety of mechanical pencils from the nineteenth century (clock-wise): brass tablet pencil, green celluloid with inlay magic pencil, gold telescoping pen/pencil combination, wooden figural pencil in the shape of an egg, pen/pencil combination in black hard rubber, mother-of-pearl magic pencil, pencil in the shape of a deer's hoof, niello gravity pencil, sterling silver cable-twist magic pencil.

Multi-purpose pencils (left to right): watch-chain pencil, telescoping ruler pencil, letter-opener pencil, pen/pencil, etc. (see John Sheldon, Chapter 8), pen-pencil-knife combination, pencil-toothpick-blade-button hook combination.

Figural pencils (see Chapter 5).

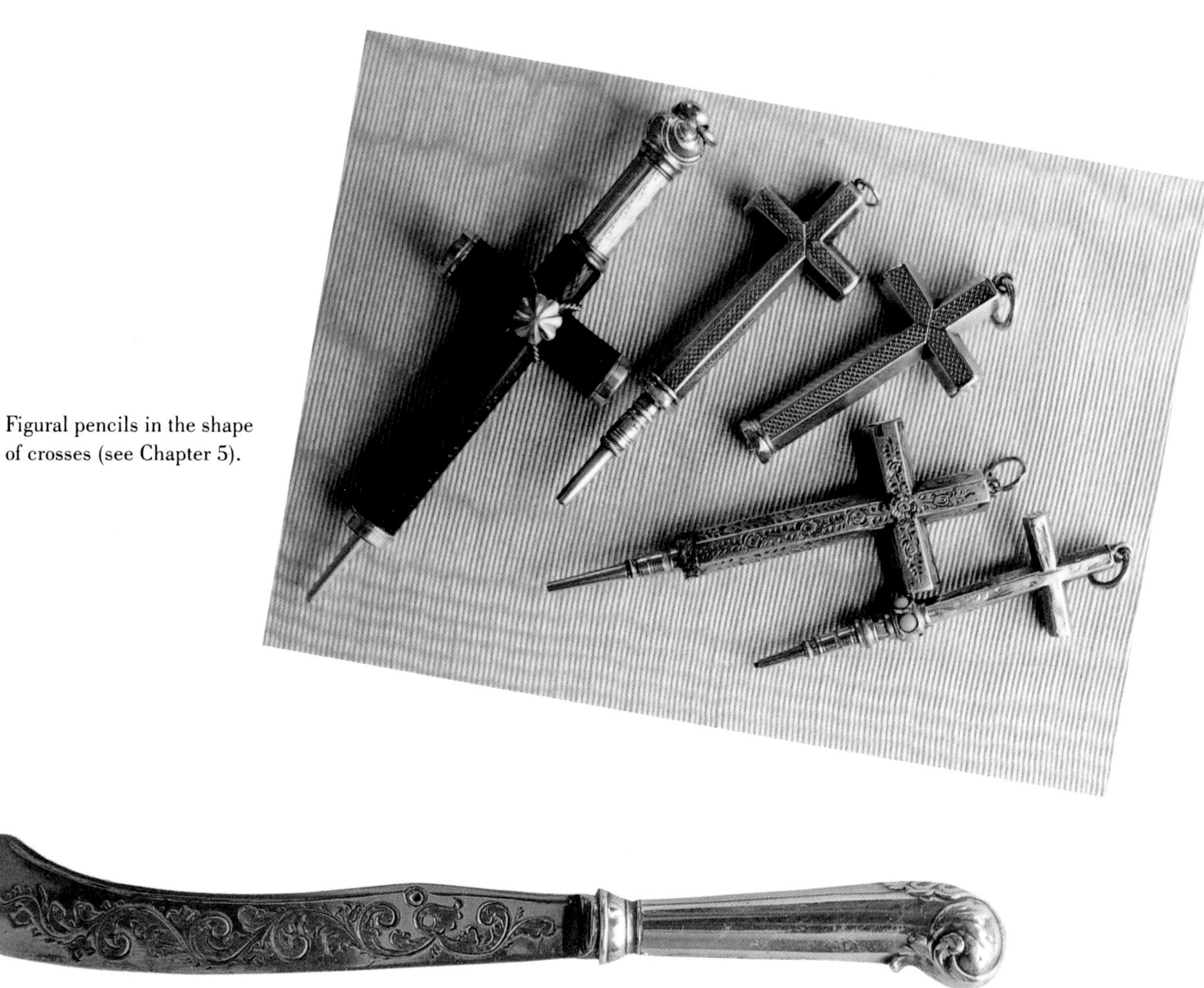

Figural pencils in the shape of crosses (see Chapter 5).

Sterling silver pencil in the shape of a butter knife, made by S. Mordan & Co. (see Chapter 5).

Sheath, vest-pocket or scabbard pencils, pencil extender, porte-crayon, aide mémoire (see Chapter 7).

Aikin, Lambert vest-pocket pencil: "Wine, Women, Song." (Enlarged to show detail.)

A variety of sterling silver or sterling silver and gold magic pencils.

Pricing

Please note that price ranges may vary by location (within the United States as well as in other countries). As with any collectible, rarity and replacement value should also be considered in determining what to pay for a particular pencil. Marked pencils, made by particular makers generally are of greater value (for example: Aikin, Lambert & Co., Mabie Todd & Bard, Edward Todd, S. Mordan & Co., John Sheldon, etc.). Pencils made of precious metals (as well as tortoise shell, ivory, and piqué inlay) or set with precious or semi-precious stones have a higher value. Figural pencils command high prices (although the mechanism is very important, people may be more willing to have a figural pencil repaired, if all the parts are intact. Repair is very expensive, and, so far, pencil "mechanics" are very hard to find). Combination pieces are usually pricier. Both figural and combination pencils have crossover value. (Knife collectors, for example, may be interested in pencil-pen knife combos. People who collect dog-related antiques may be intrigued by a pencil in the shape of a dog.)

The value ranges listed in the captions are for what the pencil would be worth if it were in excellent condition (with a working mechanism), and are based on what these pieces actually sell for. During the past two years, there has been a marked increase in the prices of nineteenth century mechanical pencils, as with Victoriana in general.

Excellent condition: the pencil is not dented or scratched, there is no plating wear (brassing), chasing is crisp, black hard rubber is not oxidized, and any enamel is completely intact. Importantly, the pencil mechanism must be in perfect working condition for a pencil to be considered to be in excellent condition. Excellent condition reflects 100% value.

Very Good: the pencil-case might exhibit some wear, but the mechanism works. Very good condition reflects 80% value.

Good: the pencil shows wear and the mechanism is sluggish, or it comes apart but is intact. Good condition reflects 60% value.

Fair: the pencil may be interesting, but may not be in good condition. The pencil-case might be dented or otherwise damaged, and the mechanism may not work. Fair condition reflects 40% value.

Poor: the pencil may be considered as a source of parts for more valuable pencils. Poor condition reflects 15% value.

For example, if a pencil is valued at $125 in excellent condition, it would be worth $100 in very good condition, $69 in good condition, $37.50 in fair condition, and $18.75 in poor condition.

The author offers these value ranges as a guide only. Prices may vary considerably, according to demand.

Chapter One

History

Before the discovery of graphite and the invention of the pencil[1] various materials were used for marking. There was, of course, pen and ink. Styluses made of lead or silver were used, as well as charcoal—but nothing matched the versatility of graphite. In order to understand the mechanical pencil, it's important to learn a bit about the discovery and mining of graphite as well as the processes used to mix graphite powder with other materials in order to extrude it in particular sizes and degrees of hardness. We take this all for granted, and considering how abundant it is today, are surprised to find that graphite was at one time considered somewhat precious. Discovered in Borrowdale, Cumberland (England), in the sixteenth century, it was originally called *wadd, plumbago,* or *black lead.* When the properties of the material proved useful in marking various materials, particularly vellum or paper, people began using it to write with. By either wrapping graphite with string or pushing a thin strip of it into a hollow reed, they learned that they could use it without getting it on themselves. By the early seventeenth century, pencils made of black lead and wood (similar to pencils still being made today) were being sold in London. It was discovered early on that the marks it made could be erased with crumbs from freshly baked bread. Black lead was exported to the continent, where it quickly became popular. In 1789, plumbago was renamed "graphite" ("to write") by A. G. Werner, after K. W. Scheele had proven it to be a form of carbon in 1779. It's difficult to imagine that the chemical composition of graphite was misunderstood until the time of the American Revolution, but somewhat easier in that context to see why we still refer to it as "lead."

Although graphite had been discovered elsewhere, Borrowdale graphite was unexcelled in its quality. Initially, the graphite was mined and cut into blocks, then into thin strips that were encased in wood to make pencils. This process led to considerable waste, with many pieces of the black lead being unsuitable because of their size or shape and with much powder created during the cutting. Due to these factors, and because Borrowdale plumbago eventually became totally unavailable outside of England, many attempts were made to try to find ways of mixing the graphite powder with other substances. In hopes of making it go further, to make it possible to mold and extrude, and in order to control how soft or hard the lead would be, shellac, sulphur, wax, gum tragacanth, etc., were added to graphite powder. However, it wasn't until 1794 that Nicolas Jacques Conte (of France) successfully devised a process whereby the powdered graphite could be mixed with clay and fired as a ceramic material, yielding a workable substance. (Josef

Ivory plummet fitted with a conical piece of white lead (not graphite), circa 1790-1800, 2-1/2". The kind of mark made by a leaden plummet is shown on the white paper. $50-$75.

Porte-crayon made of ebony and metal, circa 1820. Implements like this were still being used by artists into the mid-twentieth century, and date back as early as the 1700s. $20-$25.

Hardtmuth also made this discovery simultaneously in Vienna.) In 1843 another process was discovered when William Brockendon (of England) found that if a sufficient amount of pressure were applied to the graphite powder in a vacuum, binding agents would be unnecessary.[2]

In England as late as the 1820s, some of the black lead for use in the recently invented "ever-pointed" pencils was cut by hand. The pieces were cut into square strips as narrow as possible, usually approximately one inch long. They were then hand-drawn through an octagonal opening cut in ruby, then through progressively rounder openings until the graphite was the appropriate size for the nozzle of the pencil. Pencil-case makers also sold leads made by the Brockendon or Conte processes.[3] Refills for early mechanical pencils were packaged in glass tubes or elaborately designed brass holders, with individual slots for each piece of black lead. (See illustration, p. 26.)

In the United States, Joseph Dixon is credited with being one of the earliest Americans to develop a substantial business in graphite. An inventor and manufacturer, Dixon was born in Marblehead, Massachusetts, on January 18, 1799. Although he didn't have much of a formal education, he invented a machine for cutting files, was clever enough to learn how to carve wooden type and to cast metal type, and became adept at wood-engraving and lithography, as well as at the photo-lithographic printing process. He found that some impure graphite mined in New Hampshire could be used to make crucibles for melting metals at very high temperatures without disintegrating or melting, and that the graphite could also be used to make pencils. In an interesting anecdote, Dixon took pencils he had made by hand to Boston to sell. He had lithographed labels for the bundles of pencils but had apparently misspelled Salem by leaving out one of the letters. When this was discovered in Boston he was told that the only way he could sell his pencils with the misspelled label was to add the word "foreign" to it and pretend he was from another country. It is said that he became so aggravated by this suggestion that he swore he'd never make another pencil again.[4] This is odd coming from a man who later became one of the foremost producers of pencils in the United States.

In 1827, Dixon obtained a factory in Salem where he manufactured crucibles, pencils (he designed most of the machinery for this process), and stove polish. At this point, he began importing graphite that was more pure than what was available in New England. He is also noted for discovering a process to grind lenses using powdered graphite, inventing a way to print currency in color (to prevent counterfeiting) and a galvanic battery. Dixon Crucible Company was incorporated in 1868, with its factory in Jersey City, New Jersey. There they made graphite crucibles, stove polish, graphited oil, lubricating graphite, machine, gear and belt grease, and silica-graphite paints as well as Dixon's Fine Office and Drawing Pencils. (Even into the twentieth

Trade card showing a porte-crayon. $5-$10.

The character from the "Puck" masthead of July 20, 1881, holds a porte-crayon.

Ivory handled mechanical pencil, circa 1830. $75-$125.

Close-up of a porte-crayon and the ivory mechanical pencil. Notice the similarity between the method of holding the lead in place. The ivory pencil is somewhat more refined, and evenly spaced prongs that are tightened when the metal tip is screwed into the threaded point hold the graphite. This type of pencil does not allow the graphite to propel or repel.

Ivory handled tablet pencil with brass propelling mechanism, 7", circa 1840. $75-$125.

century, they used the term graphite, plumbago, and black lead interchangeably in their letterhead.)

Importantly, Dixon also recognized the value of the graphite mines in Ticonderoga, New York. In addition to the mineral itself, the location also provided water power with its proximity to the outlet of Lake George into Lake Champlain. According to information provided in a Dixon pamphlet dated 1888, this combination gave them "an advantage over any establishment in the world in the Graphite trade, and it enables the Company to produce highly perfected grades of Graphite suitable for every branch of the mechanical arts of a quality never before known at reduced prices."[5]

In 1846 the Borrowdale mine was still producing, in the six weeks a year that it was mined, £30,000 to £40,000 worth of graphite. However, because of the protective measures taken to limit the amount of plumbago removed from the mine each year (which clearly limited its availability) other sources were sought and developed. Another major location of natural graphite was found in Siberia. On the packages of leads from the nineteenth century, the origin of the material is often used as a ploy to enhance the desirability of the product. The other quality that was also mentioned in advertising and packaging was the hardness of the lead. The degrees were noted by different codes. The designation VS, for example, could either describe *very soft* graphite (a more typical designation for a piece which would, because of its softness, be fairly fat) or *very small* (which would indicate that it was hard). Soft graphite generally leaves a darker line and is used up more quickly than hard graphite, which is stronger but leaves a lighter line. The nozzle on pencils made by manufacturers like Sampson Mordan & Co. were often marked with a letter as well. A nozzle with the letter "H," for example, would have a small diameter opening for a piece of hard graphite. Other marking systems were done by combinations of numbers and letters (4 H would be very hard) or a series of the same letter (BBB would be very soft, or very black).

While without graphite the mechanical pencil as we know it would never have evolved, the focus of this book is primarily on the designs and mechanisms used in creating the

Brass compass with mechanical pencil, mid-nineteenth century. $75-$125.

wonderful pencils of the Victorian era. With that in mind, we'll now proceed to the next chapter in the history of *Victorian Pencils: Tools to Jewels*.

Notes on Chapter One

[1] "The History of the Manufacture of Pencils," Eric E. Voice. The Newcomen Society Transactions, Volume XXVII.
[2] Much has been written on the discovery of graphite in England during the mid-sixteenth century. In his definitive book on the wooden pencil entitled *The Pencil: A History of Design and Circumstance* Henry Petroski not only describes the discovery, but how the graphite was put to use.
[3] *The Pencil*. Author unknown.
[4] "Dixon Lead", undated.
[5] Dixon Crucible Company pamphlet, undated.

Ebony and ivory pencil, with thread used to repair a split in the barrel. Pencils of this style were made from the early nineteenth century until the early twentieth. $20-$30.

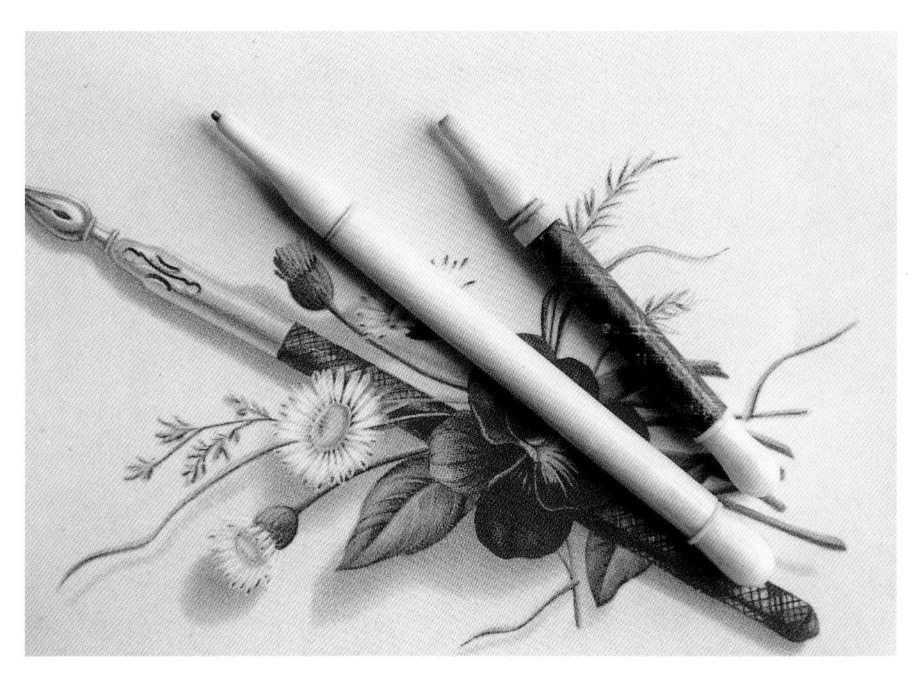

Ivory (4"), ivory and wood with tartan (2-3/4") pencils shown on an illustration of a dip pen embellished with a tartan plaid. Tartan became popular with Queen Victoria's interest in her Scottish estate Balmoral. $75-$125 each.

Bone and brass pencils, mid-nineteenth century, 2-1/2" and 3-1/2". $20-$25.

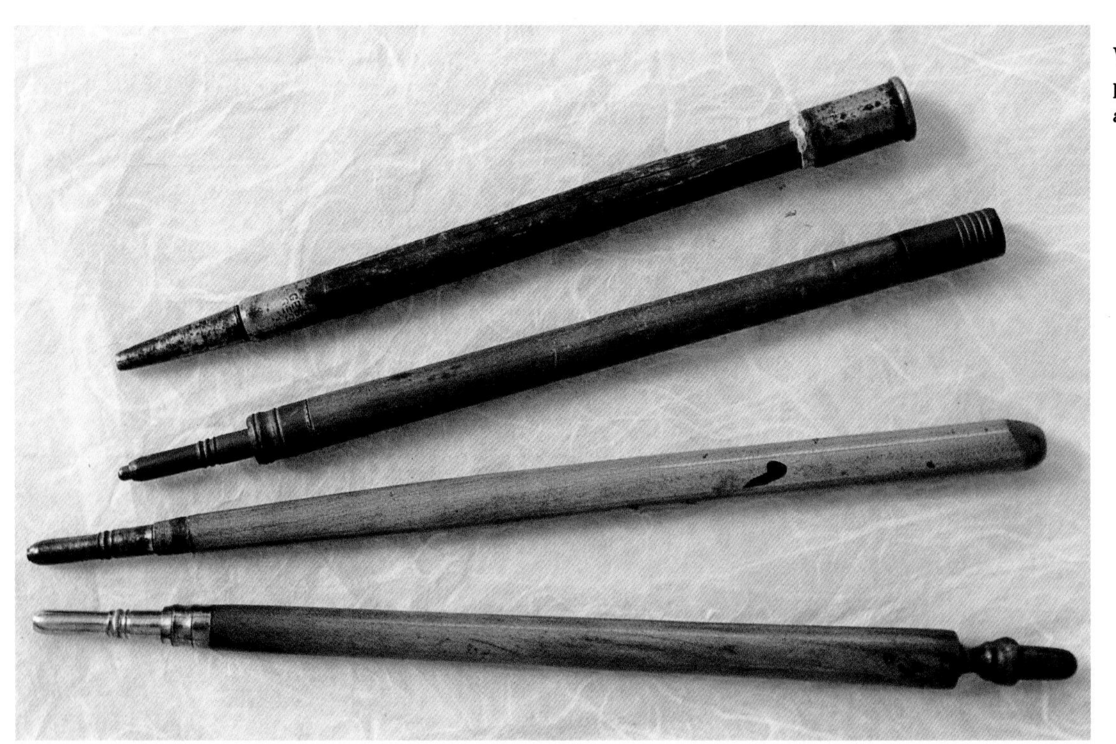

Wood and brass tablet pencils, nineteenth century, and approx. 5"-6". $20-$25.

Turned ivory and brass pencil, with turquoise set in the mid-section, 6". $125-$175.

Lottery tickets advertising "$50,000 in Greenbacks, Returnable to the Patrons of 'the Pen and Pencil' " with a pen/pencil combination in the background. $25-$35.

Portrait of Joseph Dixon, founder of the Dixon Company.

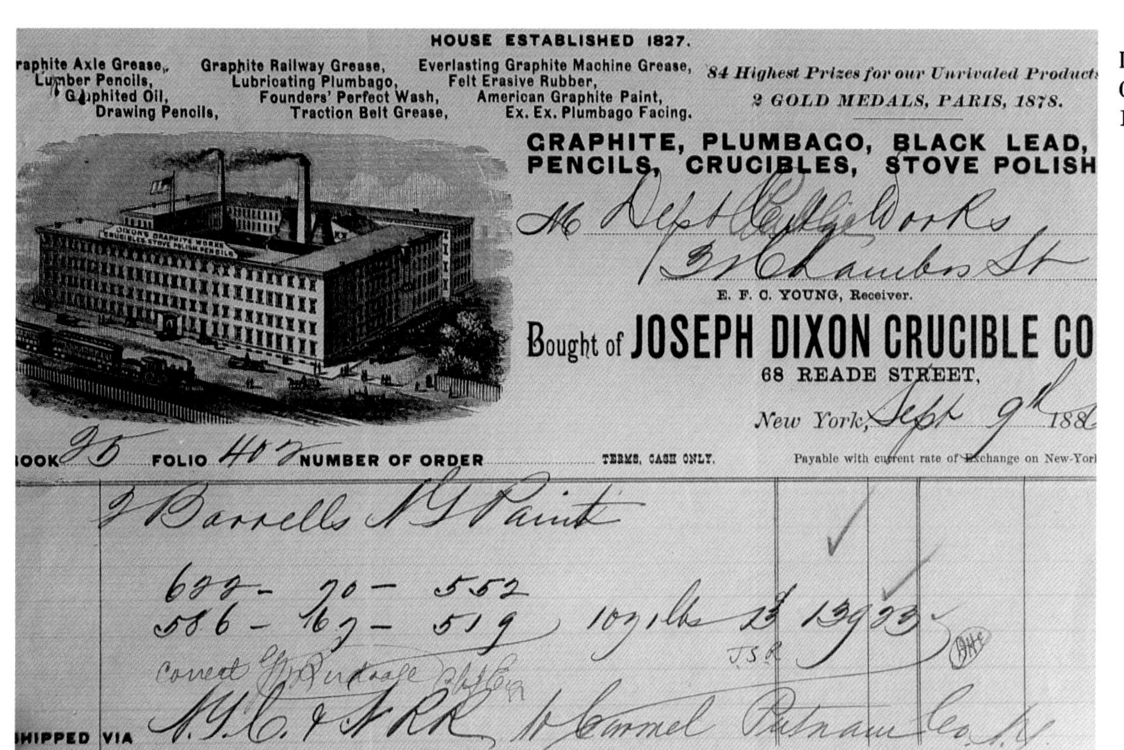

HOUSE ESTABLISHED 1827.

raphite Axle Grease, Lumber Pencils, Graphited Oil, Drawing Pencils, Graphite Railway Grease, Lubricating Plumbago, Founders' Perfect Wash, Traction Belt Grease, Everlasting Graphite Machine Grease, Felt Erasive Rubber, American Graphite Paint, Ex. Ex. Plumbago Facing.

84 Highest Prizes for our Unrivaled Products
2 GOLD MEDALS, PARIS, 1878.

GRAPHITE, PLUMBAGO, BLACK LEAD, PENCILS, CRUCIBLES, STOVE POLISH

E. F. C. YOUNG, Receiver.

Bought of JOSEPH DIXON CRUCIBLE CO
68 READE STREET,
New York, Sept 9th 188

BOOK FOLIO NUMBER OF ORDER TERMS, CASH ONLY. Payable with current rate of Exchange on New-York

SHIPPED VIA N.Y.C. & H.R.R.

Letterhead from Joseph Dixon Crucible Co., September 1886. $25-$45.

Advertisement flier from Joseph Dixon Crucible Co., 1893. $35-$50.

Illustration of Dixon stable and storage sheds, main works and offices in Jersey City, New Jersey, and the Graphite Mines in Ticonderoga, New York.

Illustration from a pamphlet on Dixon Crucible Co., with a woman "straightening the leads" from coils, in preparation for firing.

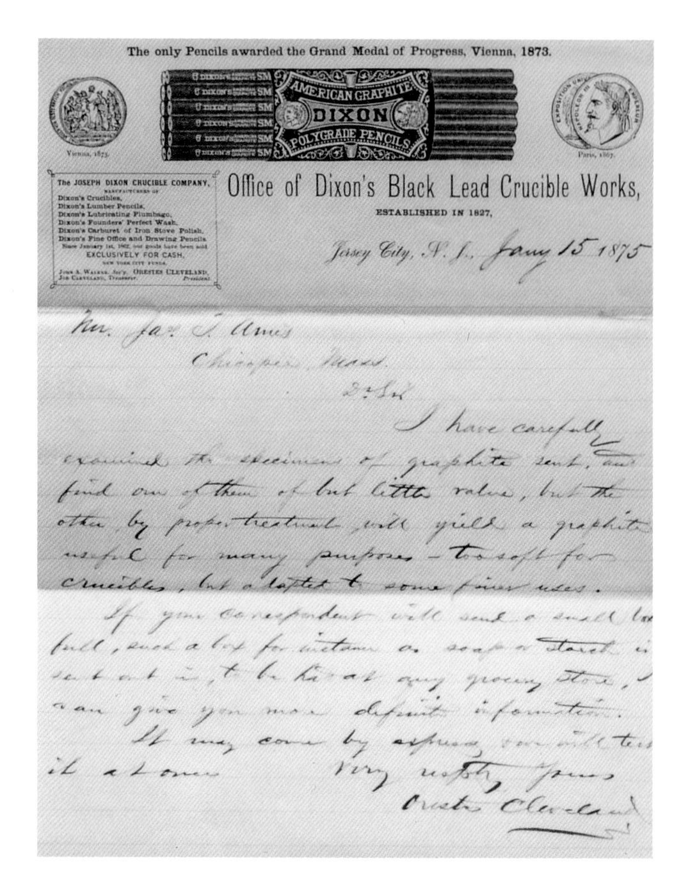

The only Pencils awarded the Grand Medal of Progress, Vienna, 1873.

AMERICAN GRAPHITE DIXON POLYGRADE PENCILS

Vienna, 1873. Paris, 1867.

The JOSEPH DIXON CRUCIBLE COMPANY,
MANUFACTURERS OF
Dixon's Crucibles,
Dixon's Lumber Pencils,
Dixon's Lubricating Plumbago,
Dixon's Founders' Perfect Wash,
Dixon's Carburet of Iron Stove Polish,
Dixon's Fine Office and Drawing Pencils
Since January 1st, 1862, our goods have been sold
EXCLUSIVELY FOR CASH,
NEW YORK CITY FUNDS.
John A. Walker, Sec'y. Orestes Cleveland,
Jos. Cleveland, Treasurer. President

Office of Dixon's Black Lead Crucible Works,
ESTABLISHED IN 1827,

Jersey City, N. J., Jany 15 1875

Mr. Jas. T. Ames
Chicopee Mass.
Dr Sir

I have carefully examined the specimens of graphite sent, and find one of them of but little value, but the other by proper treatment will yield a graphite useful for many purposes – too soft for crucibles, but adapted to some finer uses.

If your correspondent will send a small box full, such a box for instance as soap or starch is sent out in, to be had at any grocery store, I can give you more definite information.

It may come by express, you will tell it at once.

Very respectfully Yours
Orestes Cleveland

Letter written by Orestes Cleveland (president Dixon Crucible Co., and son-in-law of Joseph Dixon) describing two samples of graphite he tested in January 1875. $50-$75.

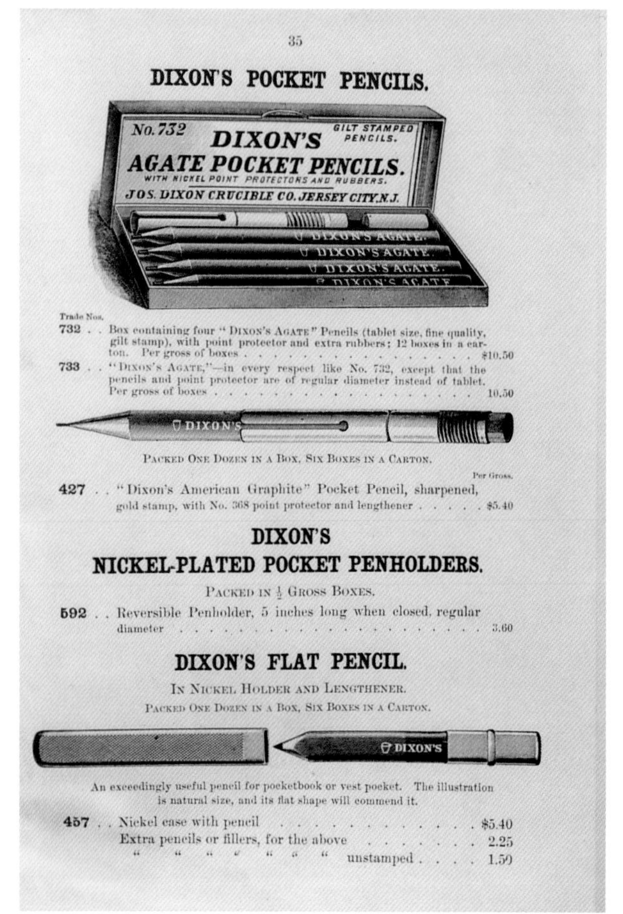

35

DIXON'S POCKET PENCILS.

Trade Nos.

732 . . Box containing four "Dixon's Agate" Pencils (tablet size, fine quality, gilt stamp), with point protector and extra rubbers; 12 boxes in a carton. Per gross of boxes $10.50

733 . . "Dixon's Agate,"—in every respect like No. 732, except that the pencils and point protector are of regular diameter instead of tablet. Per gross of boxes 10.50

Packed One Dozen in a Box, Six Boxes in a Carton.

Per Gross.

427 . . "Dixon's American Graphite" Pocket Pencil, sharpened, gold stamp, with No. 368 point protector and lengthener $5.40

DIXON'S NICKEL-PLATED POCKET PENHOLDERS.

Packed in ½ Gross Boxes.

592 . . Reversible Penholder, 5 inches long when closed, regular diameter 3.60

DIXON'S FLAT PENCIL.

In Nickel Holder and Lengthener.
Packed One Dozen in a Box, Six Boxes in a Carton.

An exceedingly useful pencil for pocketbook or vest pocket. The illustration is natural size, and its flat shape will commend it.

457 . . Nickel case with pencil $5.40
Extra pencils or fillers, for the above 2.25
" " " " " " " unstamped 1.50

Left: Page from a Dixon catalog, circa 1910.

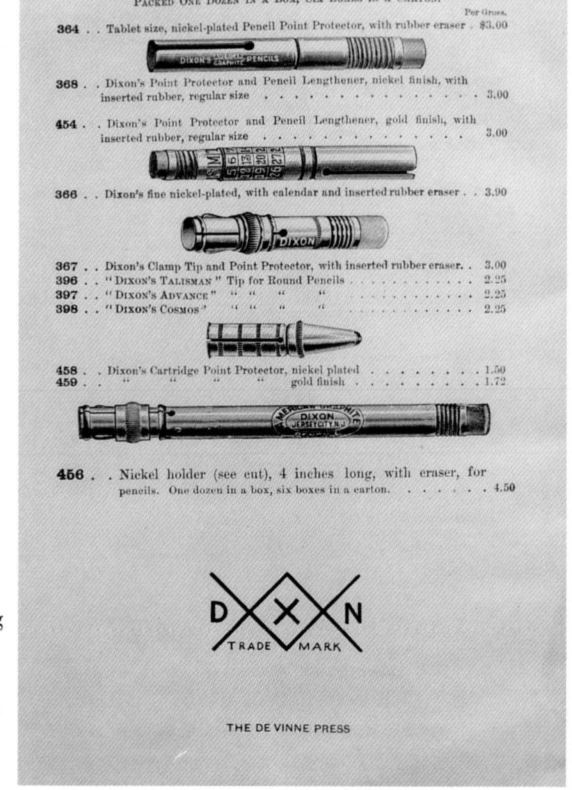

42

Dixon's Pencil Point Protectors.

Packed One Dozen in a Box, Six Boxes in a Carton.

Per Gross.

364 . . Tablet size, nickel-plated Pencil Point Protector, with rubber eraser . $3.00

368 . . Dixon's Point Protector and Pencil Lengthener, nickel finish, with inserted rubber, regular size 3.00

454 . . Dixon's Point Protector and Pencil Lengthener, gold finish, with inserted rubber, regular size 3.00

366 . . Dixon's fine nickel-plated, with calendar and inserted rubber eraser . . 3.90

367 . . Dixon's Clamp Tip and Point Protector, with inserted rubber eraser. . 3.00
396 . . "Dixon's Talisman" Tip for Round Pencils 2.25
397 . . "Dixon's Advance" " " " " 2.25
398 . . "Dixon's Cosmos" " " " " 2.25

458 . . Dixon's Cartridge Point Protector, nickel plated 1.50
459 . . " " " " gold finish 1.72

456 . . Nickel holder (see cut), 4 inches long, with eraser, for pencils. One dozen in a box, six boxes in a carton. 4.50

THE DE VINNE PRESS

Right: Page from a Dixon catalog showing point protectors, one incorporating a perpetual calendar, as well as the Dixon trademark.

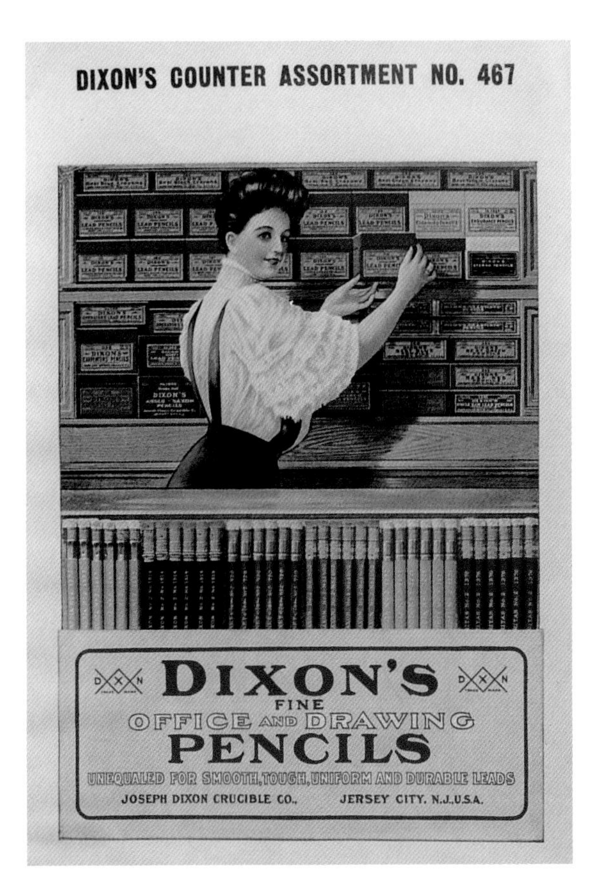

Full color page from the same Dixon catalog, showing their "Fine Office and Drawing Pencils."

Front cover of an advertising flier for Jos. Dixon Crucible Co., circa 1880, $45-$50.

Back cover of above, showing Dixon Pixies holding a pencil, and emphasizing that they were an "American Industry: American Materials, American Capital, American Machinery, American Brains, American Labor."

DIXON'S
Pencil Assortment No. 1240

A substantial tube with solid wood at the lower end, in imitation of a sharpened pencil and supported at the back by a decorated easel.
Containing one half gross of pencils of good quality, in round and hexagon shape with tips and rubbers, assorted finishes.
Packed in separate holding box, for shipment.

Price per box . . $1.72

JOSEPH DIXON CRUCIBLE COMPANY
JERSEY CITY, N. J.
NEW YORK PHILADELPHIA CHICAGO SAN FRANCISCO BOSTON ST. LOUIS PITTSBURGH BALTIMORE BUFFALO ATLANTA.

Dixon "Pixies" were often used in their advertising. Here they're shown displaying a wooden tube in the shape of a pencil, which held "one half gross of pencils of good quality."

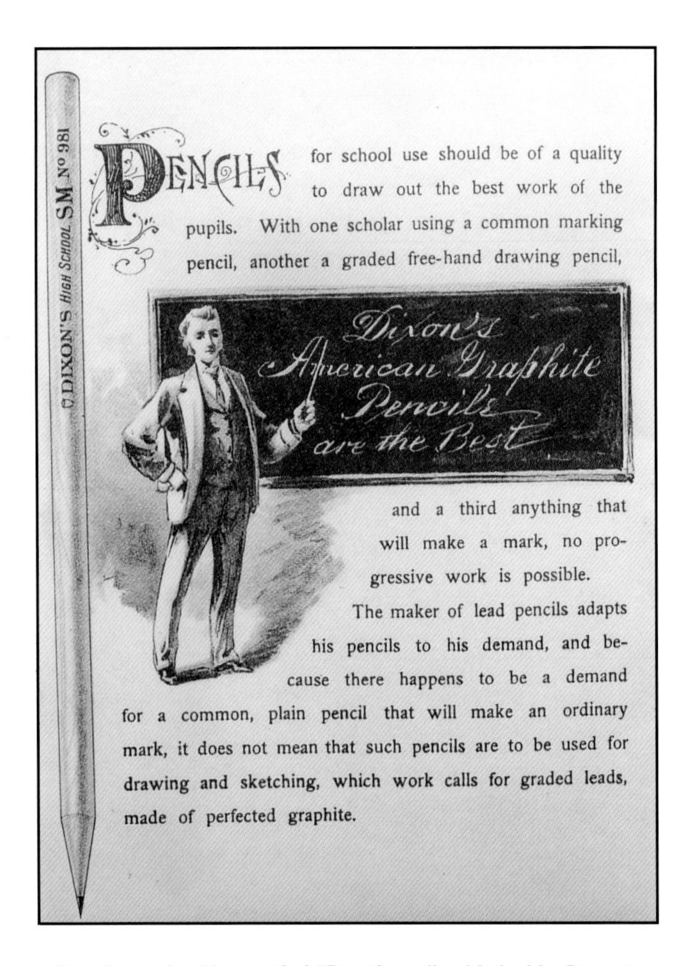

Page from a booklet entitled "Pencilings," published by Dixon in 1898. Notice the crucible trademark on the pencil.

Map from a Dixon booklet entitled "Pencil Geography For Boys and Girls of All Ages." Published in 1904, it shows Ticonderoga and "Pencilville" (Jersey City). $25-$35.

Imitation Dixon pencil, called "Dickens," made by American Pencil Co., late 1800s. $3-$5.

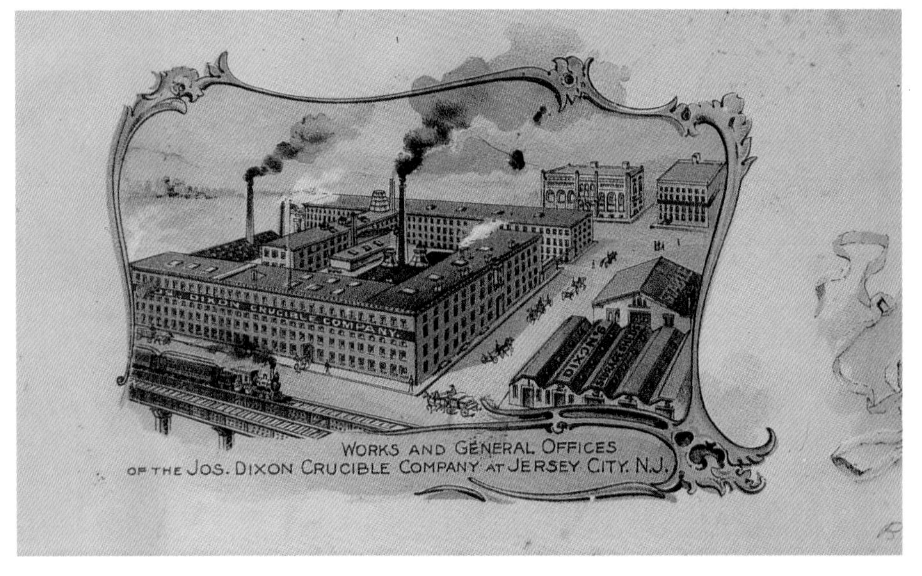

Color illustration of the Jos. Dixon Crucible Co. works and general offices in Jersey City. Note the proximity of the train tracks. Pamphlet $20-$25.

Flier advertising Dixon's penholders and pencils, 1915. Double-sided, $7-$10.

Cover of Dixon catalog, 1910. 42 pages, $75-$95.

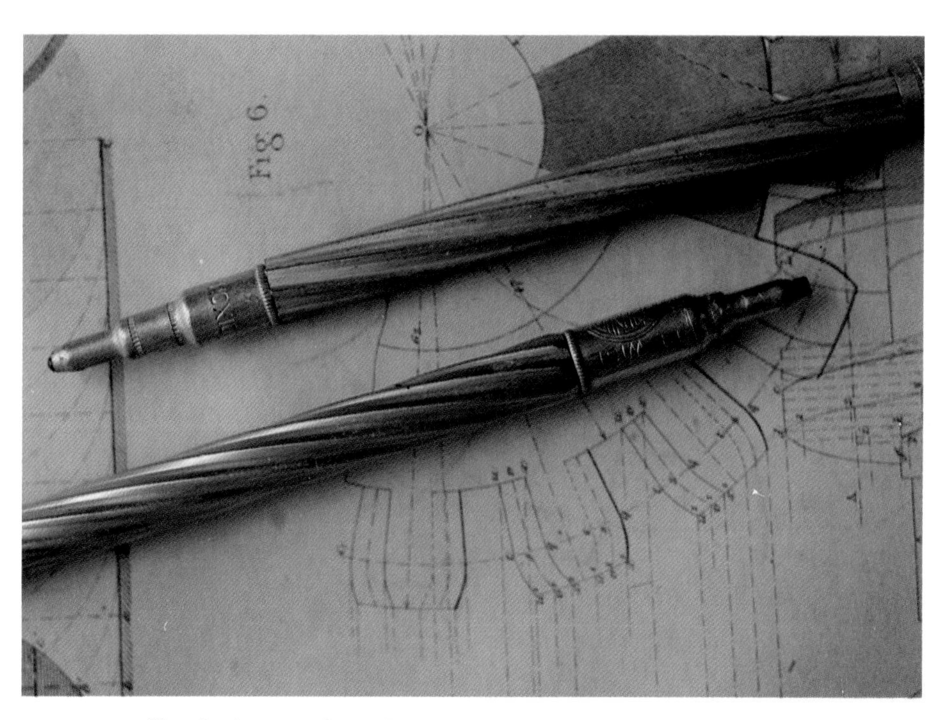

Two aluminum mechanical pencils made by Dixon, one advertising for Aetna Insurance Co. "Be wise, Aetna-ize," 1912, 5". $25-$35.

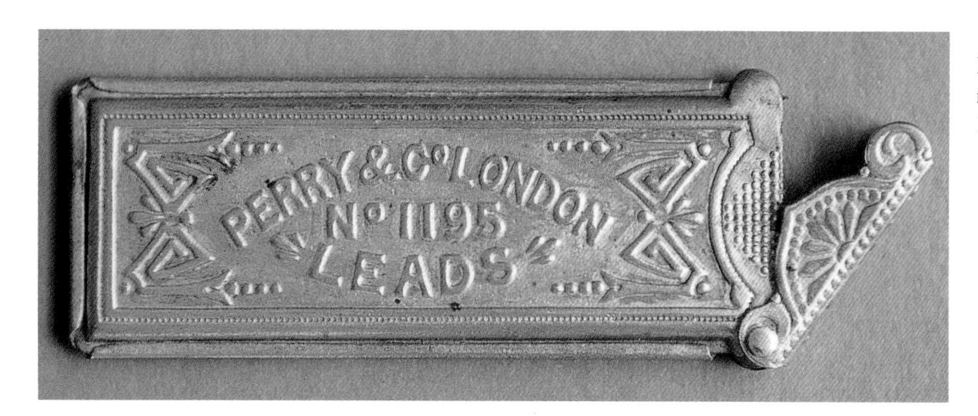

Elaborate brass holder for six pieces of graphite, made by Perry & Co., London. Mid- to late 1800s, 2-1/2" x approx. 1", $35-$50.

Patent Perry & Co. holder for six leads, approx. 2" x 1". $25-$45.

Wood and cardboard boxes to hold British compressed graphite, made by L. & C. Hardtmuth "Koh-i-noor," circa 1880, $25-$35.

Box filled with wooden containers for "Everpoint Lead" sold by William S. Hicks, New York. Circa 1880-1900, $5-$7 (each container).

Wooden holder for "Everpoint Leads," sold by Mabie, Todd & Co., circa 1880-1900. Each piece is about 9/16" long and 1/16" wide. They are not all straight, which probably created problems when trying to fill the pencil. $5-$7.

Wooden cylinder with graphite, sold by B. S. Cohen, London. $5-$15.

Cardboard box with wooden holders for "refined Everpoint leads" made by Joseph T. Mears, successor to Wm. Jackson. $15-25.

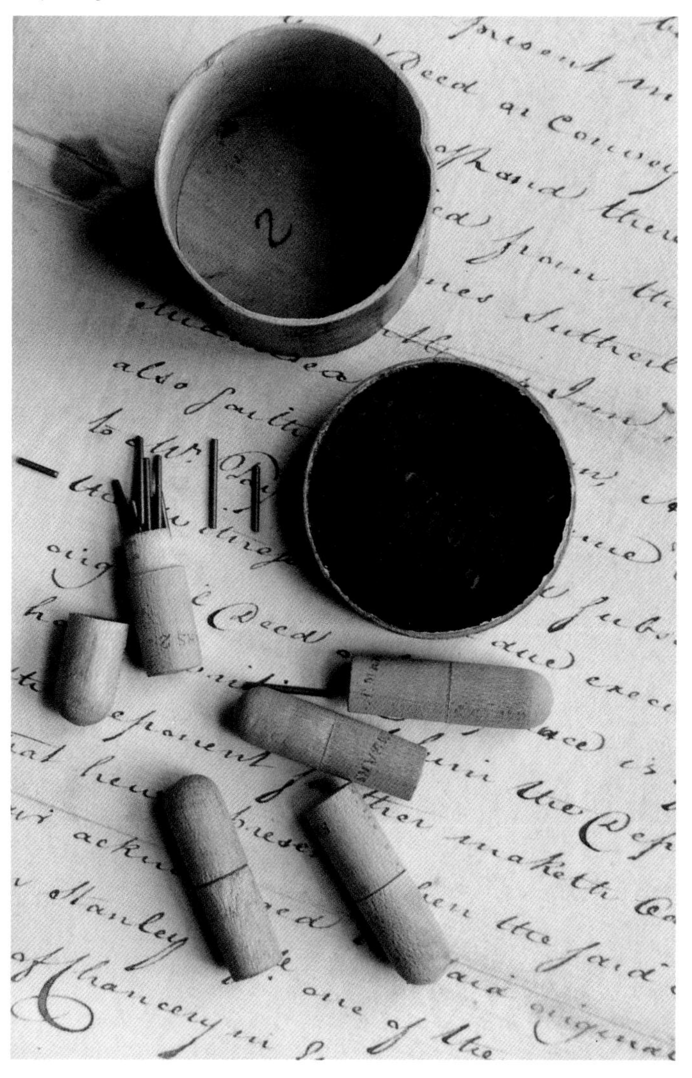

Cardboard holder for Siberian graphite, sold by A. W. Faber, circa 1870, $10-$20.

Cardboard box holding blue leads made by A. W. Faber, "for the new Everpointed Pencils," and a wood and brass pencil, circa 1870. $35-$45.

Cardboard box for six "Eagle Artist Leads for Automatic Pencils," 1879, 3-3/4" x 1-1/2", $10-$15.

Composition case holding five "A. W. Faber Polygrades" showing how the various degrees of hardness were designated, circa 1870. $50-$75.

Cedar, bone, and metal mechanical pencil with Siberian Graphite from the Alibert Mine. Approx. 6", made by A. W. Faber, Germany, circa 1870-1880. $10-$20.

Chapter Two

Materials

Because there are similarities between the materials and techniques used by jewelers, silversmiths and pencil-case makers, developments in one area often led to progress in all of these areas. Techniques used in making jewelry were often the same as those used to make objects of vertu (pencils, match boxes, etc.). The effects of the Industrial Revolution were profound. While many of the artisans at the beginning of the nineteenth century had received traditional training as apprentices, by mid-century this was no longer the case. Most pieces were created entirely by hand at the outset of this period, but by the end of the century many of these items were made (at least partially) by machine and only finished by hand.

As noted, many of the materials that were used to make jewelry were also used to make propelling and metal cased pencils. More expensive pencil-cases (in 1861, for example, Tiffany offered pencils that cost up to $269) could be made of gold (in different karats, colors, and finishes), rolled gold, silver (oxidized or white), ivory, tortoiseshell, or they could be carved out of tiger-eye. They could be decorated with mother of pearl, abalone, or enamel. Some makers embellished them with precious stones. Less expensive pencils could be gold plated or made of brass, black hard-rubber, bone, horn, or celluloid. (Celluloid was created in 1860, originally as an imitation of horn.) Novelty pencils might be made of base metal, wood, or papier-mâché, and decorated with artificial stones, paint, or even fur.

While craftsmen still employed traditional metal working techniques, new methods were also developed and refined during this period. The discovery of galvanic battery plating (or *electroplating*) was one such process. The development of this inexpensive plating system is one of the reasons why jewelry and other related items (pencils, for example) became more affordable. It was instrumental in the transition between a time when decorative arts were available only to an elite group and the time when they became accessible to the working classes.

Earlier forms of plating had required the metal-smith to fuse a thin sheet of gold or silver onto a base metal. *Rolled gold*, for example, was made in this way. Although this made metal-ware more affordable than it had been, it was still expensive because of the labor involved and the amount of gold used. Electroplating, however, used far less precious metal, required less work, and could be done in vast quantities. Electroplating could be done on pieces that had already been fabricated. The implementation of this process is one of the reasons why Providence, Rhode Island, became a center for jewelry making in the 1800s. This productive atmosphere enabled companies like A. T. Cross, one of the major manufacturers of writing implements in the nineteenth century, to create thriving businesses.

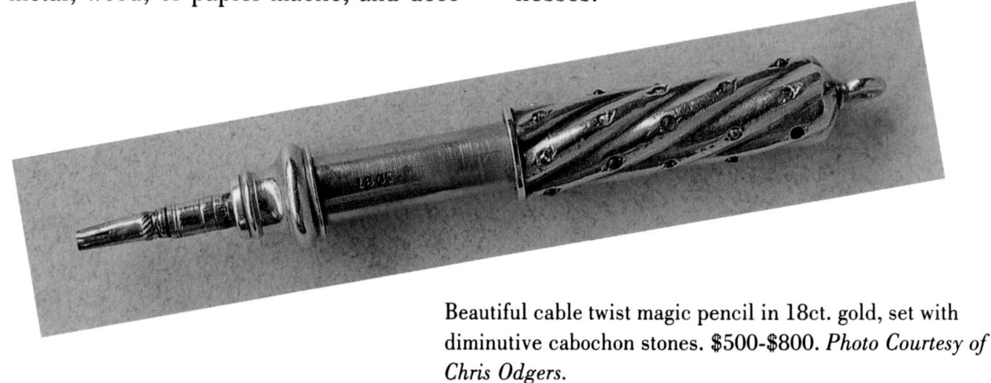

Beautiful cable twist magic pencil in 18ct. gold, set with diminutive cabochon stones. $500-$800. *Photo Courtesy of Chris Odgers.*

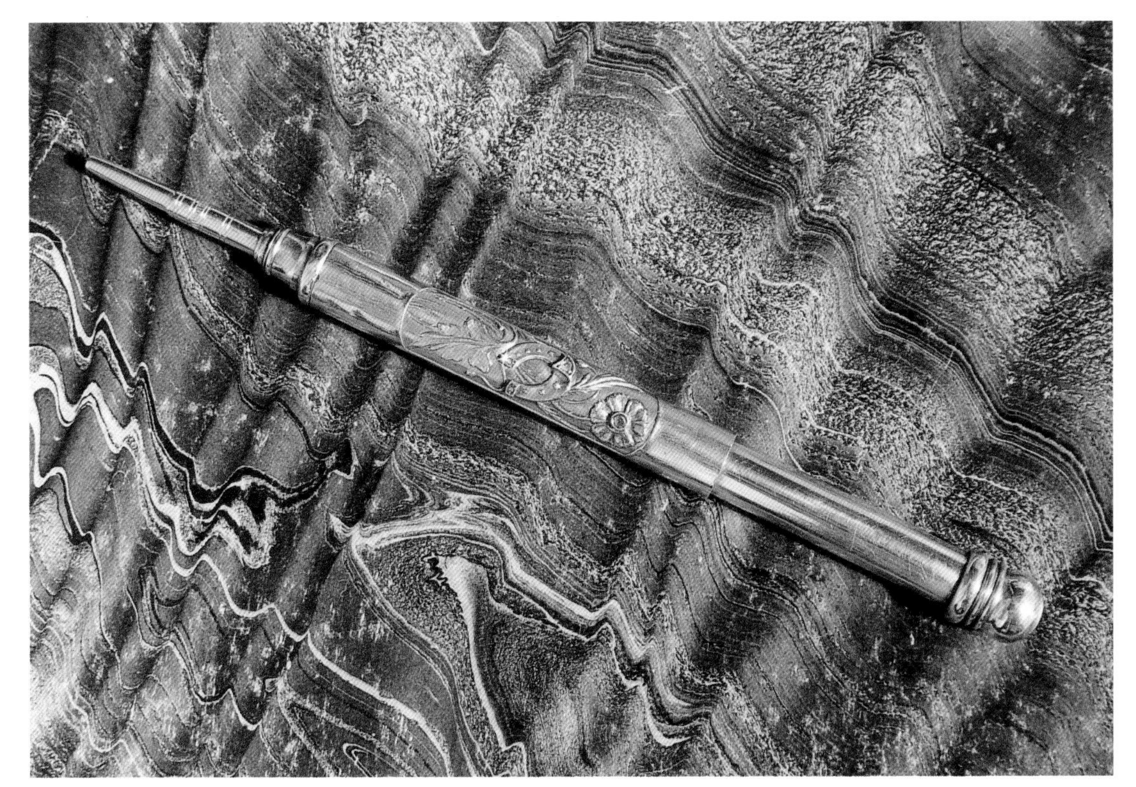

Magic pencil in rolled gold, with small mine-cut diamonds set in the repoussé flower and horseshoe. $225-$275.

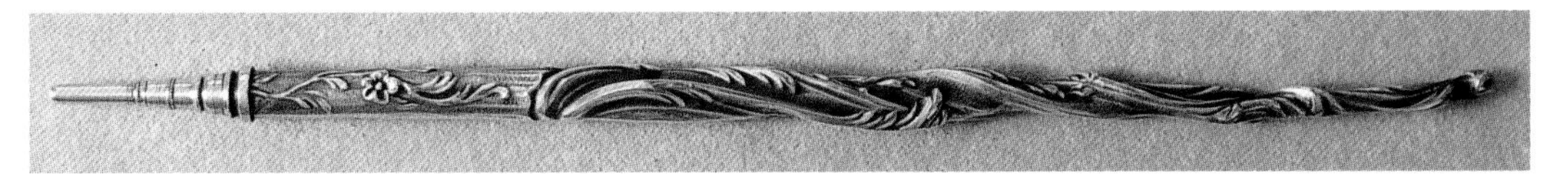

Silver desk pencil, with a French hallmark, 8". This is an unusual piece, in that nearly all nineteenth century pens and pencils are symmetrical and this one is not. The wave-like imagery of the staff stylizes plants fluttering in the wind. $150-$225.

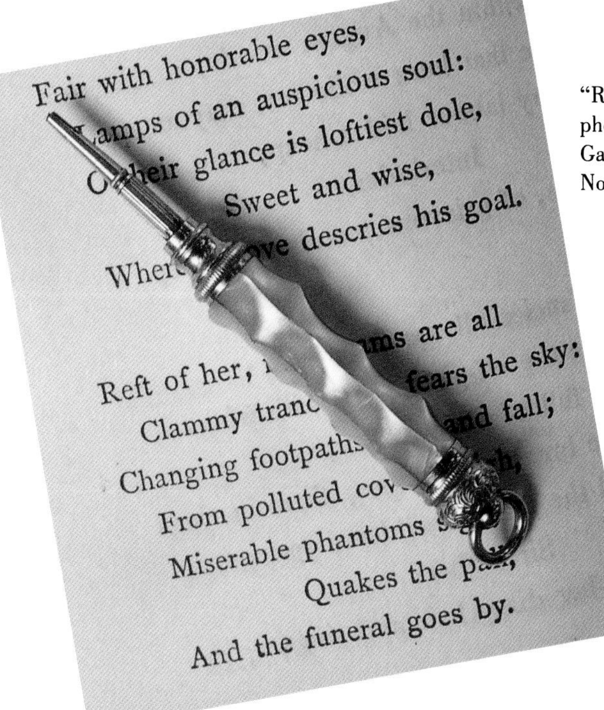

"Rustic" mother-of-pearl screw pencil, photographed on a copy of Dante Gabriel Rossetti's poem, "Love's Nocturne," both circa 1873. $75-$125.

Silver and rolled gold magic pencil with an etched design. $75-$125.

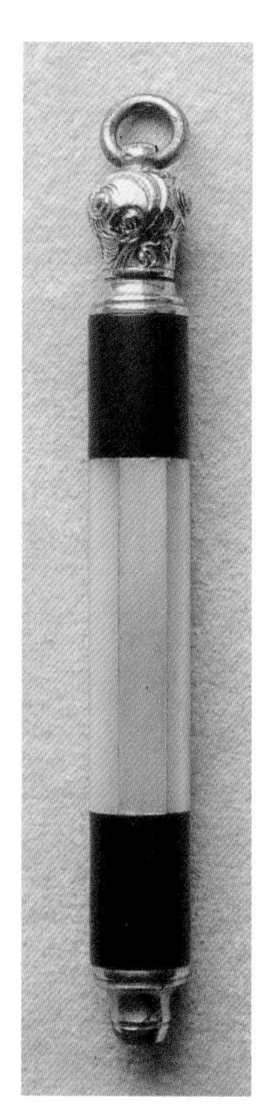

Black hard rubber and mother-of-pearl magic pencil, marked "Pearl Patent Dec. 5, '71," 2.75" closed. $75-$125.

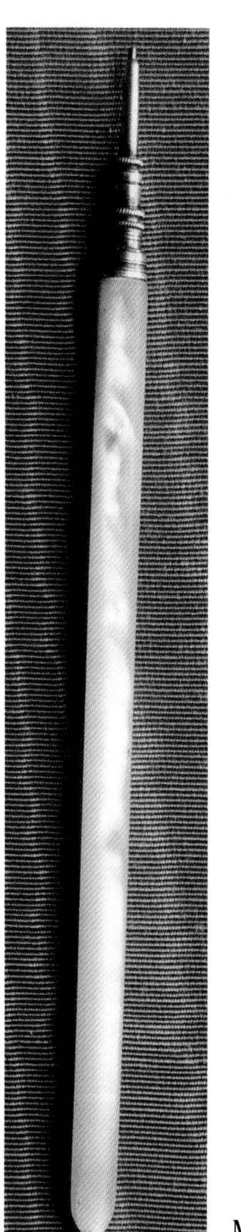

Mother-of-pearl tablet pencil, approx. 8". $45-$60.

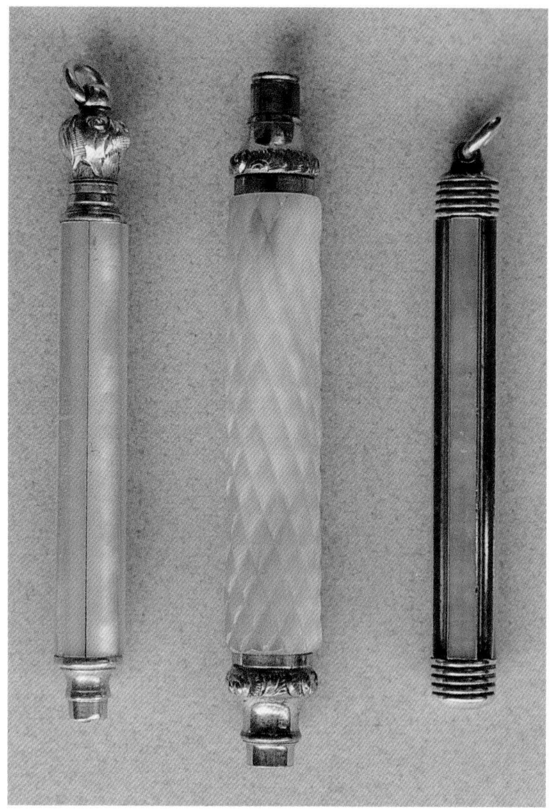

Three variations on mother-of-pearl: magic pencil with gold-filled metal and mother-of-pearl ($75-$125), gold-filled gravity or drop pencil with pearl carved in a diamond pattern ($150-$175), silver pencil with inset slabs of pearl marked with the Dec. 5, '71 patent date ($150-$175). *Photo Courtesy of Chris Odgers.*

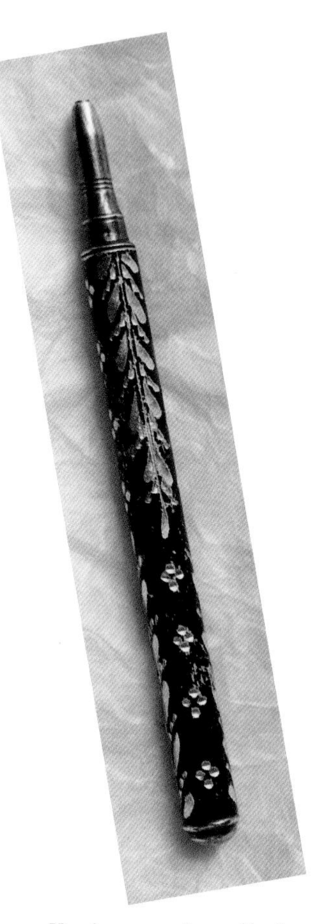

Hand-engraved, anodized aluminum, and silver mounted pencil with screw mechanism. Similarly styled fountain pens (if marked Parker or Waterman) are considered rare and valuable. Circa 1890, 3.75", $125-$150.

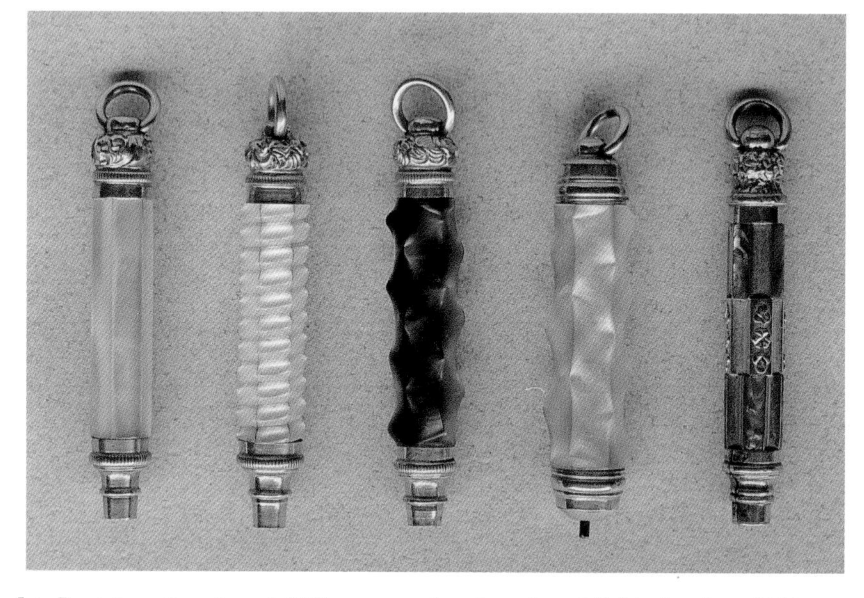

L to R: plain mother-of-pearl ($75), corrugated mother-of-pearl (Aikin, Lambert; $125 - $175), "rustic" black oyster pearl ($150-$200), "rustic" mother-of-pearl ($75-$125), abalone with bi-colored gold insets ($75-$125). Circa 1871-1890. *Photo Courtesy of Chris Odgers.*

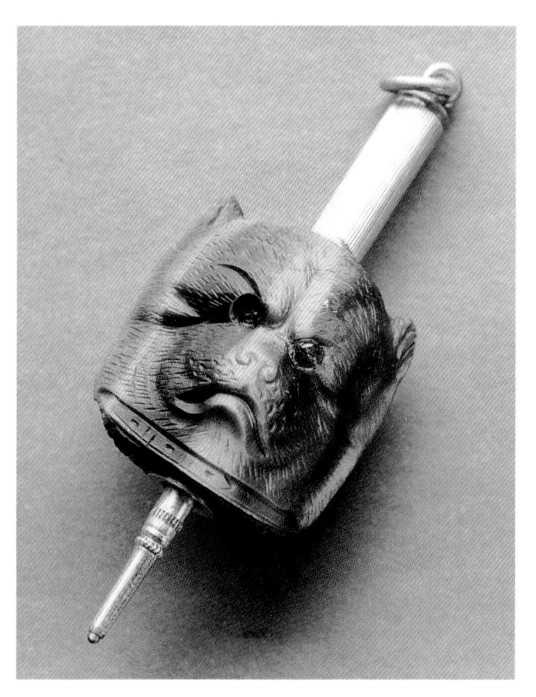

Remarkably detailed tiger-eye magic pencil, with a different breed of dog carved on each side. Because tiger-eye is so easily damaged, the excellent condition of this particular piece makes it highly desirable. Unmarked, circa 1880-1900, 1" in length when closed, 2.25" when opened, $450-$800.

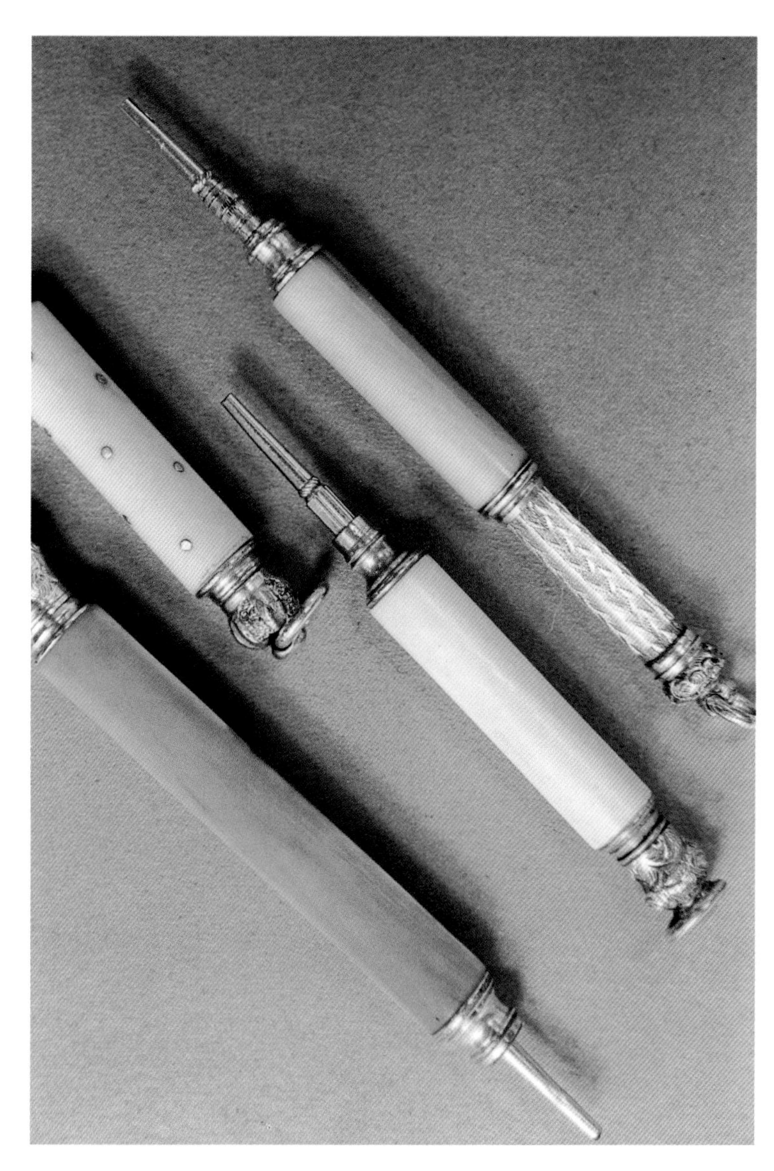

Several mechanical pencils with ivory shells. Notice the variation in the color of the ivory.

Bone and tartan pencil, with a twist mechanism. "Tartan" became a popular form of ornamentation with Queen Victoria and Prince Albert's interest in all things Scottish. Mid-nineteenth century. $75-$125.

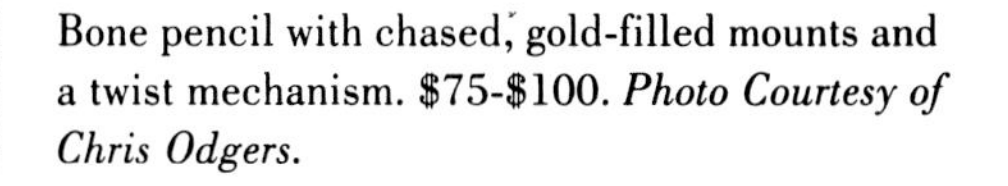

Bone pencil with chased, gold-filled mounts and a twist mechanism. $75-$100. *Photo Courtesy of Chris Odgers.*

This pencil is made of black engine-turned hard rubber, and is marked "Goodyear's Patent." Circa 1851, $50-$75. *Photo Courtesy of Chris Odgers.*

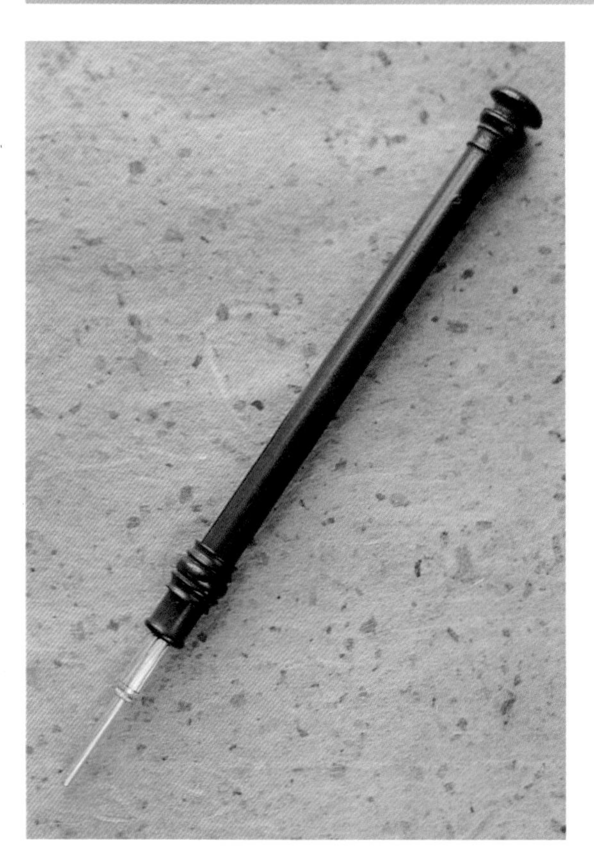

Unusual slide pencil made of black hard rubber. Marked "Goodyear," circa 1851, $50-$100. Black hard rubber oxidizes to a dull brown. The blacker the color, and the crisper the chasing (or engine turning) if there is any, the more valuable the piece.

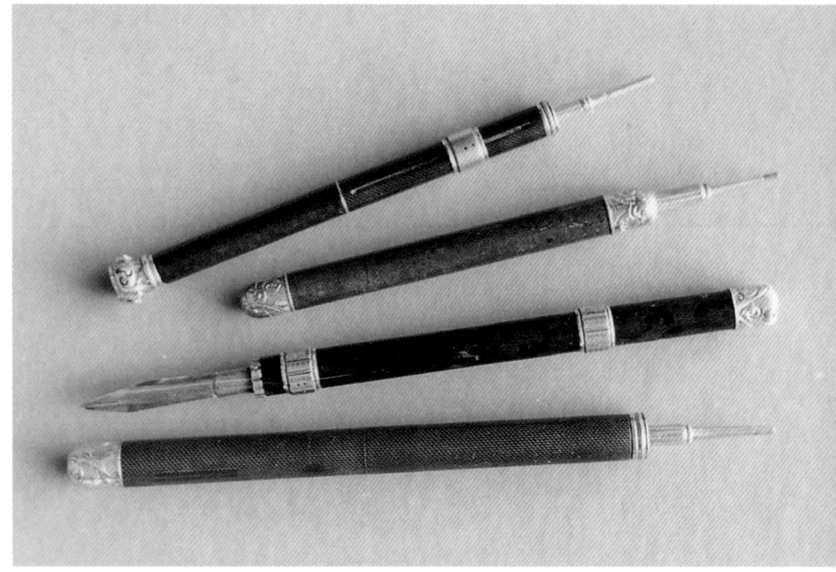

Top to bottom: Black engine-turned hard rubber combination screw pencil, slide pen ($50-$75), black engine-turned hard rubber twist pencil, marked "Goodyear's Patent, 1851" ($35-$50), engine turned black hard rubber double-slide pen/pencil combo with the nib marked "The Brilliant" ($75-$125), pen/pencil combination marked "Edward Todd & Co." as well as Goodyear ($75-$125).

Red hard rubber twist pencil, slide pen combination. Marked "Edward Todd & Co., Patent appd. for" with a number 3 Edward Todd nib. There are very few red hard rubber mechanical pencils. $175-$225.

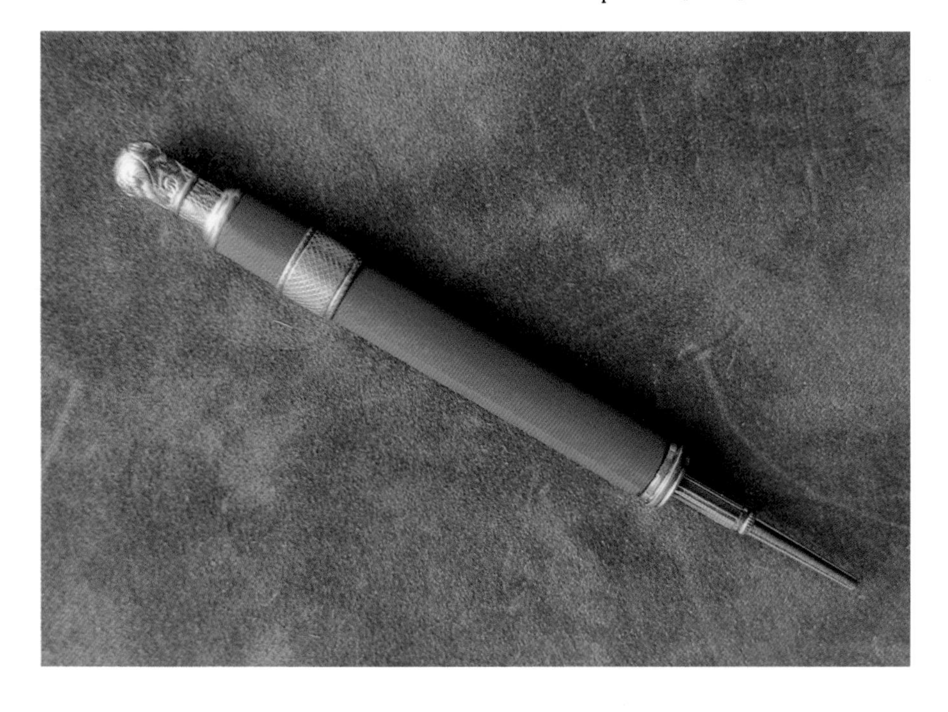

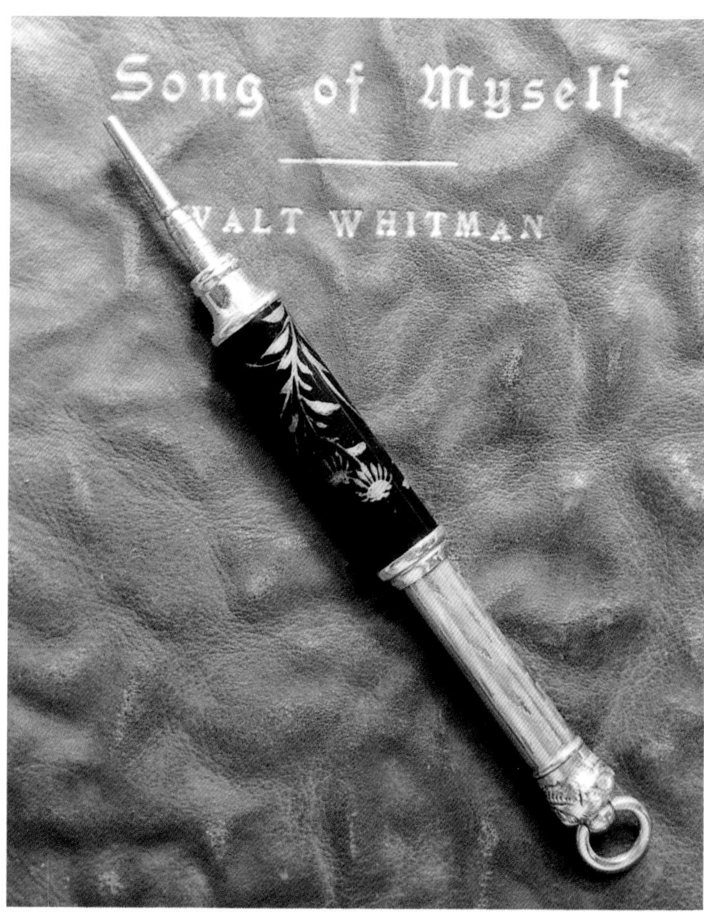

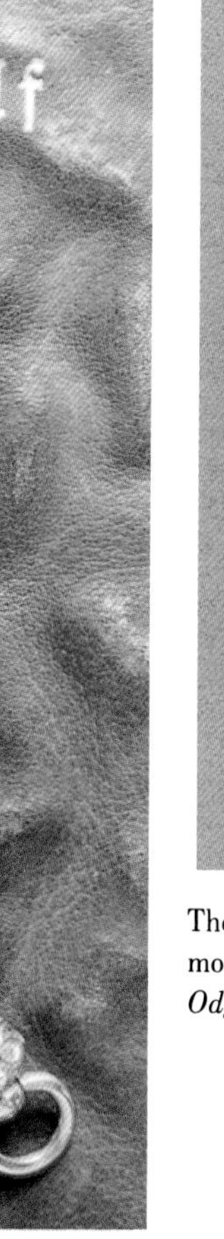

Black and gold lacquer magic pencil. Unmarked, approx. 1" closed, 3" open, $75-$125. Photographed on an edition of Walt Whitman's "Song of Myself," bound by Roycrofters in 1905.

These pencils are made of celluloid, with inlays of silver, gold, and mother-of-pearl. Circa 1880, $75-$150. *Photo Courtesy of Chris Odgers.*

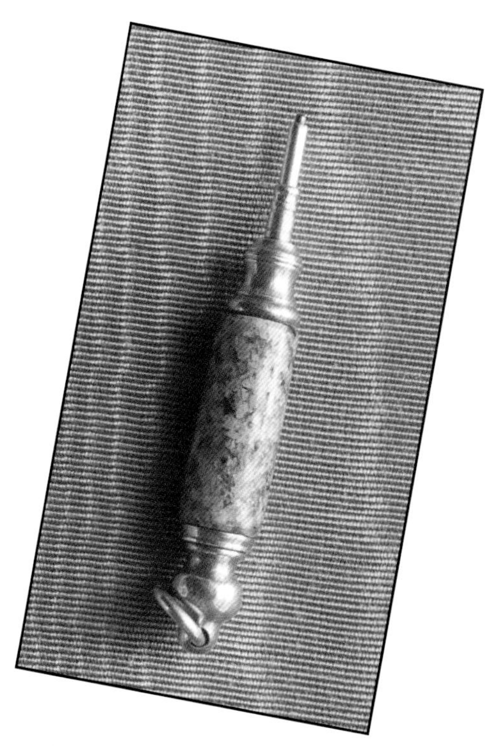

Screw pencil of pink and gray mottled celluloid, base metal trim. $35-$65.

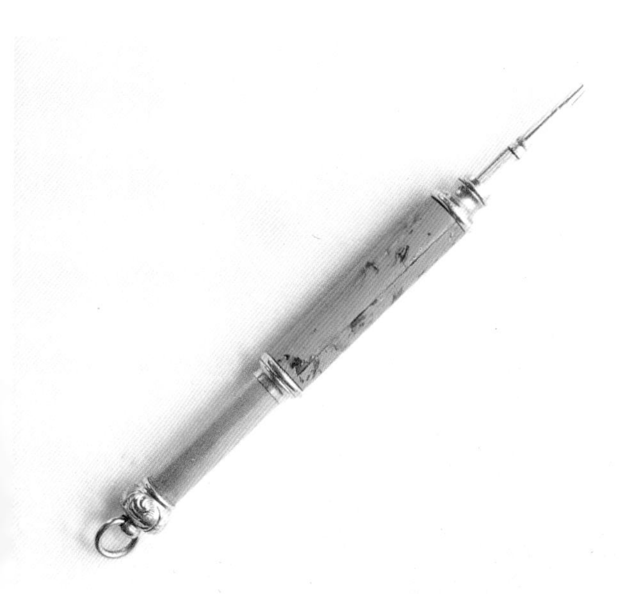

Green celluloid magic pencil with gold-filled trim, made to imitate malachite (as advertised by Aikin, Lambert in the early 1880s). $75-$85.

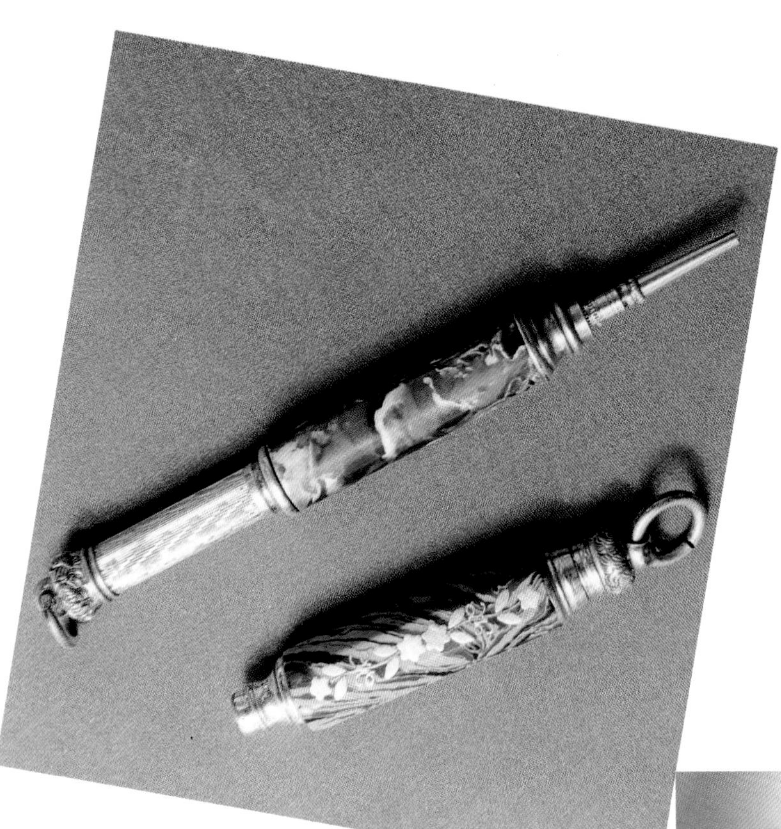

Left: mottled green, brown, white and black celluloid magic pencil with chased gold-filled trim ($75-85), "malachite" celluloid magic pencil marked W. S. Hicks, $75-$150.

Toffee colored, striated celluloid, gold-filled trim pencil. Marked "John Holland, patent May 23, 1882." ($40-$50)

Clockwise: Agate seal, agate magic pencil with gold-filled chain ($150-$175), agate aide mémoire with ivory pages ($225-$275), faceted ribbon agate desk pencil with a twist mechanism ($150-$175), smooth agate desk pencil ($150-$175), hexagonal agate dip pen, mounted with hand-engraved silver and with a terminal seal ($175-$200).

Porcupine quill and silver slide pencil made by Sampson Mordan ($150-$225). Photographed on pencil drawing from an engineering student at Rensselaer Polytechnic Institute, circa 1913.

Tortoiseshell and silver slider with wooden pencil. Made by S. Mordan & Co. $225-$350.

CHAPTER THREE
Ornamentation and Surface Decoration

With the exception of some *tablet* pencils and a few magic pencils, most nineteenth century pencil-cases were decorated. The barrel, finial, trim, and slider were decorated using any number of techniques or combinations of techniques. For example: the finial may have been cast and hand-engraved, with a bezel-set stone mounted in the end opposite where it screws onto the barrel, the barrel may have been engine-turned and the slider may have been cast. The body (or barrel) may be gold or very heavy rolled gold and the trim may have been electroplated. Or, a pencil-case could be cast (as is the case in many of the figurals) and plated so different parts would be emphasized by different colors of metal, then partially enameled.

The decoration of pencil-cases could be said to fall into two categories. In the first, the surface decoration enhances the piece but does not interfere with its usefulness. In the second, the ornamentation disguises the pencil and sometimes makes it very difficult to actually use. The earliest mechanical pencils generally had a *reeded* barrel, which, with its longitudinal lines, elegantly reinforced the form of the pencil. This type of decoration was soon replaced, however, with more complex designs created by engine turning and engraving. Figural pencils (that could, technically, be used to write with but were in reality very awkward to use) were being made as early as the 1840s.

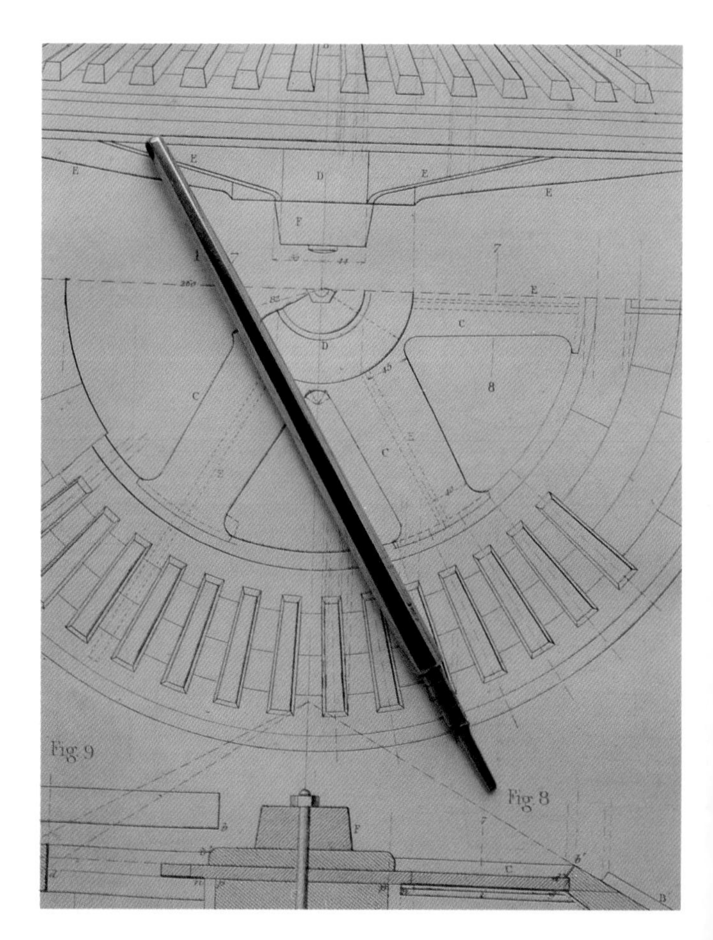

Tapered, hexagonal tablet pencil in 9 ct. gold, marked A.D. This is an example of a pencil without chasing, engraving, enamel, or applique. 5.75", $125-$150.

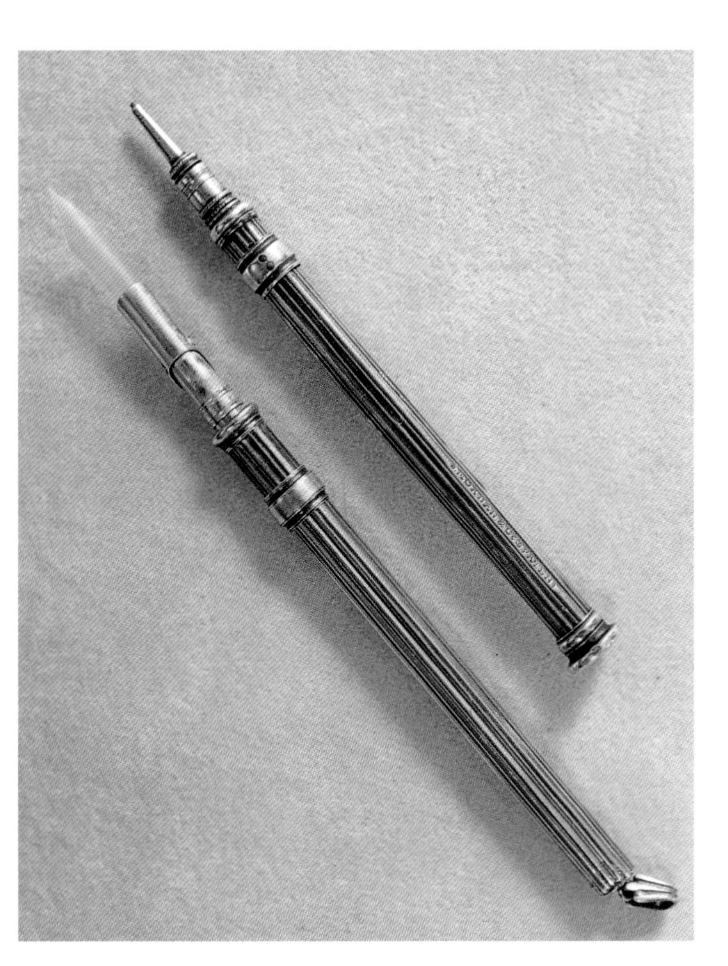

Examples of reeded barrels. Both made by S. Mordan prior to 1830. Pen with Bramah style quill holder, $225-$325. Pencil with an early Mordan mark ("S. Mordan & Co.'s Patent"), $175-$250.

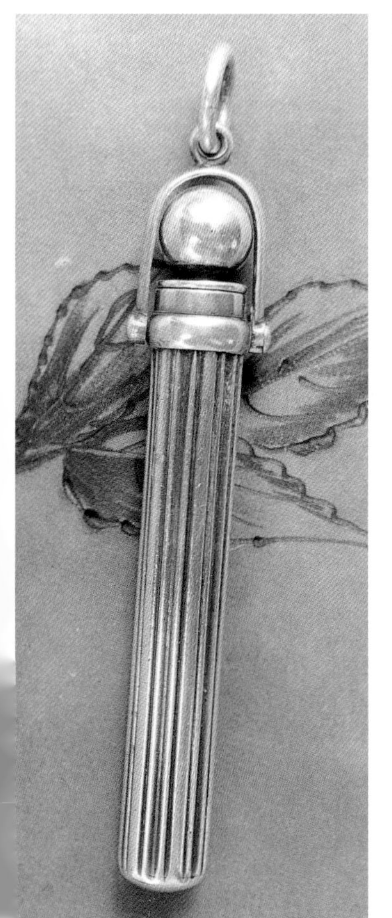

S. Mordan & Co. removable "sheath" telescoping pencil in silver. 3" with stirrup loop, circa 1890s, $125-$150.

Applique

Pieces of silver or gold (cast or fabricated separately from the pencil-case) were sometimes soldered onto the barrel of a pencil. This is called *applique*.

Rose gold-filled with green gold applique, magic pencil made by Fairchild. $100-$125.

Nickel silver pencil with applique of a figure on a horse on a fleur-de-lis background. $45-$65.

Engraving

Nineteenth-century pencils were sometimes hand-engraved. Engraving is a form of surface decoration where the metal is inscribed with a sharp tool called a *graver* or *burin*, which actually cuts away slivers of metal. A design that's especially sharp and crisp may have been *bright cut* (so-called because of the way light bounces off the sharp edges). Engraved lines can vary in depth and width. Depending on the amount of pressure applied to the *burin*, the line can go from being deep and wide to shallow and then taper down to a fine line. A line can be deeper on one side than the other.

Engraving is also a technique used in printmaking. The form of script most often emulated by nineteenth century scribes was called *copperplate*, so named because it was inspired by the formal letterforms inscribed in copper by early printmakers. Copperplate script was modified and popularized by several well-known Victorian penmen (notably Platt Rogers Spencer), who embellished their versions of it with elaborate flourishes. Many of the graceful motifs hand-engraved on nineteenth-century pencils are reminiscent of these flourishes.

Example of Spencerian flourishes from an autograph book, circa 1881.

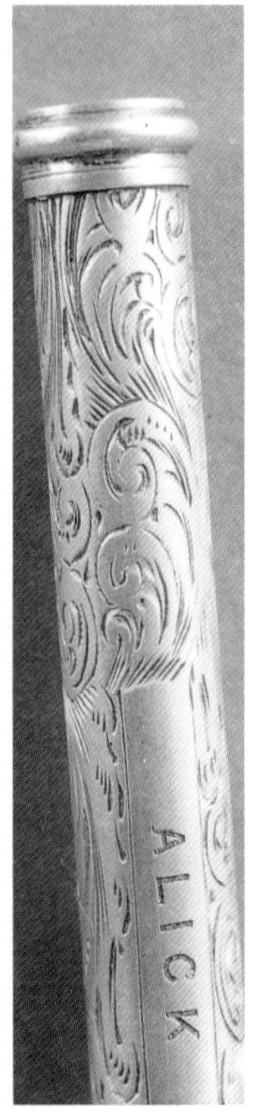

Engraved name, "Alick", on the barrel of a porte-crayon.

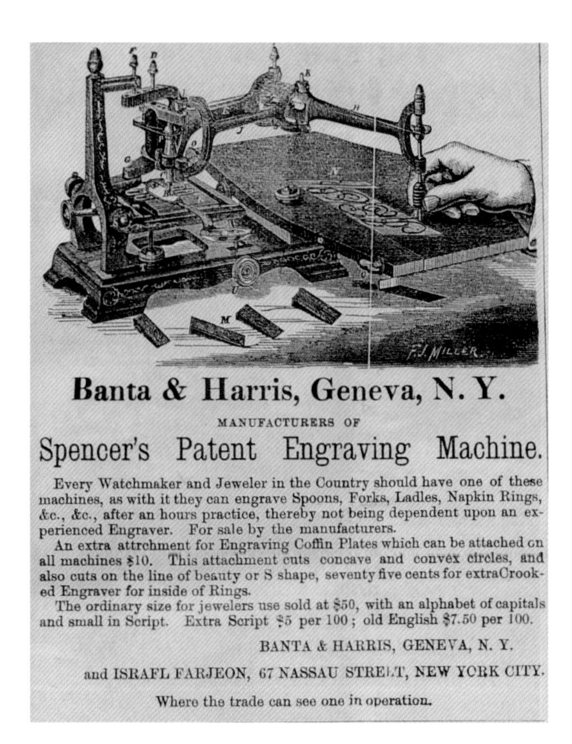

"Spencer's Patent Engraving Machine," from a late nineteenth century Ad. "Every Watchmaker and Jeweler in the Country...can engrave...after hours of practice, thereby not being dependent on an experienced Engraver..."

Hand-engraving on the blade of a pencil shaped like a butter knife.

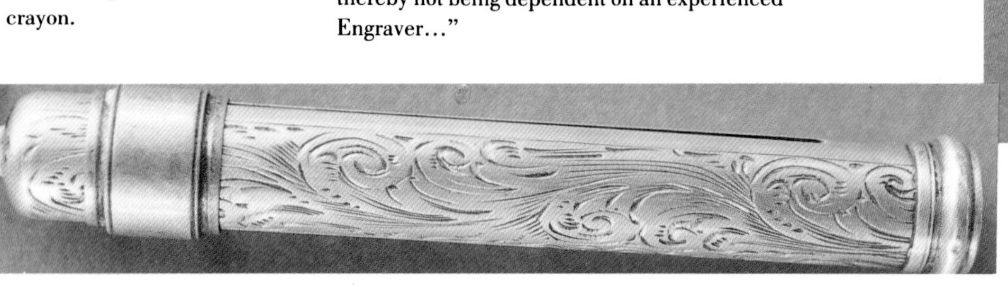

Silver porte-crayon with hand engraved barrel. 2.5", $75-$125.

Magic pencil with hand engraved panels alternated with slabs of mother-of-pearl, with engine turning on the pull. $100-$150.

Engine Turning

One of the most prevalent types of surface decoration used on pencils is called *engine turning* (the French term *guilloché* is frequently used). Also called *Rose engine turning* because it was done on a *rose-engine lathe*, it involved the utilization of *rosette mandrels* (or pattern disks). The metal cylinder, which was to become the barrels of pencils, was secured to the lathe. The pattern was inscribed as the lathe moved the metal, not spinning it as in a common wood lathe, but by conforming to the pattern of the rosette mandrel. A description of this process from Volume III of *Knight's American Mechanical Dictionary* states:

"Rose-engine Lathe (*engraving*). A lathe in which the rotary motion of the lathe and the radial motion of the tool combine to produce a variety of curved lines. The mechanism consists of plates or cams set on the axis of the lathe, or suitably rotated and formed with wavy edges or grooves which govern the motion of the cutting point toward or from the center.

Three examples showing the range of designs possible with engine turning and stamping.

"In another form, the combined radial and rotary motion is inherent in the work, the tool being stationary. In this case, the center of the circle in which the work revolves is not a fixed point, but is made to oscillate with a slight motion while the work revolves around it.

"The mandrel upon whose end the work is clutched does not rotate. In stationary standards like those of the common lathe, the oscillating motion is given to the mandrel by means of rosettes or wheels fixed on a mandrel. As the mandrel revolves, the wavy periphery of the rosettes is applied to the roller, which moves on a stationary axis and causes a vibratory motion of the mandrel as its frame moves to and fro on its axis. The mandrel contains a number of rosettes of different patterns."

Numerous designs are evident in the pencils decorated by engine turning. Some consist of a series of wavy lines, some appear to have been created from patterns that allowed flat raised areas, most often in a floral or foliate design. Early lathes were either hand or treadle operated, but even as they became powered by steam engines, they still required the expertise of a skilled craftsman.

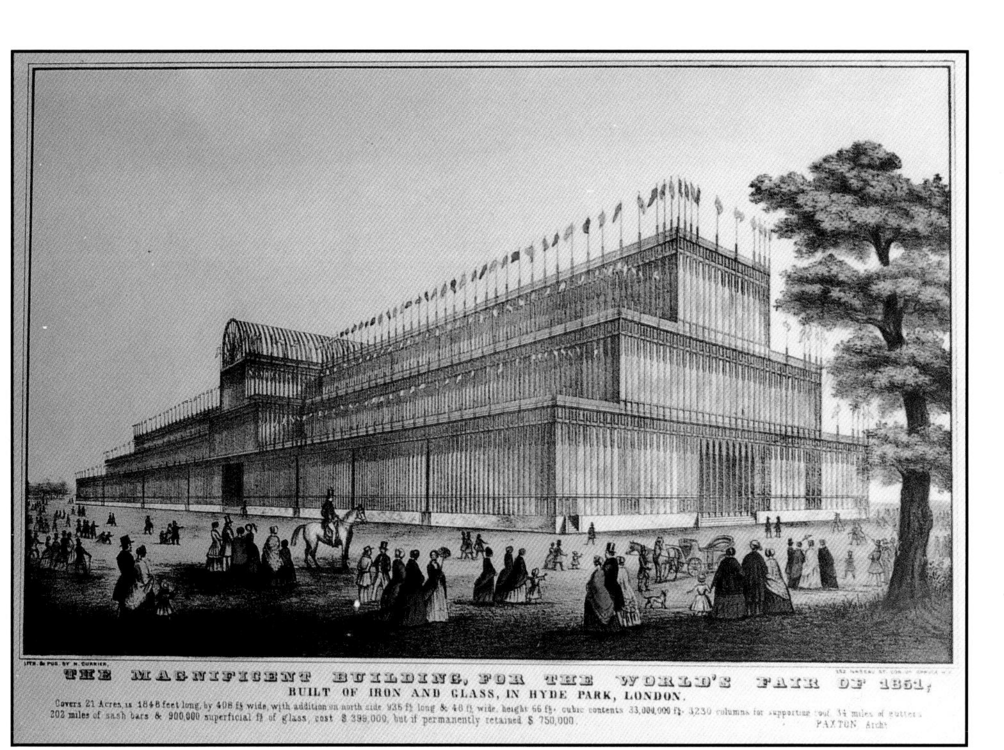

1851 illustration of the Crystal Palace by Currier and Ives.

Silver pencil with stamped motto, "Friendship." $75-$125.

One silver pencil marked Fendograph, with opaque blue enamel inlay, and four silver pencils with guilloché enamel. The engine-turned pattern is clearly visible under the transparent enamel. $150-$225.

At the Crystal Palace Exhibition in 1851, Sampson Mordan & Co. exhibited, in addition to various writing implements, a rose lathe. To what degree Mordan improved on the lathe and what the impact his improvements had on the pencil industry is not clear. It is evident, however, that the company considered the machine innovative and important enough to include it in their display at the great exhibition.

Sometimes hand-engraving is done on top of or in addition to the engine turning, giving even more dimension to a design. Another technique that is particularly handsome is when transparent enamel is applied over an engine-turned metal surface. The enamel appears darker over the recessed areas and lighter over the raised areas, adding both color and shading to the engine-turned piece. The correct term for this technique is basse taile, but it is generally referred to as guilloché.

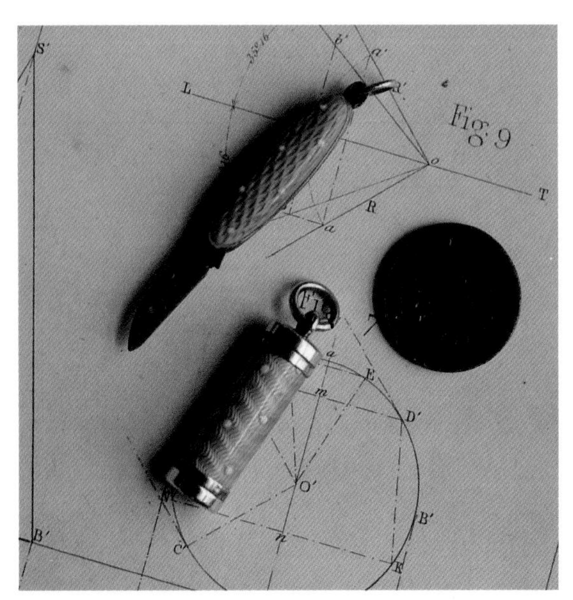

Tiny removable-sheath pencil, shown with an equally diminutive folding knife. $175-$250 (pencil), $125 (knife).

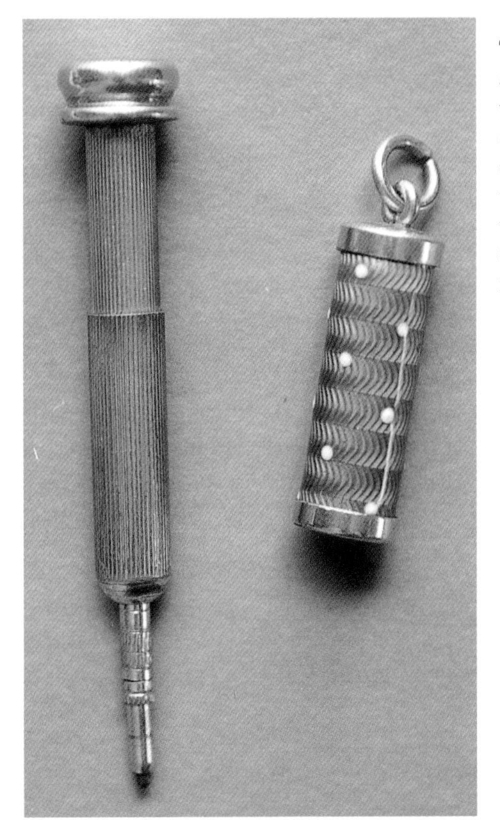

Telescoping pencil, and removable sheath with transparent enamel over an engine-turned base, with dots of opaque white enamel. $175-$250 if the enamel is not chipped.

Enamelling

In addition to *transparent enamels* (see guilloché), *opaque enamels* were also used by pencil makers to enhance the jewel-like qualities of their goods. The opacity of these enamels allowed an artisan to paint scenes or images in color on the barrel of the pencil. Also, some hand-painting was done on a thin shell of porcelain.

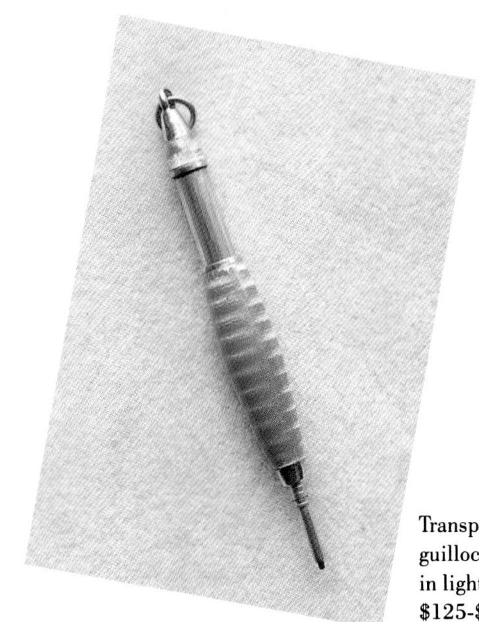

Transparent enamel, guilloché magic pencil in light blue on silver. $125-$175.

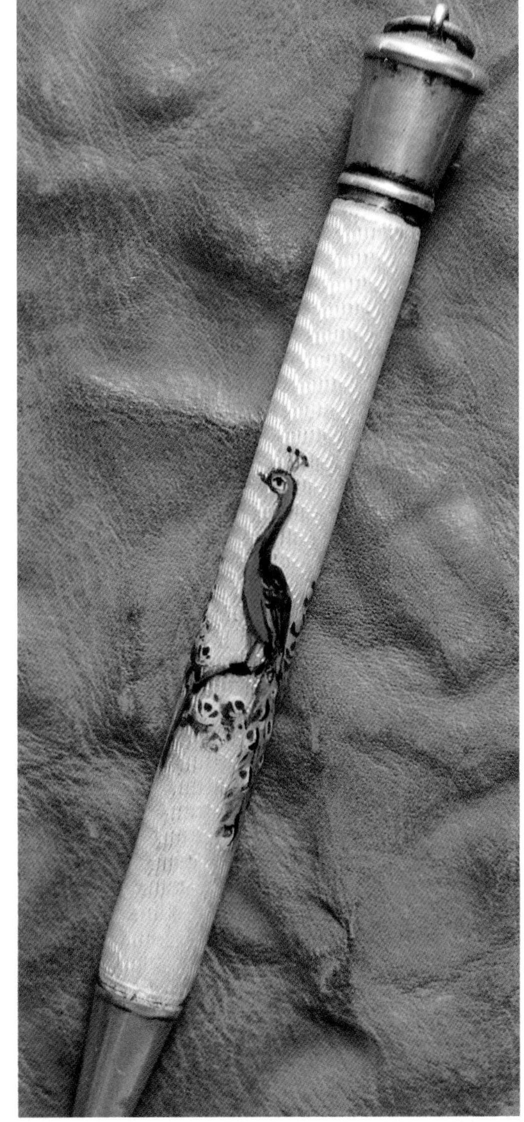

Silver pencil with transparent enamel over engine turning, with an opaque enamel peacock. Unmarked, $125-$150.

Exquisite telescoping pencil in red enamel on gold, with bands of seed pearls. $400-$800.

A rainbow selection of pencils in jewel-like enamel. $150-$225.

Sterling silver and transparent enamel pencil with a pink rose in opaque enamel. $75-$125.

Two magic pencils with guilloché enamel, and floral designs in opaque enamel. $125-$175. *Photo Courtesy of Chris Odgers.*

Five silver and enamel magic pencils in an array of rich colors. $125-$225.

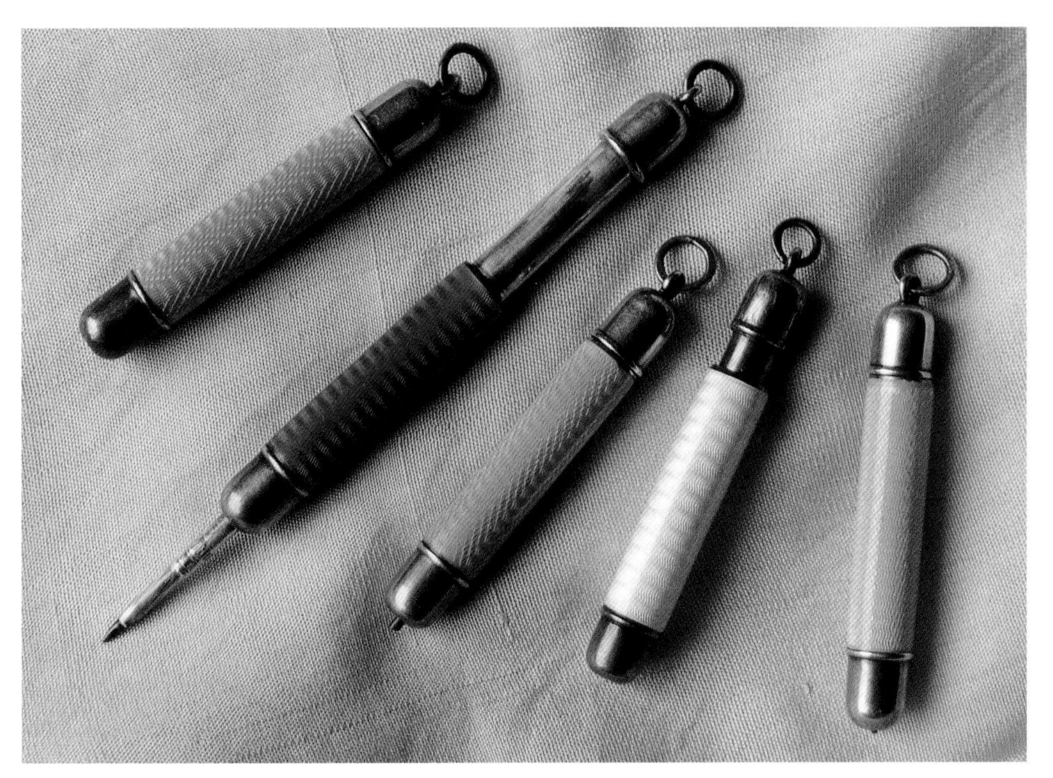

Five guilloché enameled pencils and holders for cedar pencils. Left to right: $75-$85, A. T. Cross pencil, $75-$95, $125-$175, $125-$175, $75-$90. *Photo Courtesy of Chris Odgers.*

Magic pencil with a hand-painted scene in opaque enamel. Several makers, including John Holland, offered pencils like this. $250-$350.

Magic pencil with hand-painted figure of a young girl. Unmarked, $250-$350.

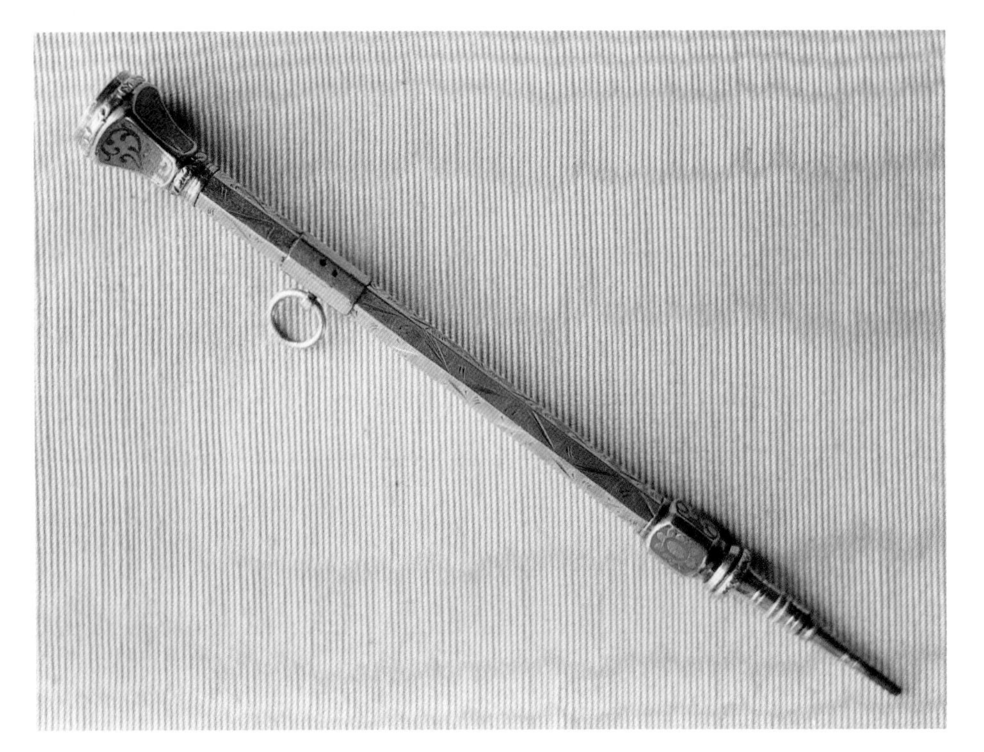

Hexagonal pencil, gold-filled, with inlaid enamel in the ring sliders, with a bloodstone set in the terminal. $150-$175.

Elegant pen-pencil combination, gold with blue enamel and diamonds. Unmarked, $3,800-$4,200.

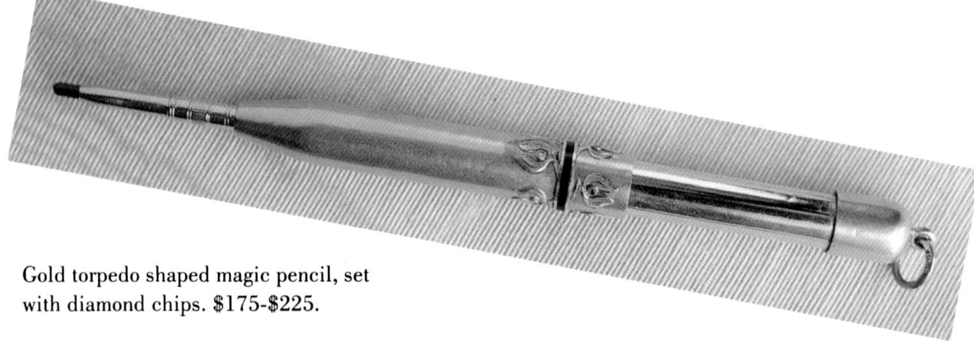

Gold torpedo shaped magic pencil, set with diamond chips. $175-$225.

Magic pencil with black and gold lacquer scene of a bird and flowers. $125-$175.

Die Stamping and Repoussé

Both *die stamping* and *repoussé* involve pushing the metal from behind. Repoussé is a hand process, allowing greater variation from piece to piece. Die stamping, whether done by hand or machine, involves the use of a die that replicates the image identically each time it's used. The raised design can have a variety of heights, curves and angles, but can't have any undercuts. If seen from behind the metal will also conform to the design, but in reverse. Some die stamped pieces have a flat backing of thin gauged metal soldered to them, which covers the negative side of the image. This is most commonly seen on well-made pieces or on pieces where a flat back would be necessary for support or function. The seam between the front and back is generally hard to find, because the solder is usually melted between two flat surfaces and the two pieces of metal are finished as one. In the example shown here, a repoussé lily graces the front of an *aide mémoire* made by Edward Todd & Co. Both the notebook and the pencil (which allows the cover to pop open when it's pulled from its holder) are marked.

Sterling silver repoussé lily aide mémoire, and pencil marked Edward Todd. $275-$350.

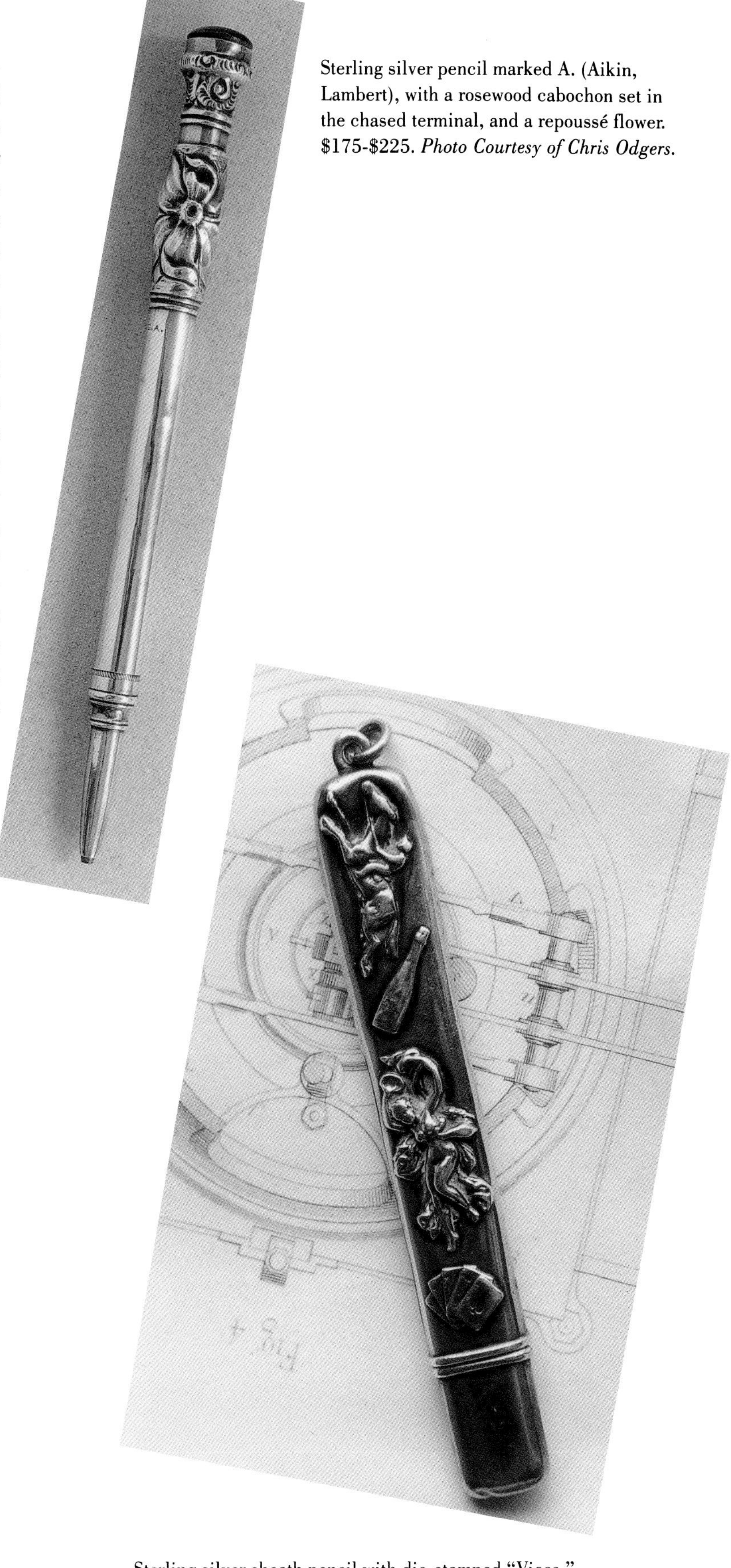

Sterling silver pencil marked A. (Aikin, Lambert), with a rosewood cabochon set in the chased terminal, and a repoussé flower. $175-$225. *Photo Courtesy of Chris Odgers.*

Sterling silver sheath pencil with die-stamped "Vices," including horse racing, drinking, women, and gambling. Probably Aikin, Lambert, $165-$225.

Sterling silver magic pencils with heavily embossed flowers. $175-$225.

Repoussé and chased flowers on a silver pencil marked H. (Heath). $175-$225.

Mechanical pencil with etched design, including a figure of a Native American with a quiver of arrows. Marked M. T. (Mabie, Todd & Co.), $175-$275.

Etching

The areas that are to remain raised on an *etched* piece are covered with a *ground* (asphaltum) that protects the metal from the acid. Different depths may be achieved by applying ground on some of the areas that have already been *bit* by the acid and by resubmerging the metal back into the acid. This technique is also used in printmaking.

Sterling silver magic pencil, marked E. T. (Edward Todd), with etched foliate design. 2.75" closed, 5" open. $150-$200.

Silver "Mabie Magazine" pencil, patented 1910, with etched floral design. $125-$150.

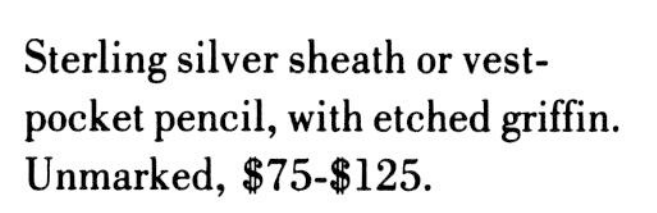

Sterling silver sheath or vest-pocket pencil, with etched griffin. Unmarked, $75-$125.

Sterling silver dip pen with etched design of a griffin, $150-$175.

Piqué (Inlay)

When silver or gold wire is inlaid into another material, traditionally tortoiseshell, the technique is called *piqué*. In pencils, the wire is pressed into celluloid or black hard-rubber made to simulate tortoiseshell or other materials.

Gold and silver pique inlay in green celluloid (imitation malachite). Made by W. S. Hicks, 1.75" closed, $125-$175.

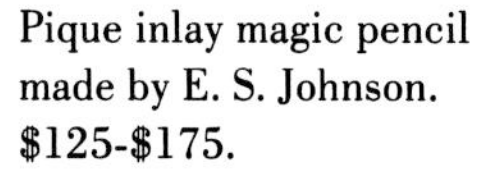

Pique inlay magic pencil made by E. S. Johnson. $125-$175.

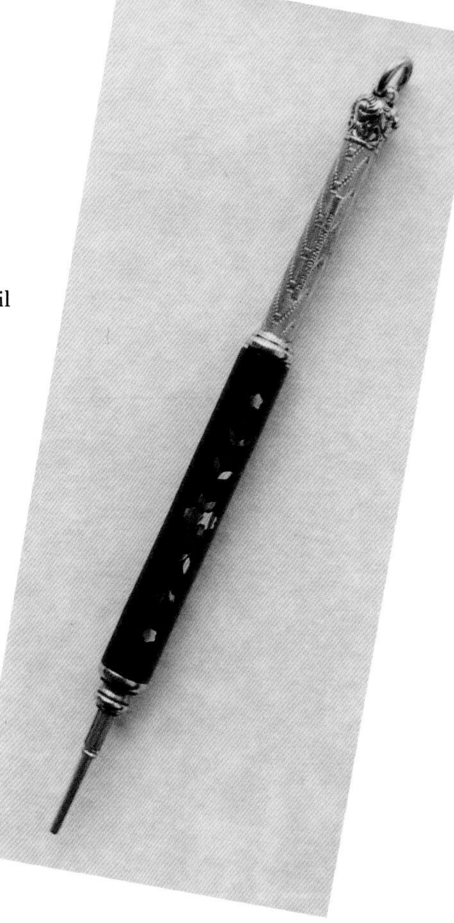

Pique inlay magic pencil, patented Feb. 6, 1872. $125-$175.

Chasing

Chasing is a method of embellishing the surface of metal by hitting it with a small chisel-like tool that is struck with a mallet. The metal isn't actually carved, but a relief design is created when the metal is pushed from one place to another.

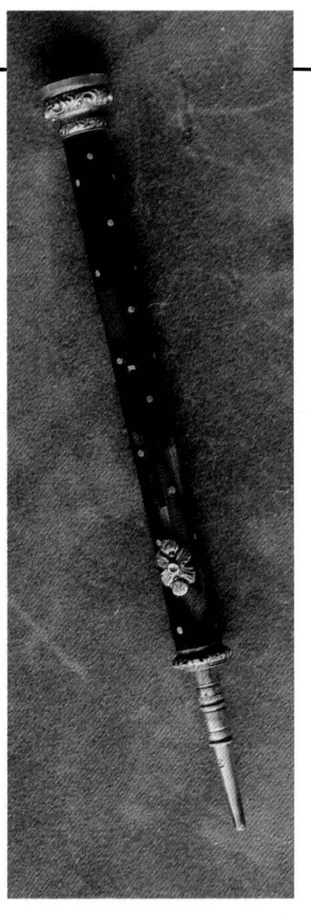

Silver inlay in tortoise shell, cast foliate slider buttons, amethyst set in terminal. Unmarked, $275-$350.

Heavily chased, sterling silver magic pencil. With acorn mark, probably W. S. Hicks. $150-$175.

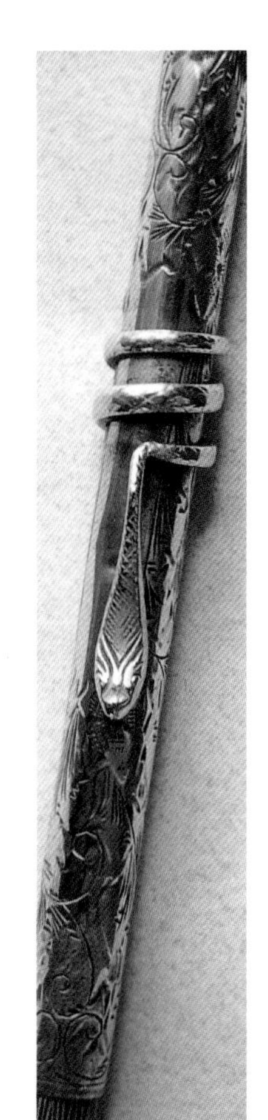

Cast snake clip on hand-engraved pencil made by Aikin, Lambert and Hardtmuth's Koh-i-Noor. $150-$225.

Tablet or desk pencils in sterling silver with chased designs. $125-$175.

Detail of above.

Desk pencil in sterling silver, with a chased pattern. $125-$175. Photographed on a pencil drawing from the mid-1880s.

Magic pencil with diamond shaped inlays of mother-of-pearl (often abbreviated as MOP), and triangular shaped abalone. $125-$175.

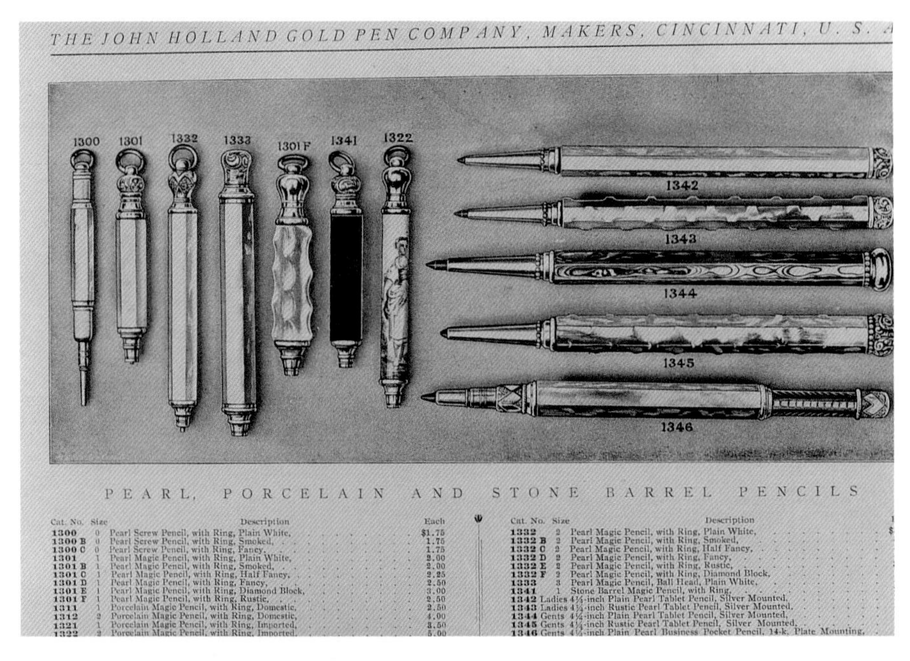

THE JOHN HOLLAND GOLD PEN COMPANY, MAKERS, CINCINNATI, U. S. A.

PEARL, PORCELAIN AND STONE BARREL PENCILS

Cat. No.	Size	Description	Each
1300	0	Pearl Screw Pencil, with Ring, Plain White,	$1.75
1300 B	0	Pearl Screw Pencil, with Ring, Smoked,	1.75
1300 C	0	Pearl Screw Pencil, with Ring, Fancy,	1.75
1301	1	Pearl Magic Pencil, with Ring, Plain White,	2.00
1301 B	1	Pearl Magic Pencil, with Ring, Smoked,	2.00
1301 C	1	Pearl Magic Pencil, with Ring, Half Fancy,	2.25
1301 D	1	Pearl Magic Pencil, with Ring, Fancy,	2.50
1301 E	1	Pearl Magic Pencil, with Ring, Diamond Block,	3.00
1301 F	1	Pearl Magic Pencil, with Ring, Rustic,	2.50
1311	1	Porcelain Magic Pencil, with Ring, Domestic,	2.50
1312	2	Porcelain Magic Pencil, with Ring, Domestic,	4.00
1321	1	Porcelain Magic Pencil, with Ring, Imported,	3.50
1322	2	Porcelain Magic Pencil, with Ring, Imported,	5.00

Cat. No.	Size	Description
1332	2	Pearl Magic Pencil, with Ring, Plain White,
1332 B	2	Pearl Magic Pencil, with Ring, Smoked,
1332 C	2	Pearl Magic Pencil, with Ring, Half Fancy,
1332 D	2	Pearl Magic Pencil, with Ring, Fancy,
1332 E	2	Pearl Magic Pencil, with Ring, Rustic,
1332 F	2	Pearl Magic Pencil, with Ring, Diamond Block,
1333	3	Pearl Magic Pencil, Ball Head, Plain White,
1341	1	Stone Barrel Magic Pencil, with Ring,
1342	Ladies	4¼-inch Plain Pearl Tablet Pencil, Silver Mounted,
1343	Ladies	4¼-inch Rustic Pearl Tablet Pencil, Silver Mounted,
1344	Gents	4½-inch Plain Pearl Tablet Pencil, Silver Mounted,
1345	Gents	4¼-inch Rustic Pearl Tablet Pencil, Silver Mounted,
1346	Gents	4½-inch Plain Pearl Business Pocket Pencil, 14-k, Plate Mounting,

Page from John Holland catalog, showing "pearl, porcelain, and stone barrel pencils."

Several examples of pencils decorated with MOP.

Ad from "The Cosmopolitan" (late 1800s), showing L. E. Waterman pens with chased metal barrels.

Ad showing chased patterns on L. E. Waterman pens.

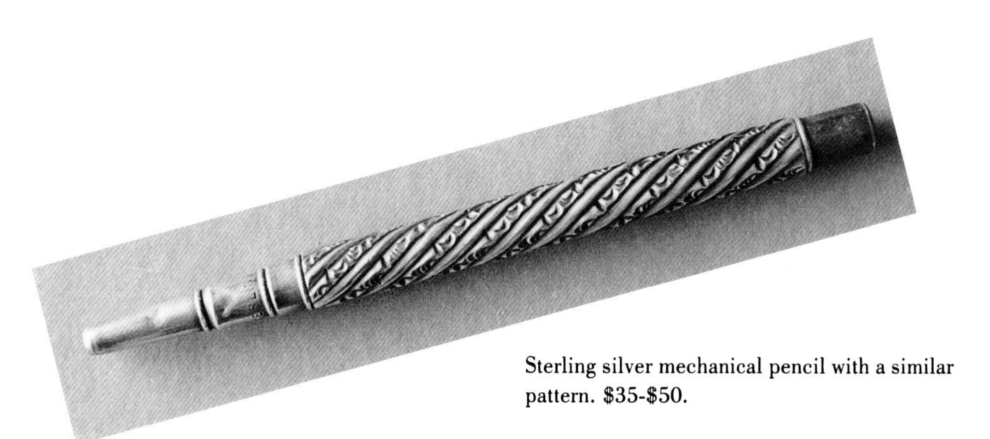

Sterling silver mechanical pencil with a similar pattern. $35-$50.

Ad showing Geo. S. Parker pens, one with a chased barrel.

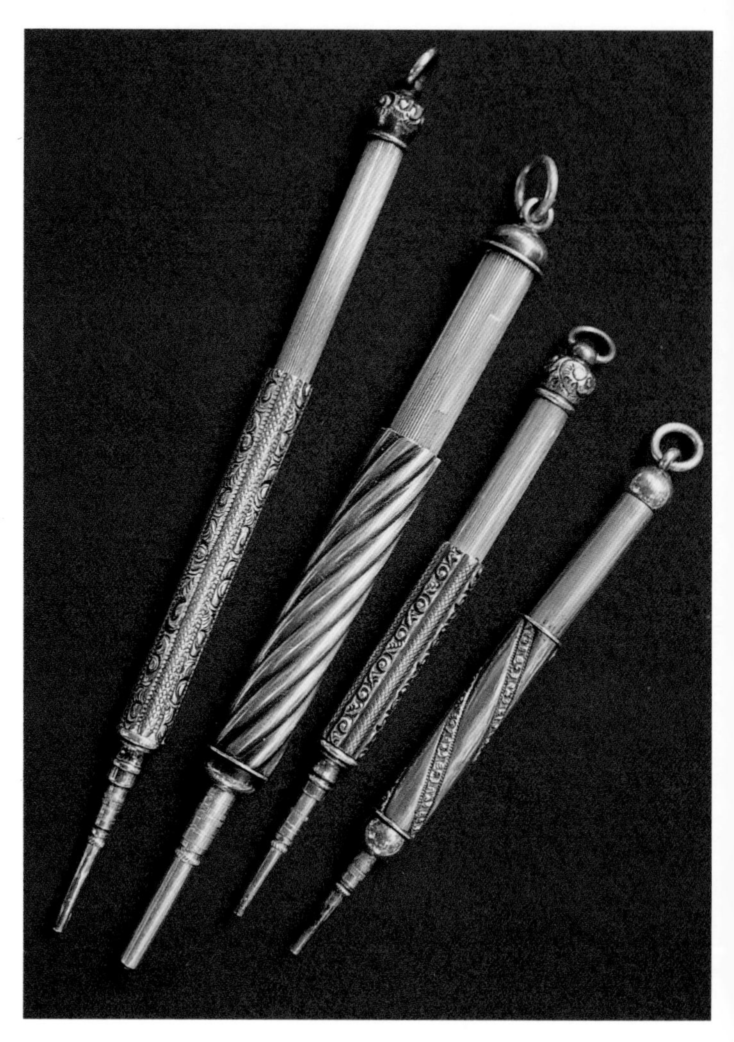

Sterling silver magic pencils with chased or cable twist barrels. $125-$150.

Magic pencil of the same era as the above ads, made of sterling silver with a chased barrel (marked "pat. apl. for"). $125-$150. *Photo Courtesy of Chris Odgers.*

Sterling silver mechanical pencil with a chased, cable twist barrel. $150-$175.

Casting

In *cire perdue* (lost wax casting) the original is sculpted of wax, either by carving or by manipulating it into the desired shape. A plaster mold is then made from this sculpture. After the wax is melted out of the mold, molten silver or gold is poured into it, creating a metal replica of the wax sculpture. This ancient process allows for considerable creativity, and the Victorians took full advantage of its potential. A great deal of detail can be captured and the only limit to the artist is that the design can't have any undercuts. Entire pencil-cases were cast, as well as sliders, finials, and metal pieces for applique.

Cast head made by Aikin, Lambert & Co. The original sculpture for this piece was made in wax, from which a plaster mold was taken. The metal was then cast in the mold, the contrasting color metal then plated over the base color, with the enamel eye being created last. 1.25" closed, 2.5" open, $500-$900.

Three examples of mechanical pencils employing niello decoration. Because the contrasting black is created by purposeful oxidation, these pieces should *not* be polished. $225-$250 for the larger drop action pencils, $135-$155 for the smaller, twist action pencil.

Niello

Niello is created when silver that is mixed with sulphur (which darkens the silver) is inlaid into common silver. Patterns are built by alternating colors of metal.

Finishes

Many different *finishes* were used on both gold and silver. A *Roman* finish on gold was matte, and was created by a quick acid bath. Lesser carat gold might be plated with a thin layer of higher carat gold, or perhaps a different color of metal (gold could be yellow, rose, or green) was applied to some areas. Silver was *oxidized* to make it black, to increase visual contrast between various parts. It was also *gilded* with a wash of gold. Sometimes it was left the color it was before the buffing stage, more white than gray.[1] Another popular finish was called *gun metal*, which represented the deep Payne's gray color of steel.

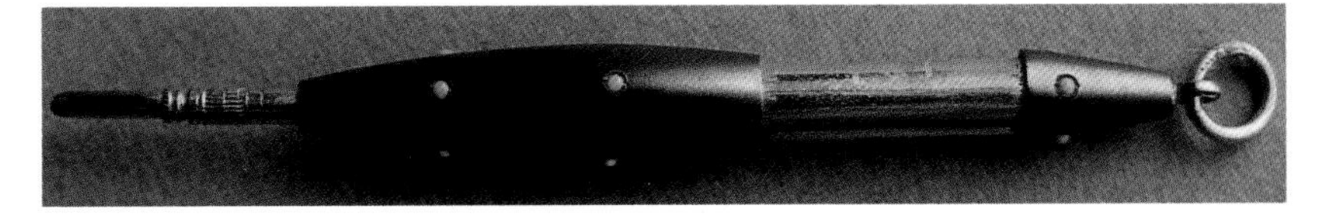

Torpedo shaped magic pencil in gunmetal finish, with rose gold, and set with turquoise. $125-$150.

Telescoping gold-filled pencil in detachable sheath made of gunmetal and gold, set with turquoise. $150-$175.

Torpedo shaped magic pencil in gunmetal and rose gold. $125-$150.

Patterns

In addition to reeded barrels, there were some that were *fluted, stamped,* and some that were left plain.

Several patterns (line and dot, barleycorn, cable twist, snail, chased barleycorn, etc.) were used by both pen and pencil makers. Because the designs used in both are so similar, it is speculated that some of the pieces were made either in the same shop or by the same maker. Notice the use of such patterns in the examples shown on pages 53 and 54, some on pens sold by L. E. Waterman and George Parker, and some on pencils sold by other makers.[2]

Gold-filled pencil with repoussé work and small, mine-cut diamonds.

Silver telescoping pen-pencil combination made by S. Mordan, in the line-and-dot pattern, with bloodstone set in the terminal. 2.5" closed, 6.5" open (not including the length of the stirrup bail), $225-$325.

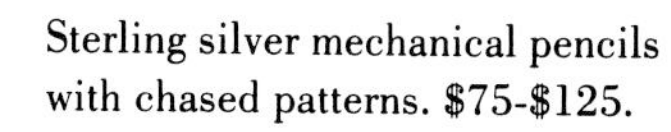

Sterling silver mechanical pencils with chased patterns. $75-$125.

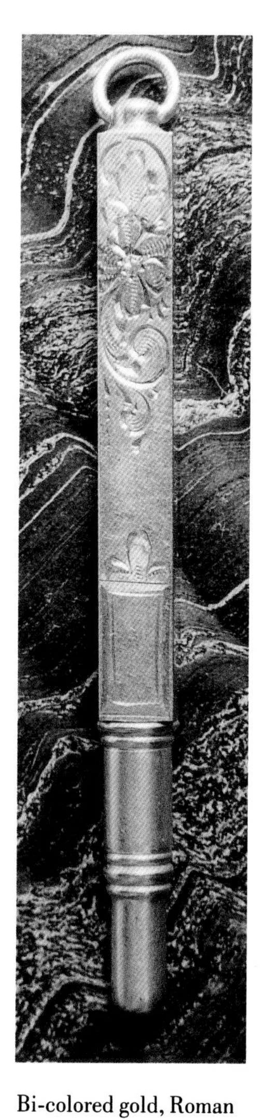

Bi-colored gold, Roman finish, square mechanical pencil. 2.5", $125-$150. The dull finish was achieved by immersing the gold briefly in acid. This finish should *not* be polished.

"Solid gold" magic pencil made by Aikin, Lambert in the barleycorn pattern. $200-$250.

Silver pencil, decorated with an unusual combination of repoussé and etching. $125-$150.

Aikin, Lambert pencil with engine-turned chevron pattern, 1905. $50-$75.

Gold magic pencil, engine-turned and mounted with small sapphires and diamonds. $225-$300.

Notes on Chapter Three

[1] Unfortunately, many of the pencils found in antique stores have been polished rather than having been left as found. In some cases the original finish has been destroyed altogether, with white finishes and oxidation having been polished away.

[2] Read the article entitled "Sterling Dip Pens" in *The PENnant*, August 1984, by Ed Fingerman.

CHAPTER FOUR
Evolution of the Mechanical Pencil

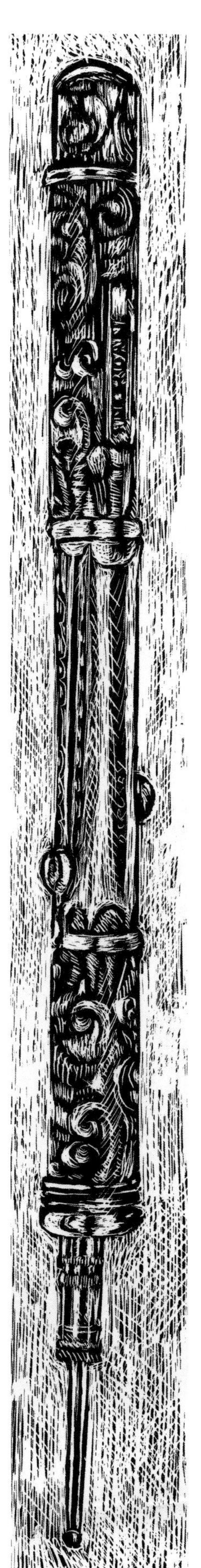

The first thing mentioned in many histories of writing implements is the wooden pencil holder illustrated by Konrad Gestner in his treatise on fossils (published in 1565). Not exactly a mechanical pencil and not quite a wooden pencil, it could be considered the forerunner of both. Essentially, it established a means of holding graphite while it was used to write or draw with, facilitating the process while keeping the hands clean.

Among the earliest pencils were *porte-crayons*. These were wooden or metal staffs with holders (sometimes on both ends) that held the graphite in place by pressure. Two half-round pieces of brass were attached to a *ferrule* (see Parts of a Mechanical Pencil, below), leaving enough space between the pieces for the graphite to be inserted. The relatively large lump of plumbago was held firmly either by the tension created by the two prongs or when a metal ring was slid up from the ferrule, pressing the two tines together around the graphite. These tools were often used by artists, but were somewhat difficult to use for fine detail because of the size of the point. The idea, however, of a lead held in place in this manner was adapted on a smaller scale and used in ivory handled architect's pencils. This concept also may have eventually given birth to the idea that a point could be pushed into and out of a metal tube.

Sampson Mordan and the Ever-Pointed Pencil

A pencil of this nature was patented in England in 1822, by one of the best known of all pencil-case makers, Sampson Mordan. Although he was perhaps not the first to create an ever-pointed or propelling pencil, his name is certainly associated with its invention. As one of the most successful of those who ever created these versatile writing instruments, "Mordan" became synonymous with pencil.

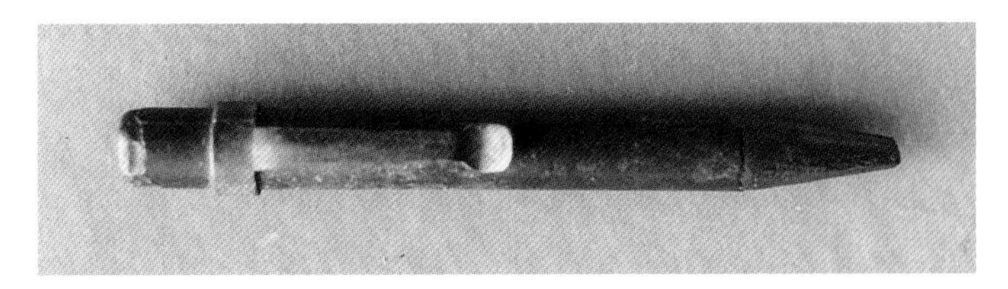

Unglamorous pencil: a piece of graphite is covered with a thick layer of lacquer, which protects the owner's hand from getting dirty when the pencil is in use, and a clip holds it in the pocket (which the lacquer also protected from being smudged, to some degree). Late 1800s, early 1900s, $2-$5.

It is believed that some form of mechanical pencil existed prior to the time when Sampson Mordan co-patented the ever-pointed pencil with John Isaac Hawkins on December 20, 1822. Some metal cased pencils are seen on chatelaines predating this patent.[1] Also, Mordan later lost a patent dispute because a variation in the mechanism he employed was said to have been in existence before his patent was granted.[2] However, Mordan was certainly one of those who best put the devise to use. (For more information about S. Mordan & Co., see Chapter 8.)

Parts of a Mechanical Pencil

The ever-pointed, propelling, screw, automatic, or magic pencil consisted of several parts. In order to understand how they worked, it's helpful to have a common language in describing the parts.

Barrel: the body of the pencil.

Bail: the metal link or loop used for hanging the pencil from a chain.

Collar: the piece that goes over the shell on the nozzle end of a magic pencil.

End Cap: the equivalent of the head or finial, but generally less ornate.

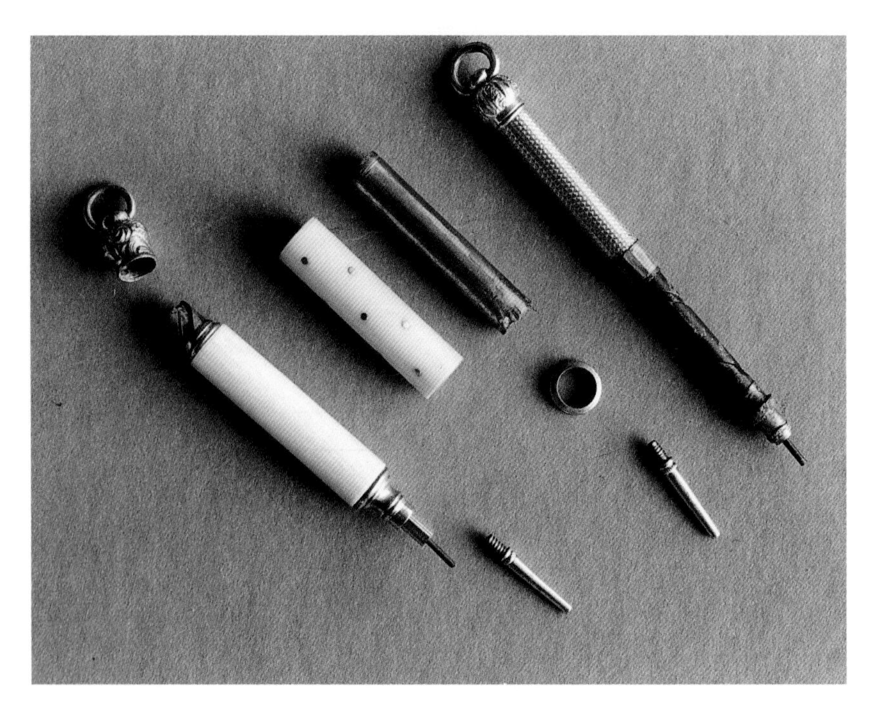

Two ivory magic pencils, disassembled to show the parts (the *action* or *mechanism*, the *pin*, the *nozzle*, *collar*, *case*, *shell*, *end cap*, and *bail.)*

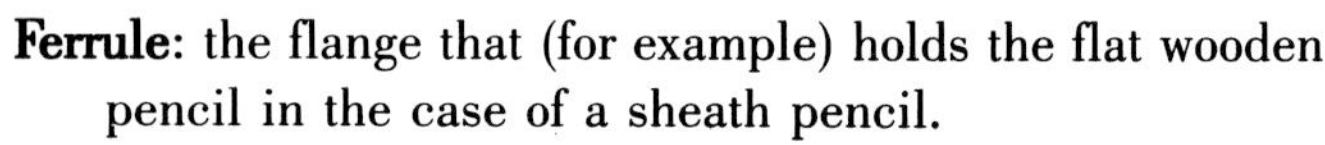

The same two pencils, reassembled to show how the parts fit together.

Ferrule: the flange that (for example) holds the flat wooden pencil in the case of a sheath pencil.

Head, Terminal, or **Finial**: the extremity opposite the writing end. It's often removable, usually by being unscrewed, and it usually covers a compartment for spare leads. Sometimes the terminal has a stone set in it (precious, semi-precious, or faux; some of the stones are inscribed and can be used as seals). Some of the earlier pencils have waffle seals in their finials.

Mechanism: the structure that allows the pencil to work. Also called the pencil action.

Nozzle: the part of the pencil holding the mechanism that holds the graphite in place.

Pen: some pencils also hold a nib, which is called a pen.

Pencil-case: the external parts of the pencil, including the barrel and terminal.

Pin: the piece within the barrel that pushes the graphite out. The nozzle fits over the pin.

Shell: a covering on the barrel which is independent of the barrel itself. It may be made of metal, ivory, porcelain, etc.

Slider button or slider pin: the knob or pin that is engaged in order to slide a pencil mechanism or cedar pencil out of the pencil-case. Some slider buttons are set with cabochon stones. Slider pins are much smaller, and are used on delicate figural pencils or telescoping pen-pencil combinations, where the pin has to be small enough to be withdrawn into one level which is telescoped into another.

Slider ring: the external ring that is used in conjunction with an internal pin to push the pencil in and out of the barrel. There are simple ring sliders, as well as more elaborate ones that are cast in a foliate relief or set with a stone.

Types of Mechanisms

Many mechanisms for propelling pencils were patented in the nineteenth century, and an in-depth study of these patents has yet to be explored. Generally speaking, however, mechanical pencils made during the 1800s and early 1900s fell into the following categories:

Automatic: a form of pencil that employs a mechanism.

Clutch: the clutch mechanism allowed a pencil to hold several different sizes of lead. After the lead was inserted into the end of the pencil, the top was turned, which "clutched" the lead in place.

Ever-pointed: so-called because the graphite in this type of pencil was small enough in diameter to not require sharpening as it was used. On account of the small size of the graphite, the holder (pencil-case) had to be designed so that the graphite would not move around while the pencil was being used.

Gravity or Drop: the graphite, either encased in a wooden pencil or held by a metal nozzle, dropped out of one end of the barrel when a button on the alternate end was pressed. (Mordan also made a "Presto" pencil, with the release button on the side.)

Magazine: a mechanism employed by Aikin, Lambert &

The *barrel* of this pencil has plain and hand-engraved alternating vertical panels. $75-$90.

Co. based on the principle that leads could be loaded into the top of the pencil (in a series of openings) and that they could fall into place as needed.

Magic: these mechanisms allowed the pencil to emerge from one end, and a handle to come out from the other end of the barrel, when one end was pulled. It was called "magic" because it didn't seem possible that such a small casing could hold so much. This type of mechanism was most often used in figural pencils, which further enhanced its mystery.

Novelty: a pencil in any unusual configuration, especially but not limited to figural.

Perpetual: the concept was based on the theory that a large supply of graphite was housed in the barrel which could be continuously fed out of the nozzle end (the holder opened to release, and closed to grasp the graphite by pushing on the opposite end and letting go when the graphite had fallen into place).

Plummet: usually a piece of metallic lead (not graphite) used for marking, which was sometimes attached to a staff of some type.

Porte-crayon: a holder for a piece of graphite, either with or without a wood casing. Some porte-crayons allowed for the pencil to be pushed into or out of the holder with a slider.

Propelling: the nozzle holding the graphite, and/or the graphite in the nozzle could be propelled out of the barrel of the pencil. Some, but not all, pencils were both propelling and repelling.

Sleeve: In the mechanism designed by John Hague and used throughout the nineteenth century, the pencil action is propelled by a sleeve of metal, as opposed to a slider in the form of a button or ring.

Slider: in this type of pencil, the action is pushed out of the barrel for use by a slider button, ring, or pin.

Screw: graphite was propelled out of the nozzle when either the nozzle or a part of the barrel was turned. Screw pencils with a joint in the center of the barrel were often used on chatelaines.

Telescopic or Collapsible: telescopic pencils could be condensed by pushing on the ends. These pencils often have a reversible end, so the point of the pencil or the nib is protected for travel.

Tablet or Desk: the length of graphite was determined by turning the nozzle. The staff, usually several inches long, remained stationary. Tablet (these pencils fit into notebooks or traveling writing cases) or desk pencils generally employed a screw mechanism. Desk pencils often had matching pens.

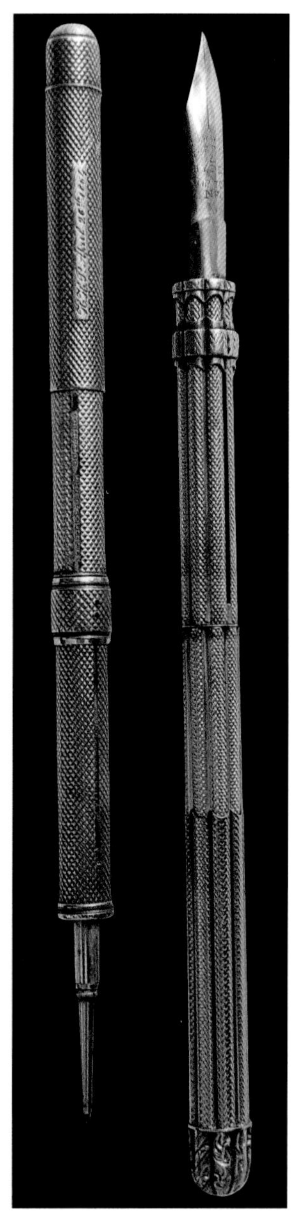

The barrel on the left is decorated with a barleycorn pattern, the barrel on the right is fluted barleycorn.

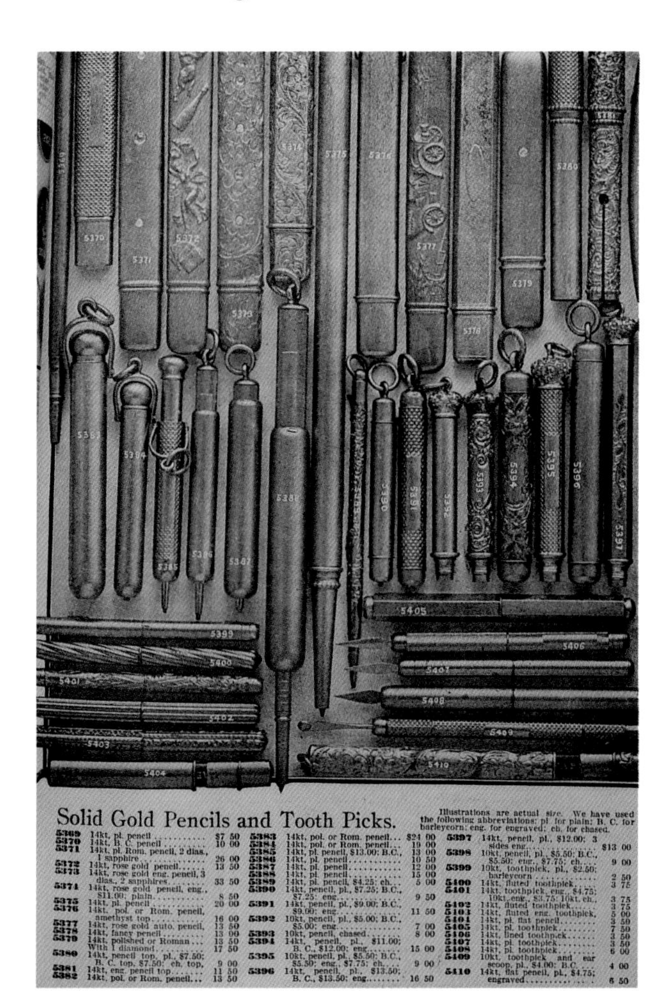

Solid Gold Pencils and Tooth Picks.

Illustrations are actual size. We have used the following abbreviations: pl. for plain; B. C. for barleycorn; eng. for engraved; ch. for chased.

5369	14kt, pl. pencil	$7 50
5370	14kt, B. C. pencil	10 00
5371	14kt, pl. Rom. pencil, 2 dias., 1 sapphire	26 00
5372	14kt, rose gold pencil	13 50
5373	14kt, rose gold eng. pencil, 3 dias., 2 sapphires	33 50
5374	14kt, rose gold pencil, eng., $11.00; plain	8 50
5375	14kt, pl. pencil	20 00
5376	14kt, pol. or Rom. pencil, amethyst top	16 00
5377	14kt, rose gold auto. pencil	13 50
5378	14kt, fancy pencil	13 00
5379	14kt, polished or Roman...	13 50
	With 1 diamond	17 50
5380	14kt, pencil top, pl., $7.50; B. C. top, $7.50; ch. top	9 00
5381	14kt, eng. pencil top	11 50
5382	14kt, pol. or Rom. pencil...	13 50
5383	14kt, pol. or Rom. pencil...	$24 00
5384	14kt, pol. or Rom. pencil...	19 00
5385	14kt, pl. pencil, $13.00; B.C.	13 00
5386	14kt, pl. pencil	10 50
5387	14kt, pl. pencil	12 00
5388	14kt, pl. pencil	15 00
5389	14kt, pl. pencil, $4.25; ch...	5 00
5390	14kt, pencil, pl., $7.25; B.C., $7.25; eng.	9 50
5391	14kt, pencil, pl., $9.00; B.C., $9.00; eng.	11 50
5392	10kt, pencil, pl., $5.00; B.C., $5.00; eng.	7 00
5393	10kt, pencil, chased	8 00
5394	14kt, pencil, pl., $11.00; B. C., $12.00; eng.	15 00
5395	10kt, pencil, pl., $5.50; B.C., $5.50; eng., $7.75; ch....	9 00
5396	14kt, pencil, pl., $13.50; B. C., $13.50; eng.	16 50
5397	14kt, pencil, pl., $12.00; 3 sides eng.	$13 00
5398	10kt, pencil, pl., $5.50; B.C., $5.50; eng., $7.75; ch....	9 00
5399	10kt, toothpick, pl., $2.50; barleycorn	2 50
5400	14kt, fluted toothpick	3 75
5401	14kt, toothpick, eng., $4.75; 10kt, eng., $3.75; 10kt, ch.	3 75
5402	14kt, fluted toothpick	3 75
5403	14kt, fluted eng. toothpick	5 00
5404	14kt, pl. flat pencil	3 50
5405	14kt, pl. toothpick	7 50
5406	14kt, lined toothpick	3 50
5407	14kt, pl. toothpick	3 50
5408	14kt, pl. toothpick	6 00
5409	10kt, toothpick and ear scoop, pl., $4.00; B.C.	4 00
5410	14kt, flat pencil, pl., $4.75; engraved	6 50

Page from S. Kind & Son catalog, 1911, illustrating several pencil styles and mechanisms.

In addition to the continued use of graphite of varying degrees of hardness, *indelible* pencils were also introduced in the 1850s. These pencils left a mark, which, supposedly, was permanent. Ironically, though one of the earliest uses discovered for India Rubber was erasing pencil marks, erasers were not commonly incorporated in mechanical pencils until well into the twentieth century.

How Mechanical Pencils Were Made

The earliest mechanical pencils were handmade by tradesmen. These individuals, generally young men, studied their craft as apprentices with master craftsmen. Some trained as jewelers or watchmakers, some as machinists and engineers (as had Sampson Mordan). Some, like Richard Cross, received their training in the pencil-case industry in England and immigrated to America with knowledge and skill. Joseph Dixon learned through experimentation. Edward Todd is known to have traveled to Europe prior to establishing his own company, perhaps to learn more about manufacturing pencil-cases. The major contributors to the industry were inventive and clever, finding many opportunities to explore the possibilities of perfecting their product and expanding their market. Huge numbers of patents were issued for a variety of advancements or improvements in propelling pencils during the nineteenth century (between 1820 and 1873, more than 160 patents were listed pertaining to mechanical pencils).

A variety of *finials* (also called *heads* or *terminals*) on silver pencils.

Pencils were first sold either by the merchant himself or by travelling salesmen, who carried sample cases of their wares. As time went on, however, the travelling salesman was gradually replaced by mail-order business created through catalogue sales. Although catalogues existed prior to the 1800s, their use increased dramatically from the 1860s on. Catalogues, using written descriptions, drawings, or photographs of what the company offered, reached far more households than the travelling salesman ever could, and certainly increased the numbers of pencils purchased by the middle classes. Tiffany's described pencils in their Blue Books as early as 1861.

Early shops would have had few employees, but as the century advanced, workshops employed more and more help. (Even as industrialization became more prevalent, the final stages of finishing the pencils were still done by hand.) An article written for an 1879 issue of *Scientific American* describes how pencils were made in the Faber gold pen and pencil factory:

> "One of the modern pencil cases, which is extended by simply pulling one end, is a marvel of compactness. Some seven or eight pieces slide one over the other. The portion drawn out carries a spirally slotted tube which engages a pin projecting from another spirally slotted tube, and revolves the tube so that it moves the lead-carrying portion of the pencil outward. The extreme end of the spiral slot of the inner tube ends in a straight or circumferential slot, which receives the pin projecting from the lead-carrying device, and prevents the latter from moving backward when pressure is exerted on the pencil point.
>
> "The various sizes of tubes required in the manufacture of pen and pencil cases are made in the pencil department. The blanks are first cut from the sheet and bent roughly into semicircular form by hammering into a grooved block. They are then drawn through a plate to bring their edges together, when they are ready to be soldered. For the internal brass tubes silver solder is used. It is applied in a finely divided state along the seam together with a little liquid borax. The soldering is accomplished by moving the tube lengthwise in a trough formed of thin firebrick under a huge roaring

Two styles of terminals on gold pencils, top: cast and hand-engraved, bottom: cast.

blowpipe flame, which is directed into the trough. The flame is urged by the blast from a bellows, and the tube becomes hotter until the particles of silver solder melt and look like little globules of mercury, an instant more and the melted solder runs into the seam, and the operation is complete. Gold soldering is quite similar, the only difference being that the gold is applied in a thin strip instead of a powder, the strip being drawn into the seam into the tube.

"After soldering, the tubes are cleaned and drawn down to the required size on a draw bench. Most of the tubes are drawn upon a mandrel to ensure equality in the internal and external diameter of the different tubes of the same nominal size.

"The tubes are cut into different lengths for different purposes by a circular saw, having a gauge for regulating the lengths. The spiral slots are formed in the internal tubes of the "magic" pencil by a very ingenious and simple device, which consists simply of a tubular guide placed diagonally across the edge of the saw, the angle formed with the side of the saw corresponding to the pitch of the spiral to be cut. The tube being inserted in the guide and brought in contact with the edge of the saw has a short diagonal slit cut in it, and it is now pushed forward and at the same time allowed to turn, when a slit will be cut, having a true pitch from end to end.

"The several operations in pen and pencil case making are carried forward by workmen who have acquired skill by long practice, and who, under the guidance of an able superintendent, make and assemble the parts rapidly. Each workman has a special piece, which he makes carefully and perfectly, so that when all the parts are brought together there is no difficulty. All of the pieces work together smoothly. The tubes forming the outer case are drawn in plain corrugated dies, and are ornamented by chasing, engraving, or knurling.

"It would be futile to attempt to describe in detail the different operations in pen and pencil case making in an article of this character, as the great variety of ways in which they are made would require an entire volume to properly describe them... " *(Scientific American,* November 15, 1879)

Identifying Marks

Many of the mechanical pencils produced in the 1800s are unmarked. Although a system of marking metal (that was at least 925 parts silver per thousand and 75 parts copper) as sterling silver had been in existence in Great Britain for several hundred years, many small pieces, perhaps made for export, were not marked. However, if a pencil is hallmarked, it's fairly easy to determine the year in which it was made (see page 84). A row of tiny impressions made by four or five stamps would indicate several things. Firstly, if it were sterling silver a lion passant would usually be stamped at the beginning of the row. Secondly, the town it was assayed in would be indicated (for example, the leopard's head was used to symbolize that the piece was made in London). Thirdly, there would be a maker's mark, and, finally, there would be a letter indicating the

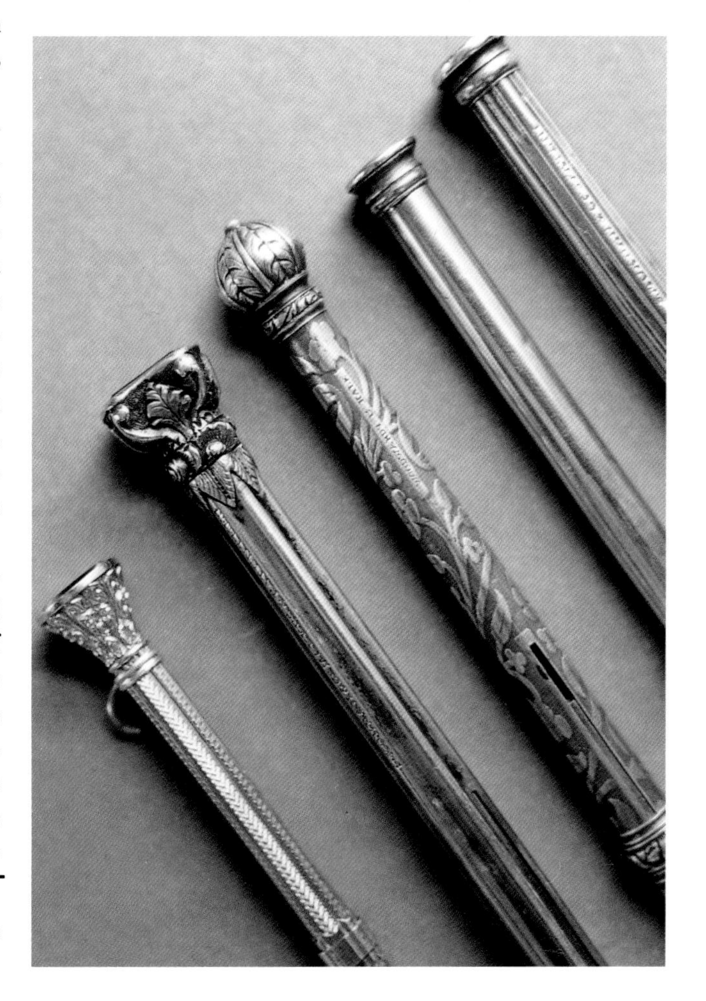

The earlier style of terminal is shown on the two pieces at the upper right.

year the piece was made. There are many books that show precisely how to interpret these marks, and it is useful to have access to one of these books. With such information, one can determine precisely when a pencil was made. Another form of marking is called the British registry, and was used from 1842 till 1883. In this system, a diamond shape has the letters *Rd* in the center, surrounded by codes that indicate what the piece is made of, as well as the day, month, and year. The designation "800" is for silver that has a larger content of alloys than 925 does. *German silver*, or *nickel silver*, looks much like the metal it imitates, but contains no actual silver.

Some American pencils are marked "silver" or "sterling silver." Sometimes, the mark is followed by a letter or a symbol. An "acorn" would indicate that the pencil was made by W. S. Hicks. The letters **A** or **E** would represent Aikin Lambert or Eagle, respectively. An **X** usually stands for A. T. Cross. Surprisingly, pencils made of gold are often not marked, although sometimes one finds a pencil marked as "solid gold" or one that is stamped with a number (14, 18, etc.) followed by *ct* or *kt*.

Marks representing particular manufacturers are listed in the chapter on pencil-case makers.

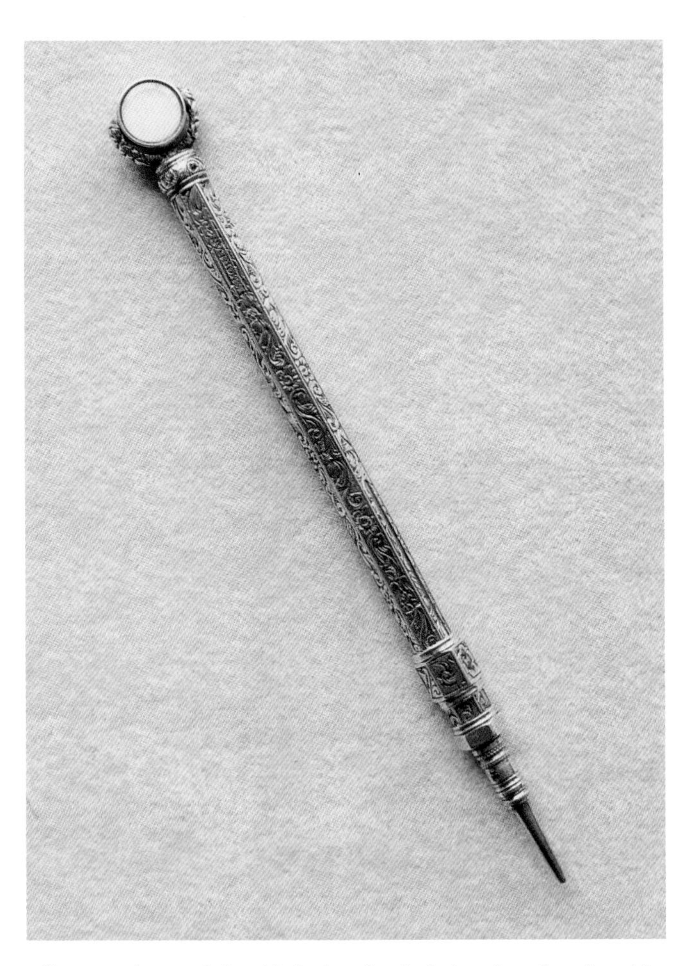

Same as photo on left, with the bezel swiveled to show the other side.

This finial employs a swivel, which allows two stones to be set in this pencil. Made by S. Mordan & Co., hand-engraved silver, hexagonal, slider mechanism, $250-$325.

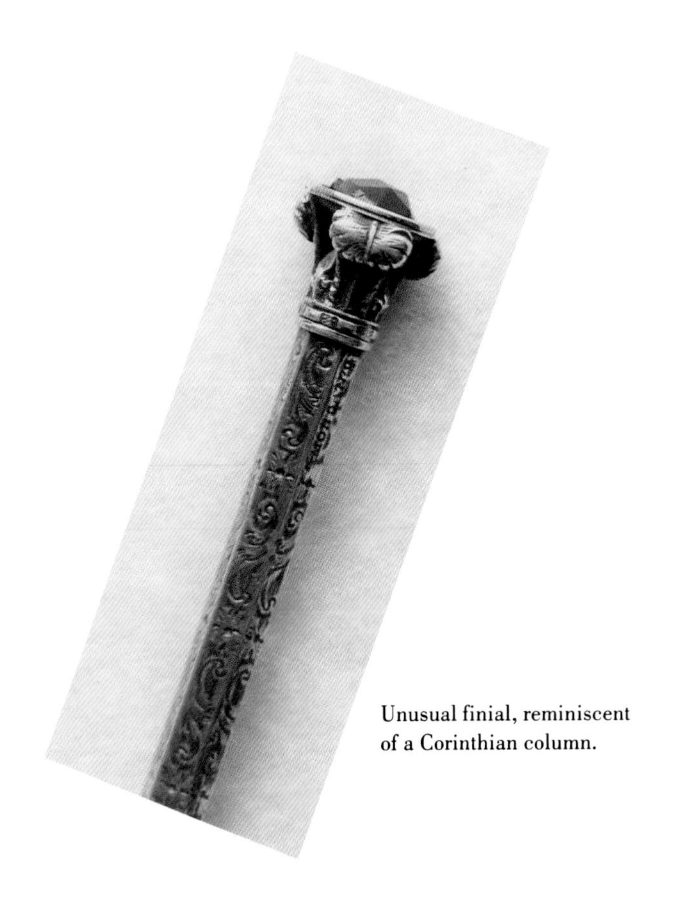

Unusual finial, reminiscent of a Corinthian column.

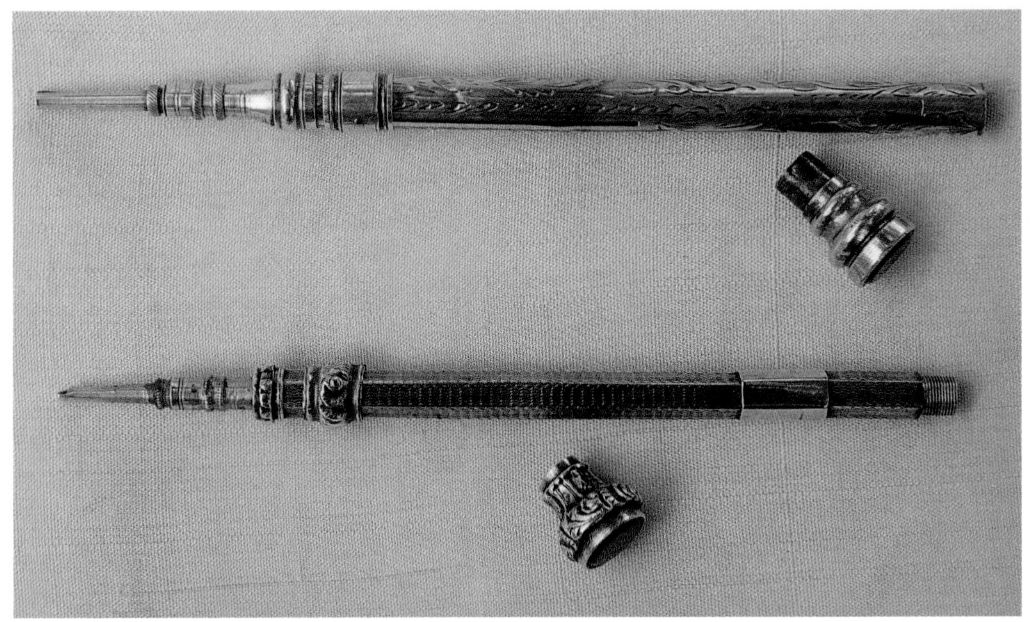

Normally, the head screws into the barrel, as in hexagonal, engine-turned pencil on the bottom. In the silver, engraved pencil on the top, however, the head or terminal is friction fit into the barrel. $75-$100.

Many different stones, some real, some faux, were bezel set into the finials. Among the semi-precious stones that were used in the nineteenth century were agate, bloodstone, carnelian, lapis lazuli,citrine, and amethyst.

The *nozzle* holds the graphite. Most of these nozzles have knurled ridges.

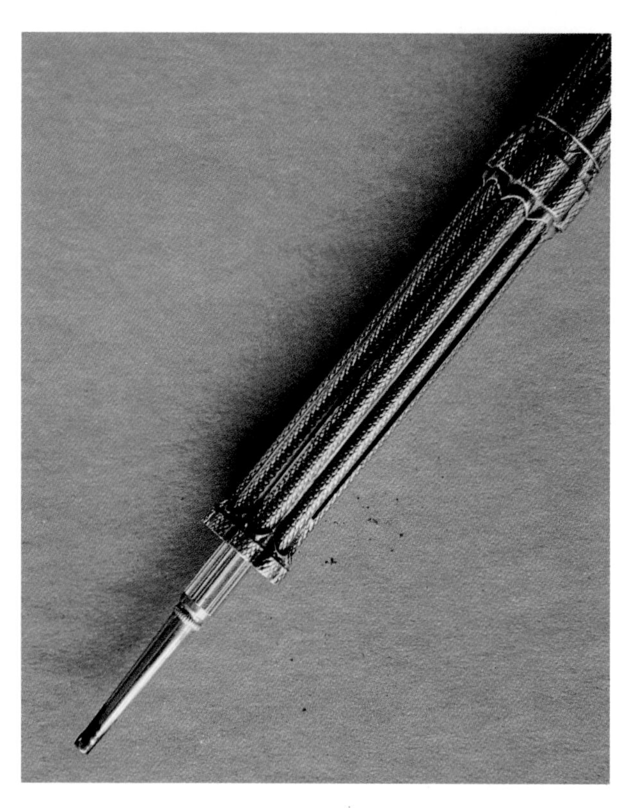

Detail showing the nozzle of the mechanical pencil as it appears when it is twisted (or screwed) out of the barrel of a Mabie patent pencil.

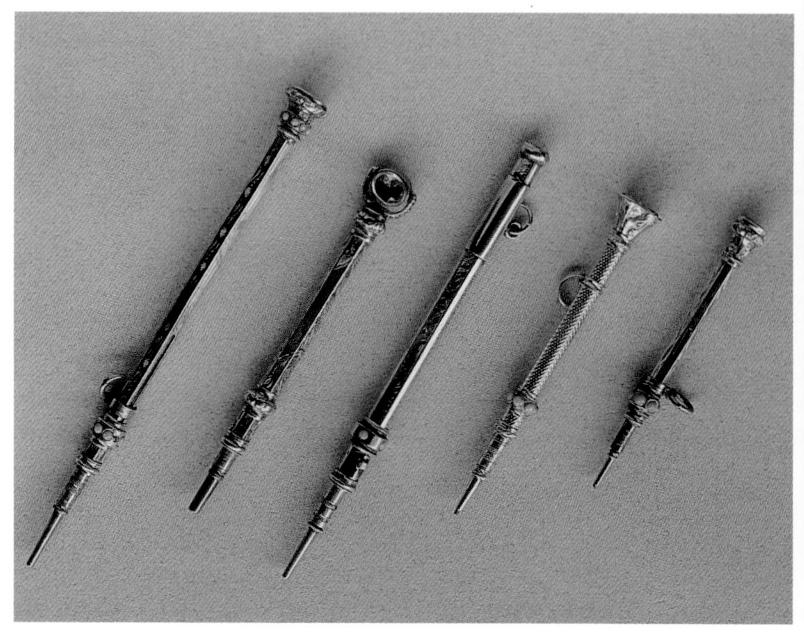

A selection of slide pencils, four having turquoise set in the *slider rings* (and one with a revolving terminal), $125-$225. *Photo Courtesy of Chris Odgers.*

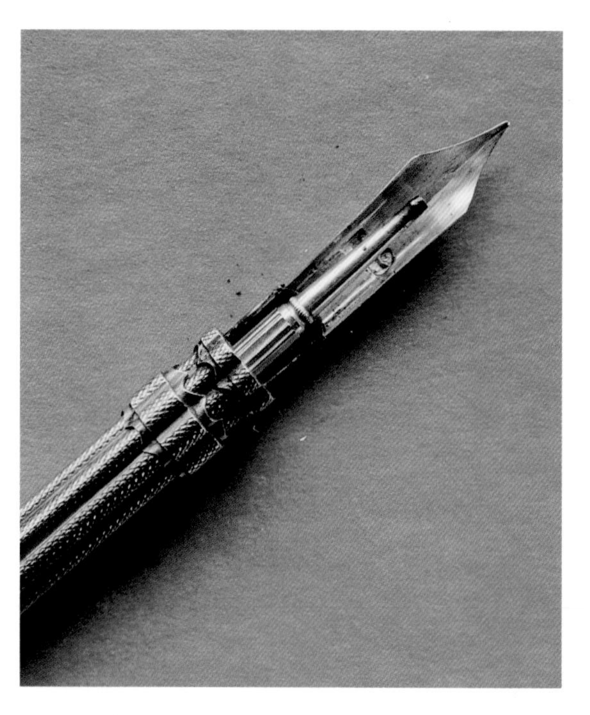

Detail of the above pen-pencil combination, showing the nib from behind. The nib is slid into its working position with a slider ring, the pencil mechanism twists out from a joint in the center of the barrel.

Gold *porte-crayon* with an engraved *slider button*. *Slide actions* were used in both porte-crayons (metal holders with cedar pencils) and mechanical pencils with metal nozzles.

Two porte-crayons with slide actions: one in silver, made by S. Mordan (hallmarked), $75-$125; the longer of the two made in gold, also by S. Mordan (note the arrow on the escutcheon plate), $175-$275.

Porte-crayons. The wooden pencil slides in and out of the pencil-case when the slider ring is manipulated along the incised groove. Bottom: silver niello porte-crayon, $150-$175. Top: silver cable-twist porte-crayon, $100-$125. *Photo Courtesy of Chris Odgers.*

Lightweight, gold porte-crayon, with loop for a watch chain. Notice the knife-sharpened cedar pencil. 2", $50-$75.

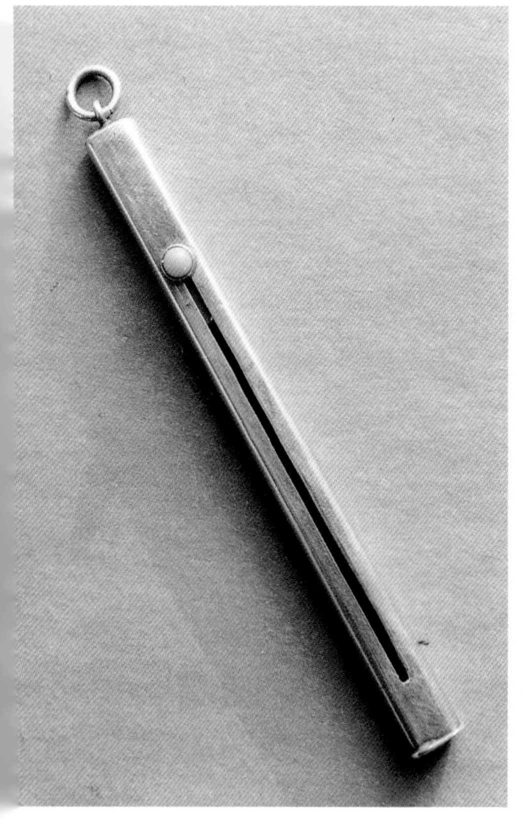

Sterling silver porte-crayon with turquoise set into the button slider, 3" long. A rectangular wooden pencil would slide in and out of this pencil-case, which is hallmarked W. M., Chester, 1836. $75-$100.

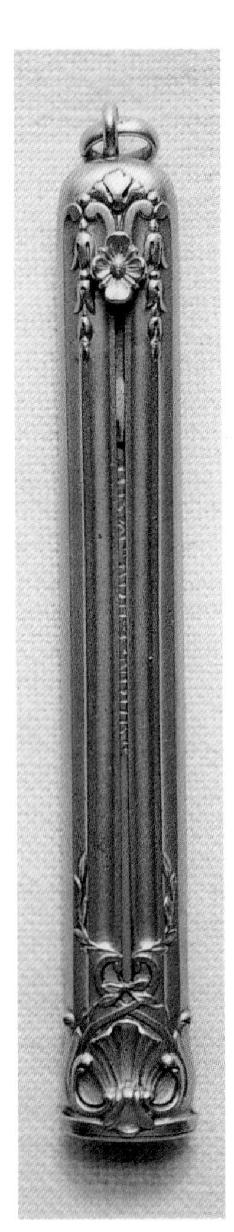

Nineteenth century, flat porte-crayon, made of gold, Roman finish, with a cast button slider shaped like a flower. $135-$150.

Hexagonal, hand-engraved slide pencil in silver. The slider ring is also hexagonal, and a bloodstone is set into the terminal. Two tiny metal pins hold the slider ring to the pencil mechanism, which turns to both propel and repel the graphite from the nozzle. 4.5" open, $75-$125.

Large rolled gold, slide pencil, hexagonal, with engine turning. $150-$175.

Eleven rolled gold, slide pencils with faceted glass "stones" set in the terminals. $75-$125.

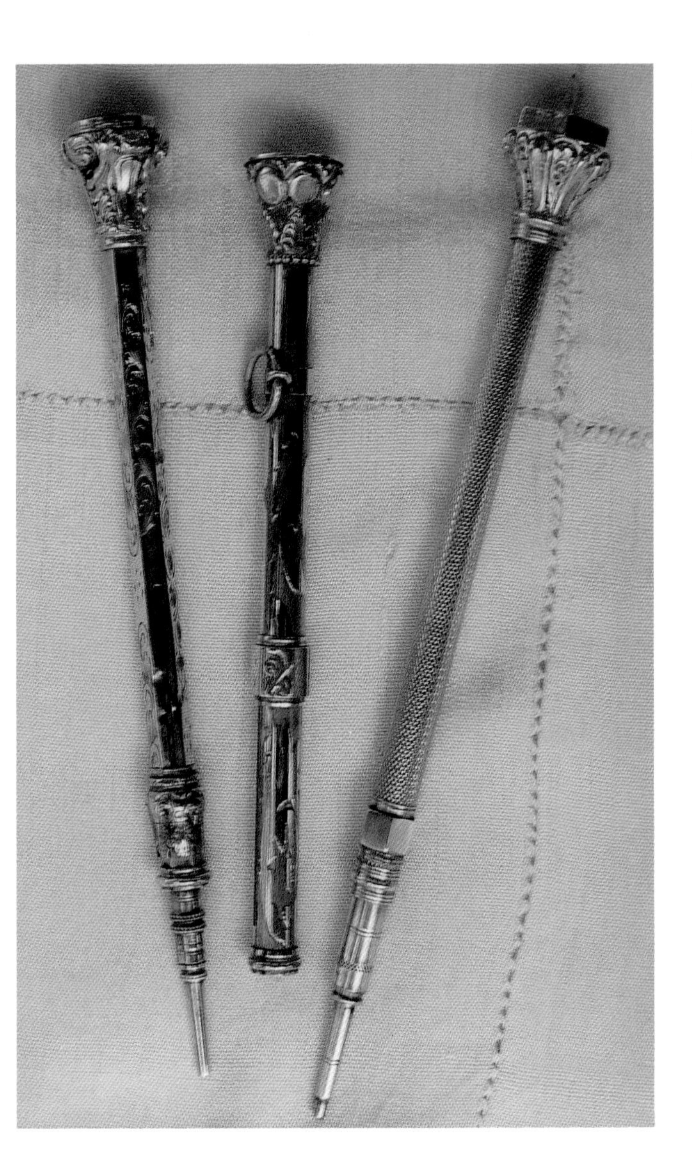

Three typical rolled gold slide pencils from the mid-1800s. Of all the mechanical pencils, these are perhaps the easiest to find, but very little advertising for them has yet to be found. $75-$125.

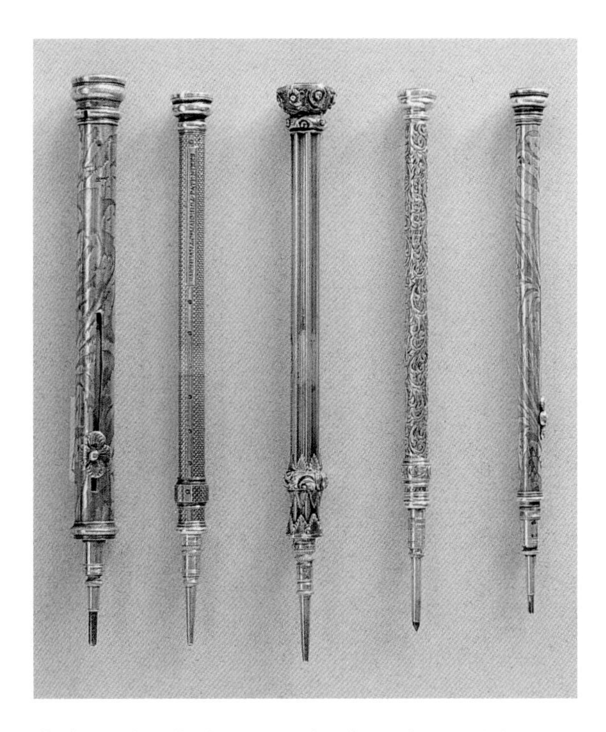

Left to right (all silver): pencil with cast button slider in the shape of a flower ($125-$150); pencil made by "S. Mordan & Co. Makers & Patentees" in the line and dot pattern (early, $175-$250); pencil with an ornate, gothic style slider, unmarked but similar to those made by Mordan ($150-$175); engraved pencil ($125-$175); pencil made by Butler ($135-$175). *Photo Courtesy of Chris Odgers.*

Tiny slide pencils, all less than two inches long, closed, and three inches, opened. The pencil on the bottom left is rolled gold, the others are silver, many have chatelaine or watch-fob rings. Notice the size comparison between the pencils and a #2 Waterman nib. $50-$75.

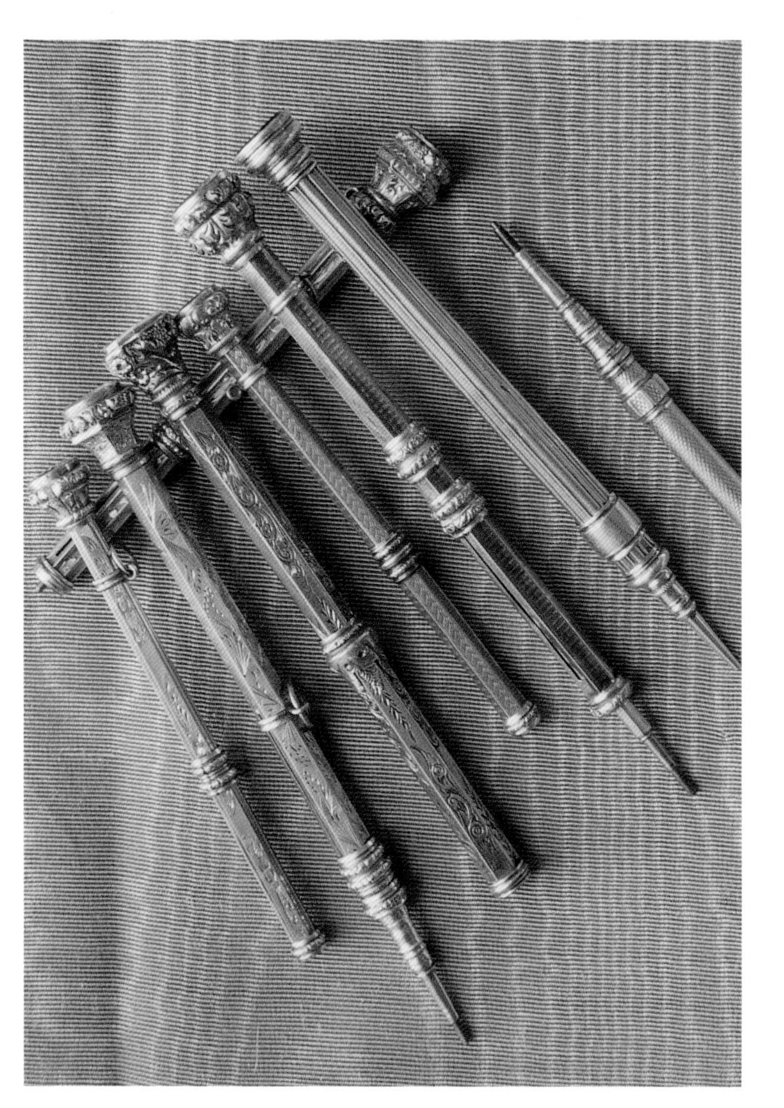

Typical slide pencils, circa 1860.

This beautiful gold pencil (hallmarked) has a slider that follows the twist of the metal, rather than going straight up and down. There is a shield shaped stone set in the terminal, with individual tubes to hold the lead in the storage chamber. 3.5" closed, 4.75" open, $250-$350.

Advertising pencil, "Hotel Vendig, Philadelphia." The sleeve mechanism patented by John Hague is utilized here. $25-$35.

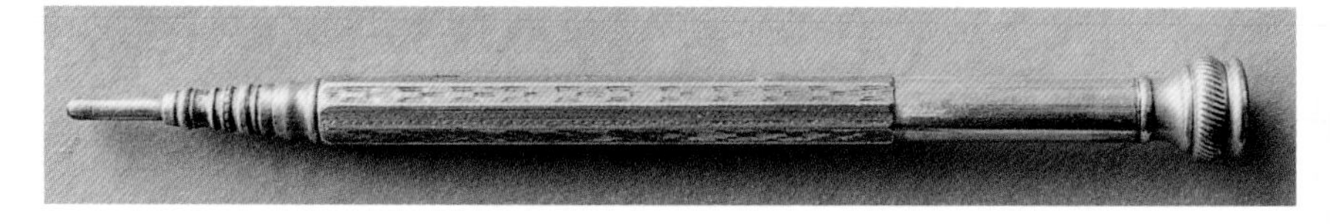

Another pencil with the sleeve mechanism, shown open. 4.5", unmarked, made of brass, $35-$45.

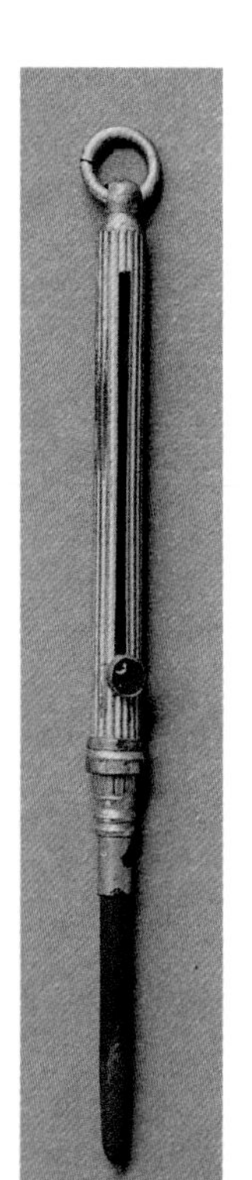

Gold filled pencil with button slider. The red pencil has been fully extended by pushing the red glass button forward. $15-$25.

Same as photo on left, but shown closed.

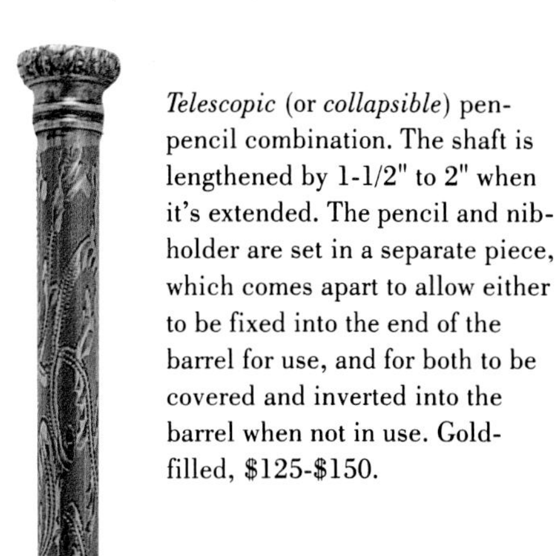

Telescopic (or *collapsible*) pen-pencil combination. The shaft is lengthened by 1-1/2" to 2" when it's extended. The pencil and nib-holder are set in a separate piece, which comes apart to allow either to be fixed into the end of the barrel for use, and for both to be covered and inverted into the barrel when not in use. Gold-filled, $125-$150.

Gold pencil with barleycorn pattern, shown closed.

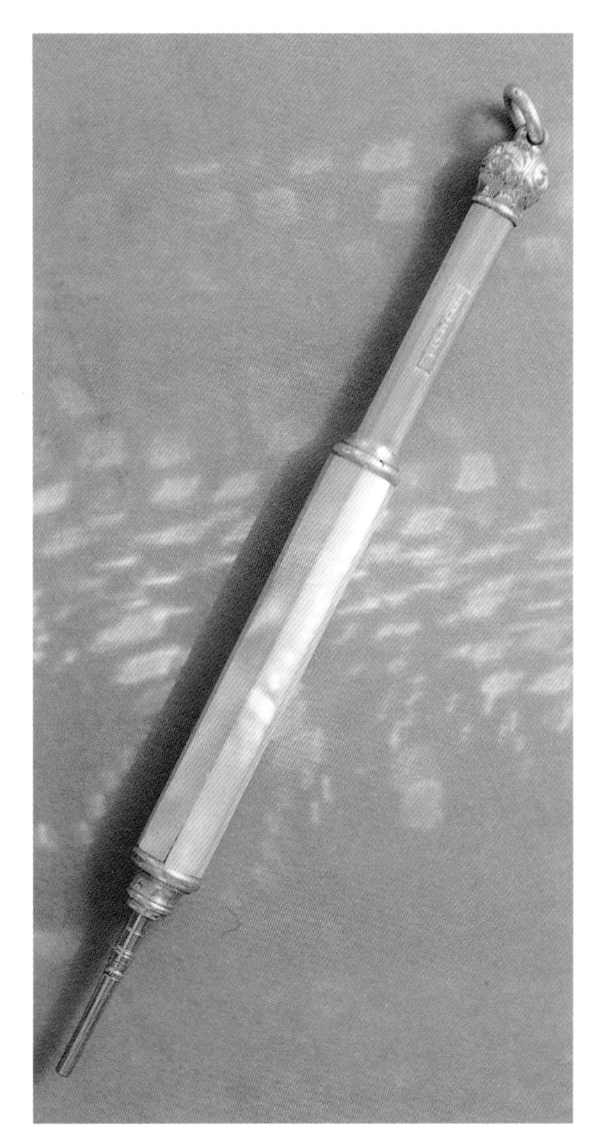

Magic pencil with mother-of-pearl, marked Edward Todd & Co., patented December 1871. 2.5" closed, 4.5" open. When the terminal end of a magic pencil is pulled, both that end and the nozzle end protrude. $125-$150.

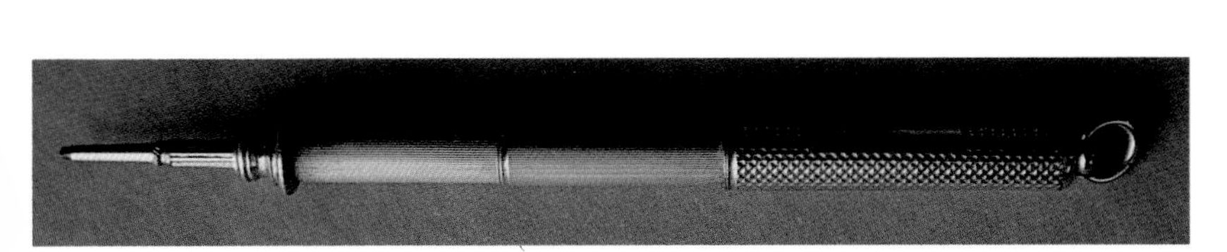

The same pencil, its length more than doubled when it is telescoped out. $150-$200.

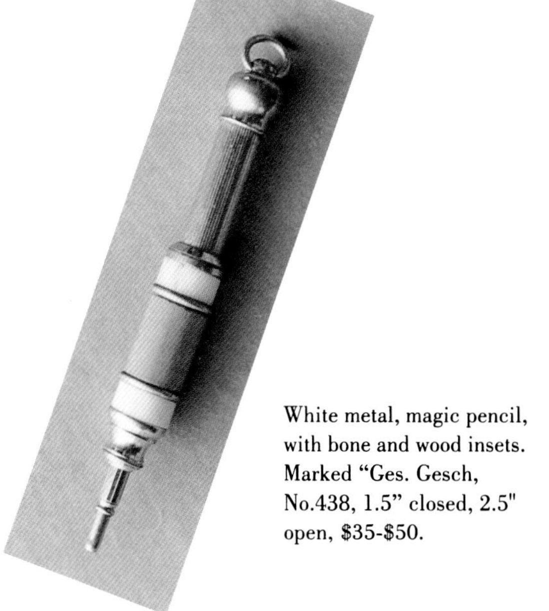

White metal, magic pencil, with bone and wood insets. Marked "Ges. Gesch, No.438, 1.5" closed, 2.5" open, $35-$50.

Sterling silver and rose gold magic pencil. Marked "L. W. F. & Co.," (Leroy Fairchild & Co.), 2" closed, 4" open, $125-$150.

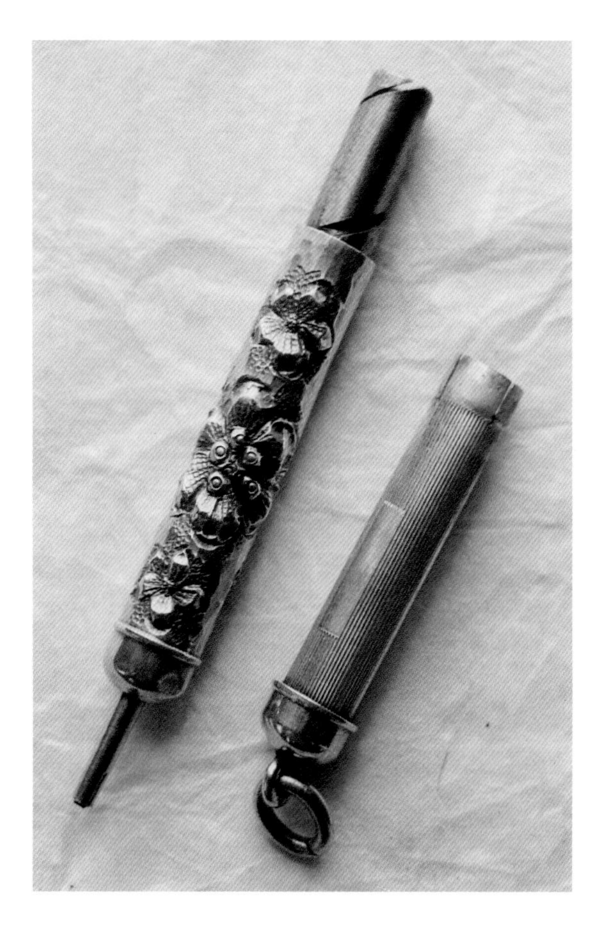

Silver pencil shown with internal mechanism visible.

Two 10ct gold pencils, one showing the internal mechanism.

Pearl pencil made by Aikin, Lambert. Note how the pin rides in the slot, allowing the action to work.

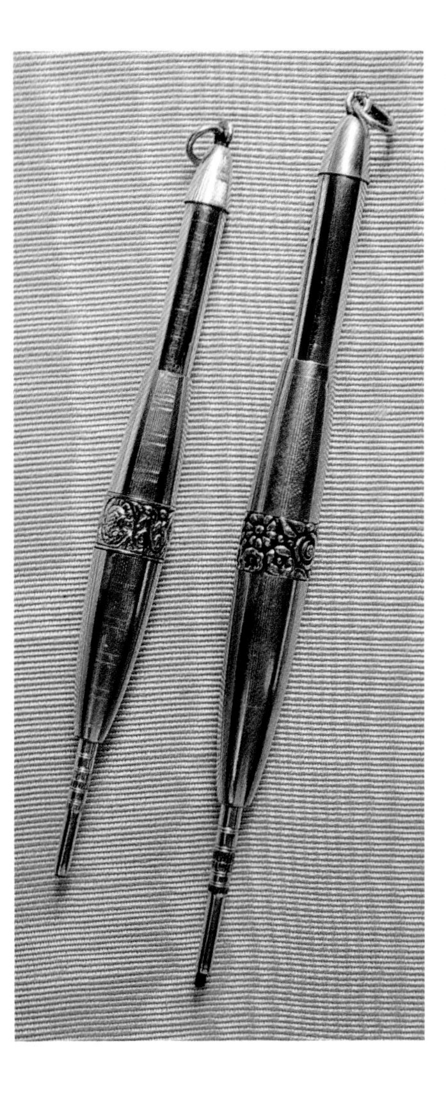

The same two, torpedo shaped gold pencils. 2.5" closed, 4.5" open, $125-$150.

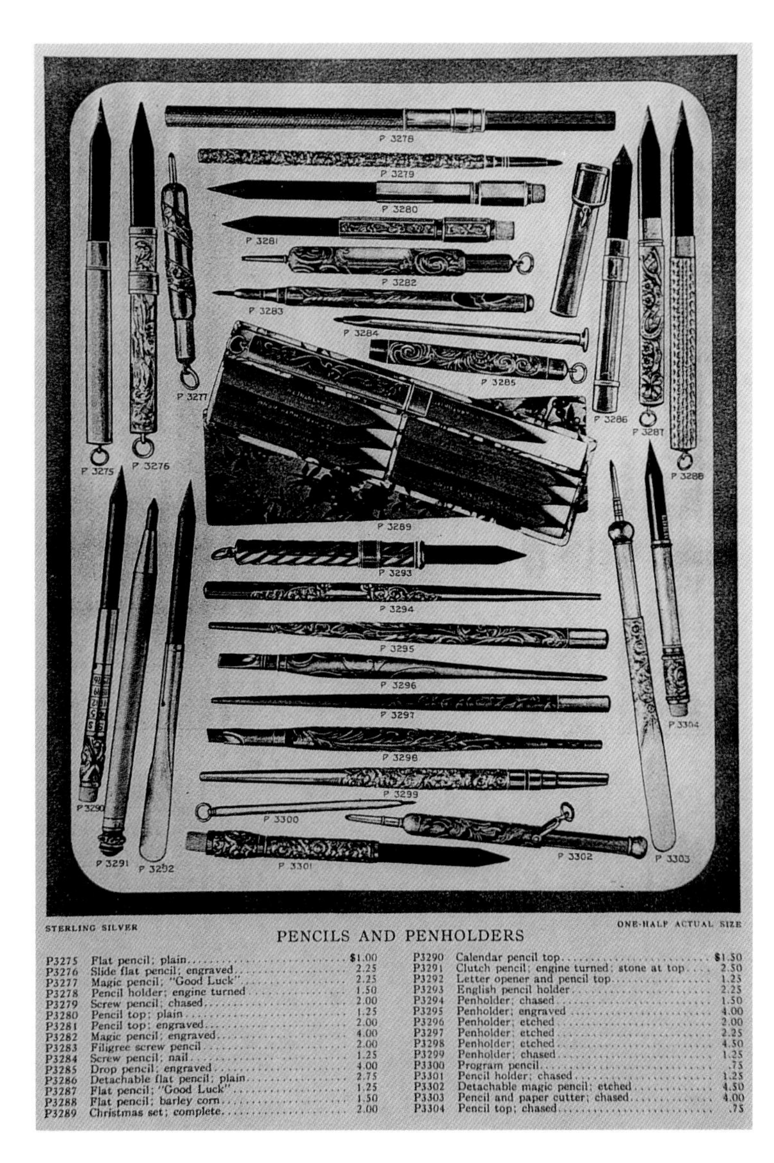

STERLING SILVER

ONE-HALF ACTUAL SIZE

PENCILS AND PENHOLDERS

P3275	Flat pencil; plain	$1.00
P3276	Slide flat pencil; engraved	2.25
P3277	Magic pencil; "Good Luck"	2.25
P3278	Pencil holder; engine turned	1.50
P3279	Screw pencil; chased	2.00
P3280	Pencil top; plain	1.25
P3281	Pencil top; engraved	2.00
P3282	Magic pencil; engraved	4.00
P3283	Filigree screw pencil	2.00
P3284	Screw pencil; nail	1.25
P3285	Drop pencil; engraved	4.00
P3286	Detachable flat pencil; plain	2.75
P3287	Flat pencil; "Good Luck"	1.25
P3288	Flat pencil; barley corn	1.50
P3289	Christmas set; complete	2.00
P3290	Calendar pencil top	$1.50
P3291	Clutch pencil; engine turned; stone at top	2.50
P3292	Letter opener and pencil top	1.25
P3293	English pencil holder	2.25
P3294	Penholder; chased	1.50
P3295	Penholder; engraved	4.00
P3296	Penholder; etched	2.00
P3297	Penholder; etched	2.25
P3298	Penholder; etched	4.50
P3299	Penholder; chased	1.25
P3300	Program pencil	.75
P3301	Pencil holder; chased	1.25
P3302	Detachable magic pencil; etched	4.50
P3303	Pencil and paper cutter; chased	4.00
P3304	Pencil top; chased	.75

Page from a catalog, circa 1900, showing *flat*, *sheath*, or *vest-pocket* pencils (see Chapter 7), *magic* pencils, *screw* or *twist* pencils, pencil *extenders* or pencil (see Chapter 7), *drop* or *gravity* pencils, *clutch* pencils, and *detachable* pencils.

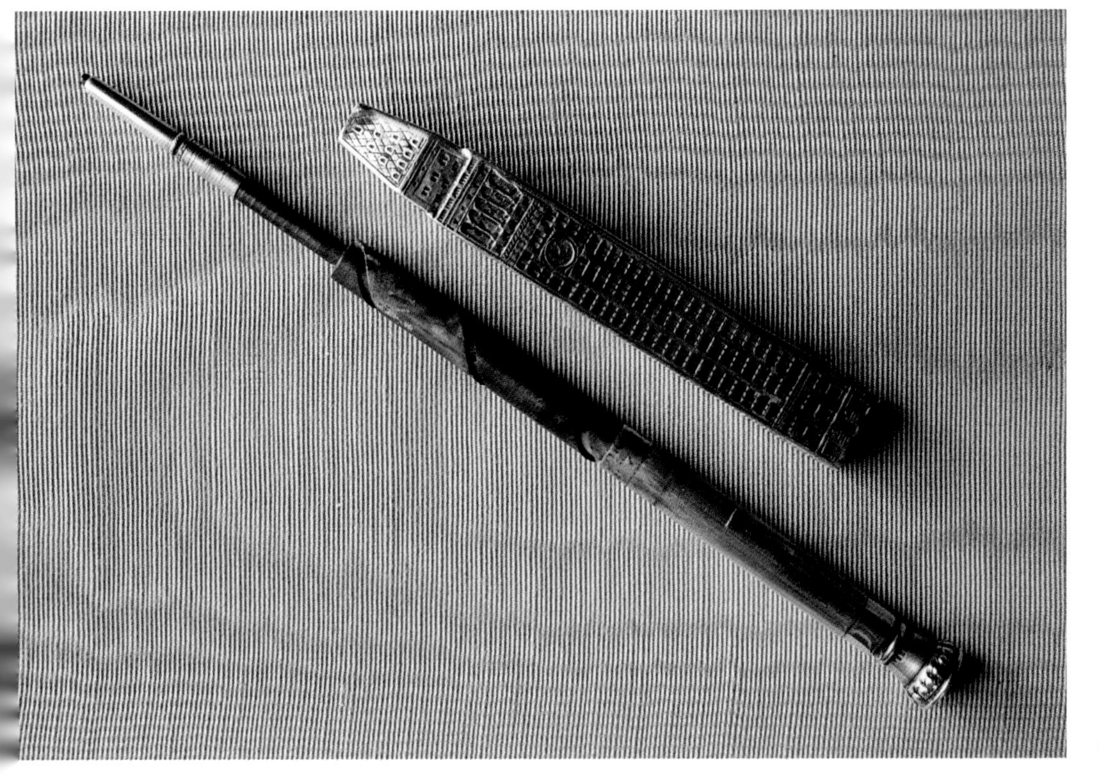

Metropolitan Life Building, figural pencil made by Tiffany's.

"Pharoah" pencil, with a portion removed to show the pencil action.

Gold "bullet" pencil, with seed pearls and detachable sheath. $300-$425.

Two silver telescopic pencils, with removable sheaths. The pencil shown at the left comes out of the top of the sheath, whereas the pencil on the right is removed for use from an opening at the bottom of its sheath. Top: $125-$135. Bottom (heavily chased sterling silver): $175-$200.

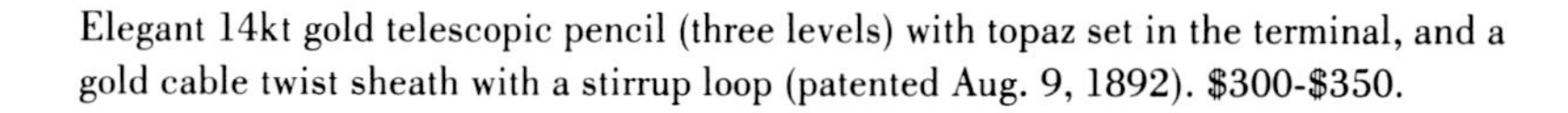

Elegant 14kt gold telescopic pencil (three levels) with topaz set in the terminal, and a gold cable twist sheath with a stirrup loop (patented Aug. 9, 1892). $300-$350.

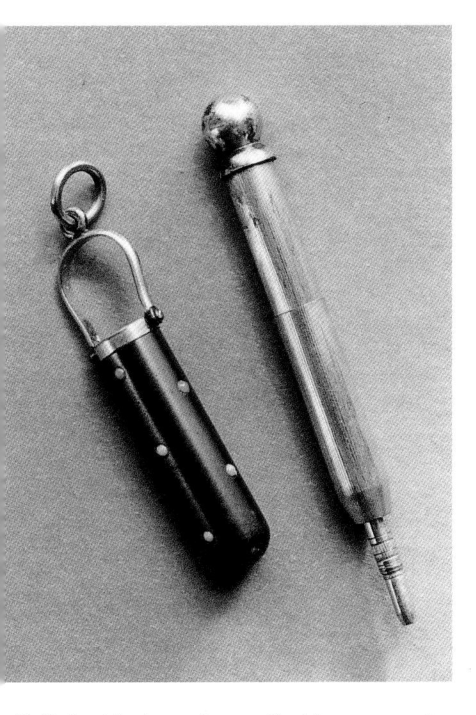

Rolled gold telescopic pencil with a gunmetal sheath, mounted with turquoise, $150-$175.

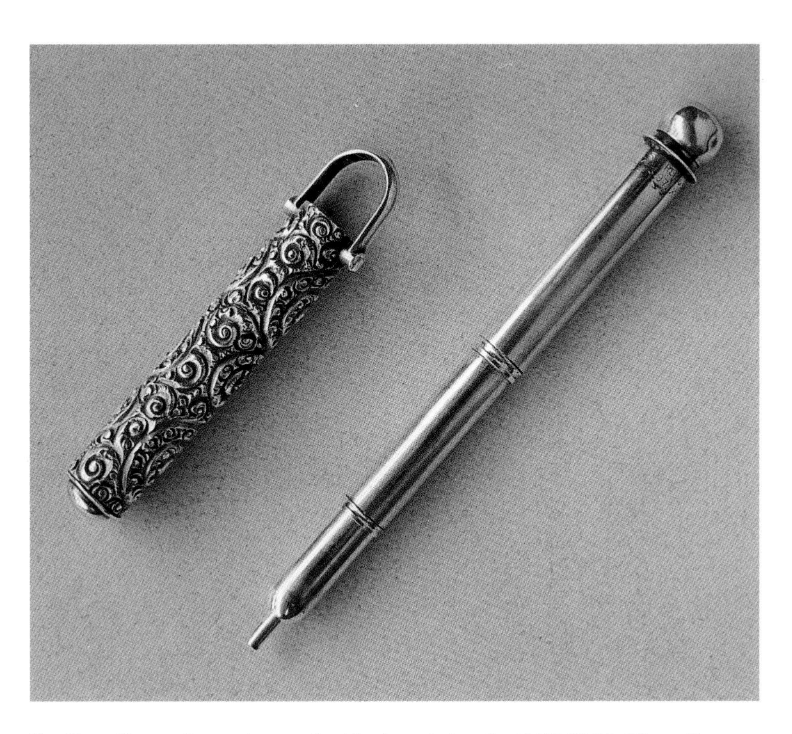

Sterling silver telescopic pencil with chased sheath, $175-$200. *Photo Courtesy of Chris Odgers.*

Collapsible pencil made by Mabie, Todd & Co., in its chased cable sheath. Marked 14kt, $300-$425.

Gravity or *drop* pencil, shown with the pencil that screws into the threads inside the metal case. (See Chapter 6.)

Mordan made a "Presto" pencil, which had the release button on the side rather than on the end. Circa 1898, shown closed.

Same pencil shown open. $300-$350.

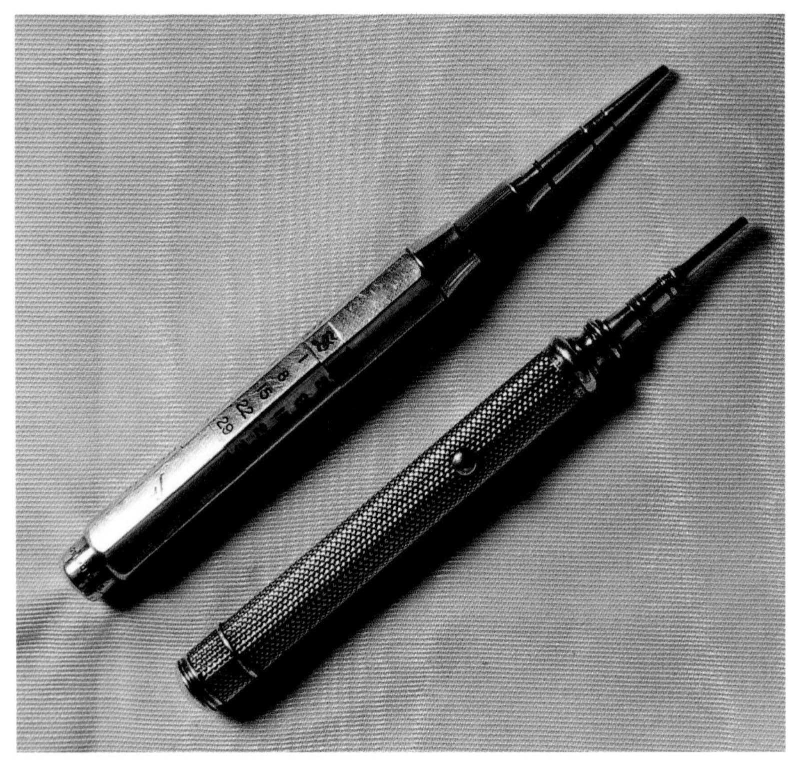

Comparison between the more common style of drop action pencil with the "Presto" pencil. Both pencils were made by S. Mordan & Co.

Cedar pencils used in gravity pencils.

Cedar pencils for drop action. Pencils made by S. Mordan & Co. $25-$40.

Three niello pencils, two with drop actions. $125-$250.

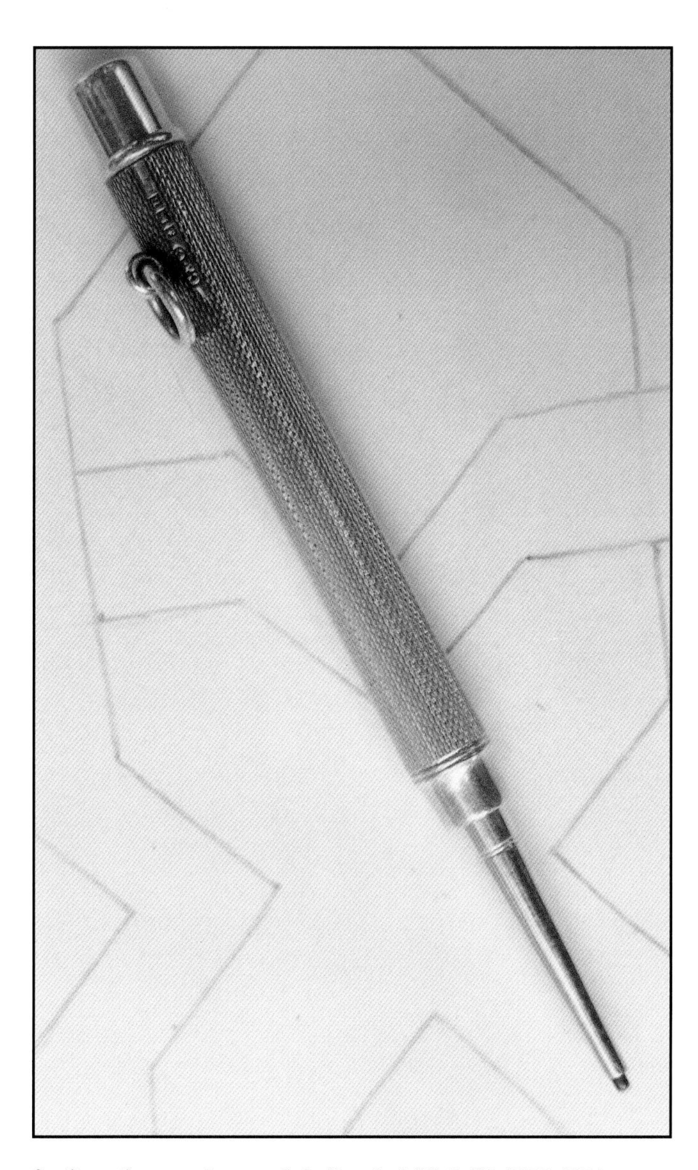

Sterling silver gravity pencil, hallmarked “C & C”, $250-$300.

Silver drop action pencil, barleycorn pattern, made by S. Mordan & Co., $325-$425.

Gold gravity pencil with metal nozzle and a perpetual calendar. $300-$500.

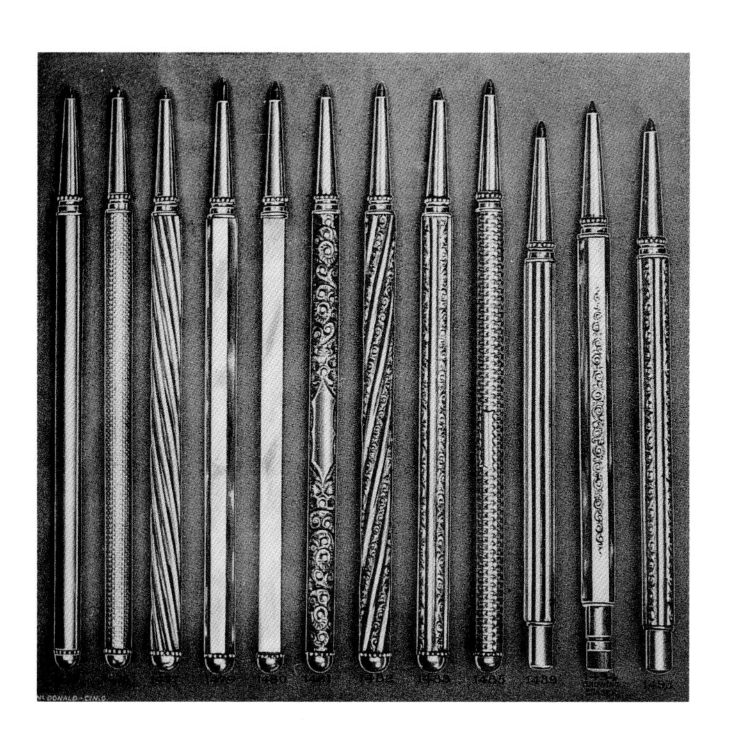

Examples of tablet pencils, also called desk pencils.

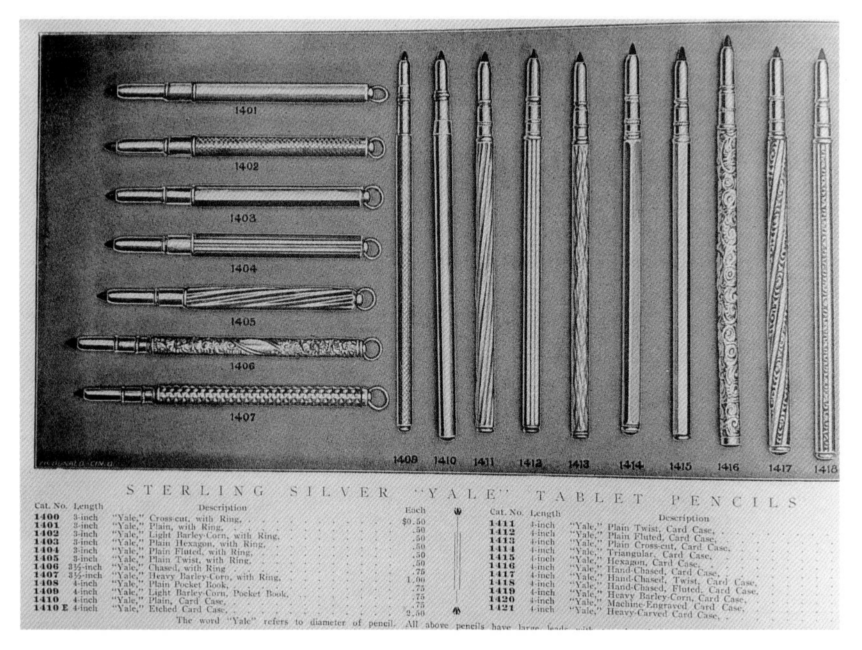

1401 1402 1403 1404 1405 1406 1407

1409 1410 1411 1412 1413 1414 1415 1416 1417 1418

STERLING SILVER "YALE" TABLET PENCILS

Cat. No.	Length	Description	Each
1400	3-inch	"Yale," Cross-cut, with Ring,	$0.50
1401	3-inch	"Yale," Plain, with Ring,	.50
1402	3-inch	"Yale," Light Barley-Corn, with Ring,	.50
1403	3-inch	"Yale," Plain Hexagon, with Ring,	.50
1404	3-inch	"Yale," Plain Fluted, with Ring,	.50
1405	3-inch	"Yale," Plain Twist, with Ring,	.50
1406	3½-inch	"Yale," Chased, with Ring	.75
1407	3½-inch	"Yale," Heavy Barley-Corn, with Ring,	1.00
1408	4-inch	"Yale," Plain Pocket Book,	.75
1409	4-inch	"Yale," Light Barley-Corn, Pocket Book,	.75
1410	4-inch	"Yale," Plain, Card Case,	.75
1410 E	4-inch	"Yale," Etched Card Case,	2.50

Cat. No.	Length	Description
1411	4-inch	"Yale," Plain Twist, Card Case,
1412	4-inch	"Yale," Plain Fluted, Card Case,
1413	4-inch	"Yale," Plain Cross-cut, Card Case,
1414	4-inch	"Yale," Triangular, Card Case,
1415	4-inch	"Yale," Hexagon, Card Case,
1416	4-inch	"Yale," Hand-Chased, Card Case,
1417	4-inch	"Yale," Hand-Chased, Twist, Card Case,
1418	4-inch	"Yale," Hand-Chased, Fluted, Card Case,
1419	4-inch	"Yale," Heavy Barley-Corn, Card Case,
1420	4-inch	"Yale," Machine-Engraved Card Case,
1421	4-inch	"Yale," Heavy-Carved Card Case,

The word "Yale" refers to diameter of pencil. All above pencils have large leads

Page from a John Holland catalog, showing tablet pencils.

Chased silver tablet pencil. $125-$175.

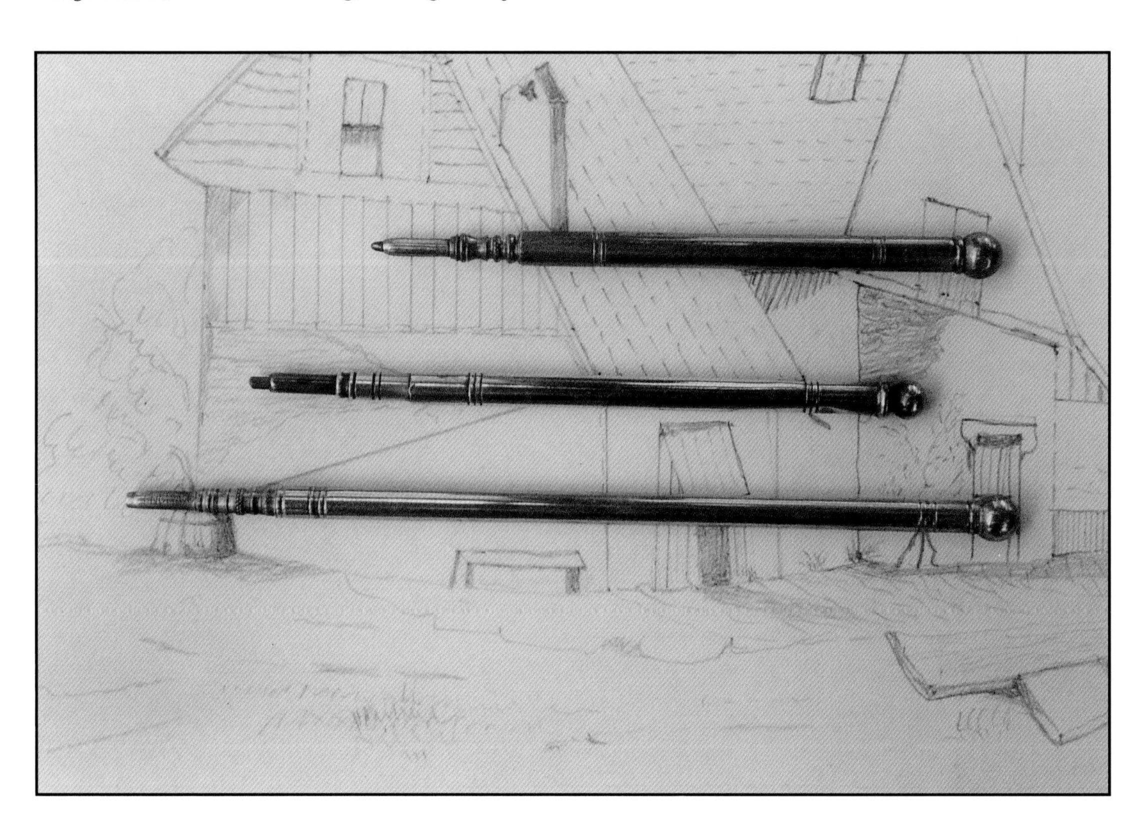

Brass *tablet* pencils, 3.5" to 5", $35-$45.

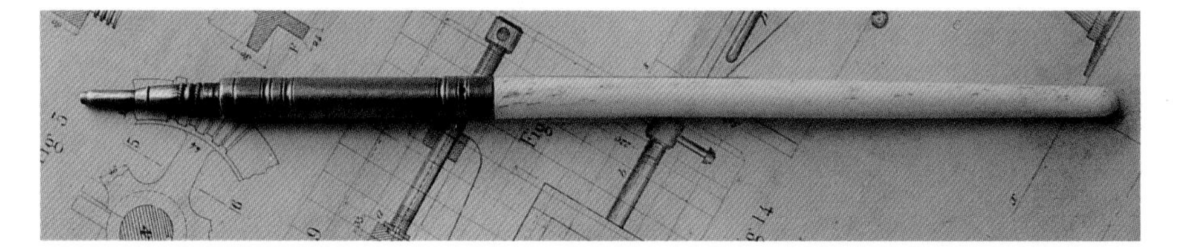

Bone and brass tablet pencil. These pencils were sometimes sold as components of stationery sets, made with leather bindings and stocked with writing paper. There are frequently loops inside the packets to hold the pen and pencil. $50-$75.

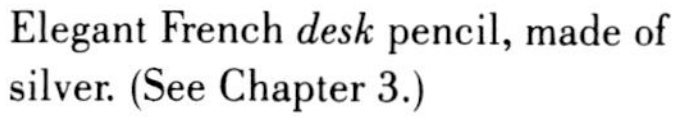

Elegant French *desk* pencil, made of silver. (See Chapter 3.)

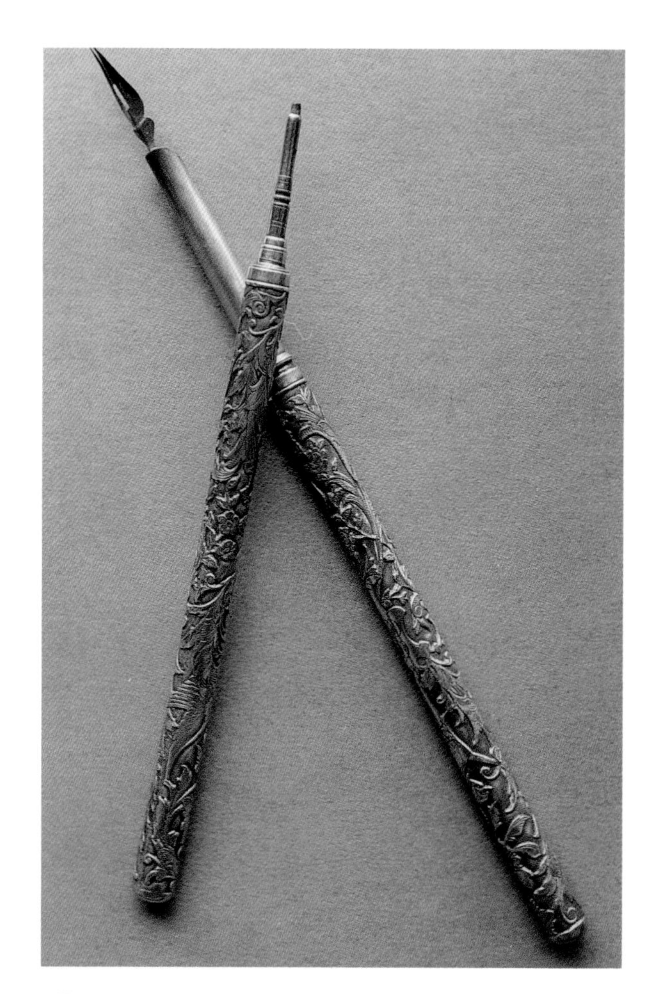

Etched floral design on a silver pen and pencil set. $250-$300.

Pen and pencil set in cable twist silver. $225-$275 set.

The Season's Biggest Novelty in The World's Biggest War

GENUINE U. S. BULLET PENCIL

What it is:

A brand new idea—GENUINE U. S. Army bullet cartridges BEAUTIFULLY finished and hand chased, and converted into a useful, practical novelty, with a REAL pencil in place of the powder.

Bound to be THE SEASON'S BIGGEST SELLER, because of the universal interest in the great European war. And because this is THE FIRST TIME the American people could obtain a genuine U. S. rifle bullet in ENDURING NOVELTY FORM.

SPLENDIDLY DISPLAYED, a dozen each on a striking card that holds the eye, tells the story, and MAKES THE SALE. ALL live jobbers and salesmen should write at once for samples and prices. There's a GOOD profit.

The war won't last FOREVER. You can't beat a bullet. GET IN WHILE THE GOING IS GOOD!

United States Army cartridge

ORDER FROM

NEW ENGLAND NEWS COMPANY

BOSTON MASS., U. S. A.

Where it's Worn:

Wherever a novel pencil looks well and will come in handy. For instance:

ON WALDEMARS

Hitch on the opposite end of the chain from the watch. Weight will keep chain taut.

VEST CHAINS

Keeps the unfastened end of the chain from falling out of the pocket. Very convenient for making memoranda.

EN SAUTOIR

Attach at end of neck chain. Mighty handy for the woman who "never can keep a pencil."

IN HAND BAGS

Not easily lost or mislaid because of the weight. Keep one in every hand bag.

Novelty pencils. Novelty pencils were made from the mid-1800s on. Figural pencils were often described as "novelties," as were pencils made of unusual materials or those that served more than one purpose (see Chapter 6). Here, pencils made from United States Army cartridges are advertised as "The Season's Biggest Novelty in the World's Biggest War."

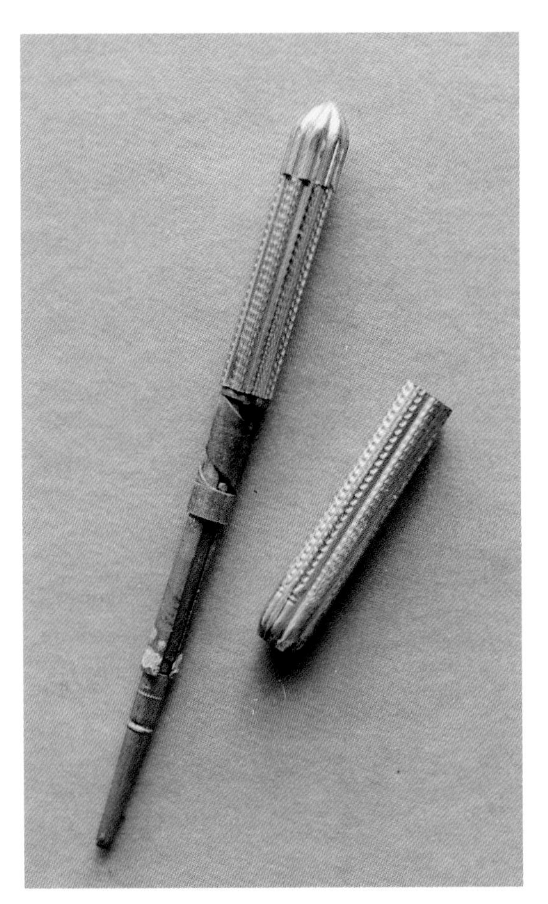

Screw or *twist* pencils were engaged by turning one end of the pencil. The mechanism is visible here, as is the original lead solder that held the case to the nozzle.

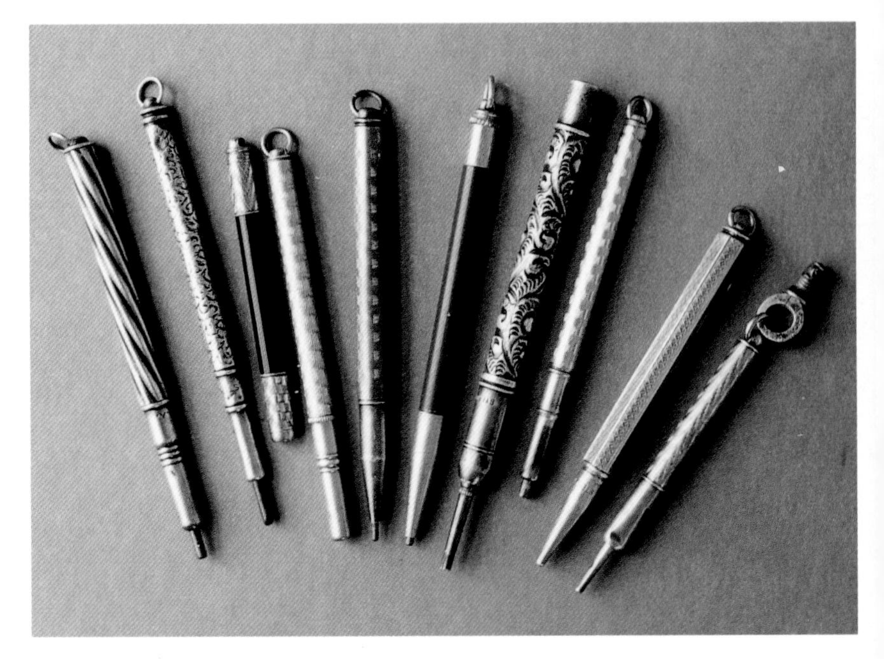

A selection of pencils, which utilize the screw or twist mechanism. $35-$60.

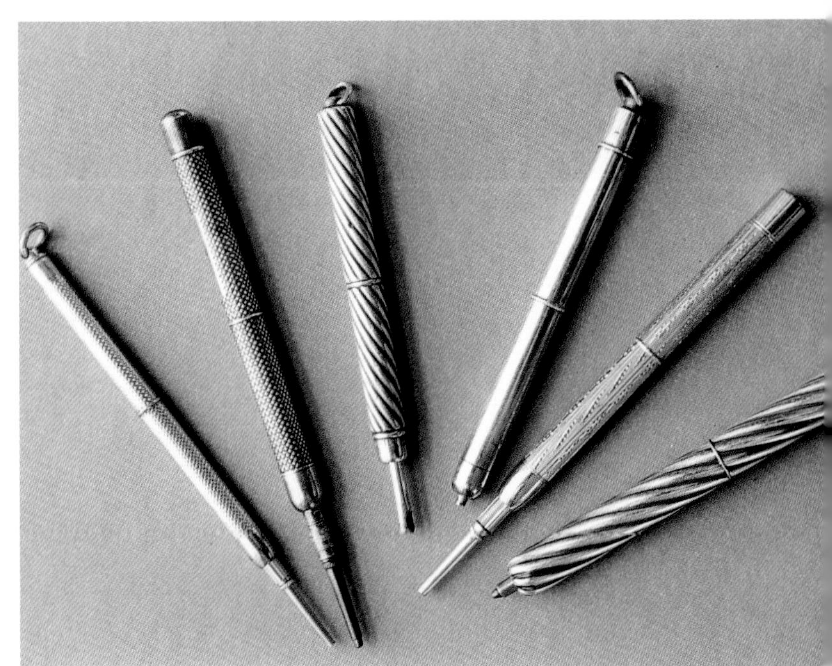

Several screw pencils with a center seam. $35-$60.

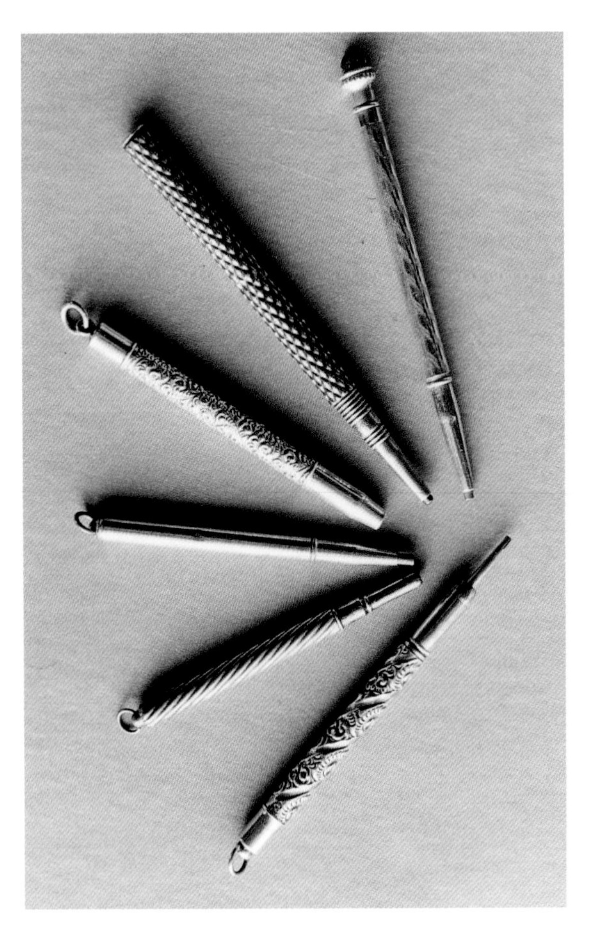

Sterling silver twist pencils, $35-$60.

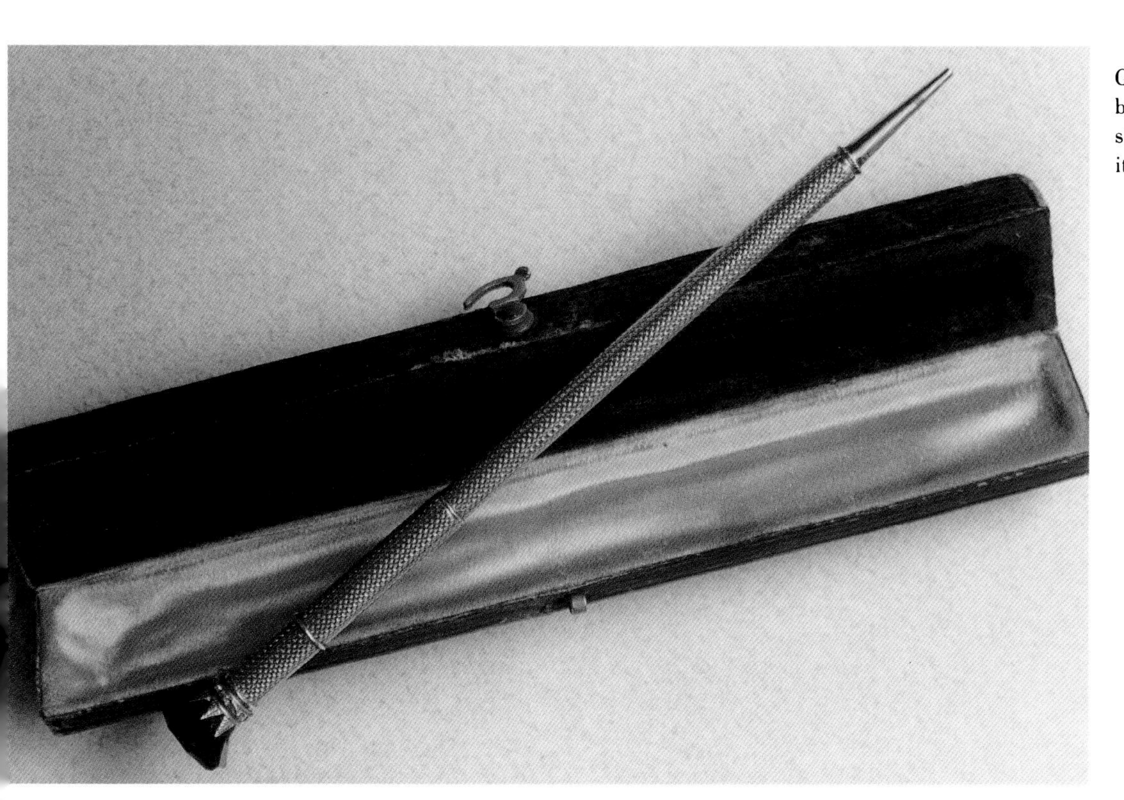

Gold screw pencil, with barleycorn pattern and a signet carved of bloodstone, in its original case, $175-$225.

Sterling silver twist pencil, hallmarked "V & J." 3", $50-$65.

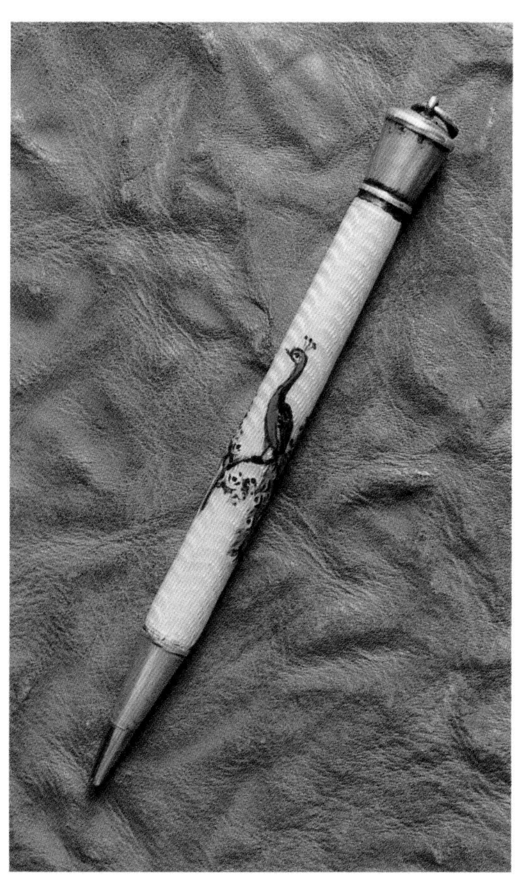

Silver and enamel twist action pencil (see Chapter 3).

The inner workings of a large screw pencil.

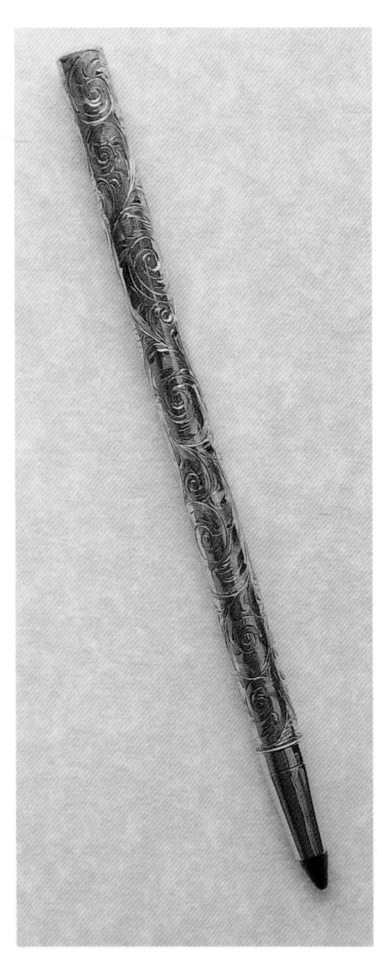

Pencils with large diameter pieces of graphite are often called *stockbroker's* pencils, and were apparently used for marking boards.

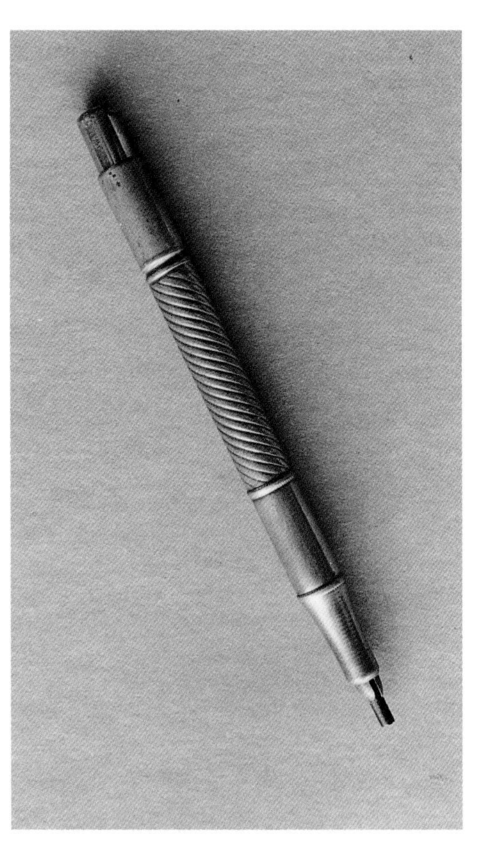

The concept behind this *automatic* pencil is that, by pushing the button on the end and tilting the pencil forward, a piece of graphite falls into place. Conversely, by pushing on the button and tilting the pencil backwards, the graphite would fall back into the barrel. $25-$45.

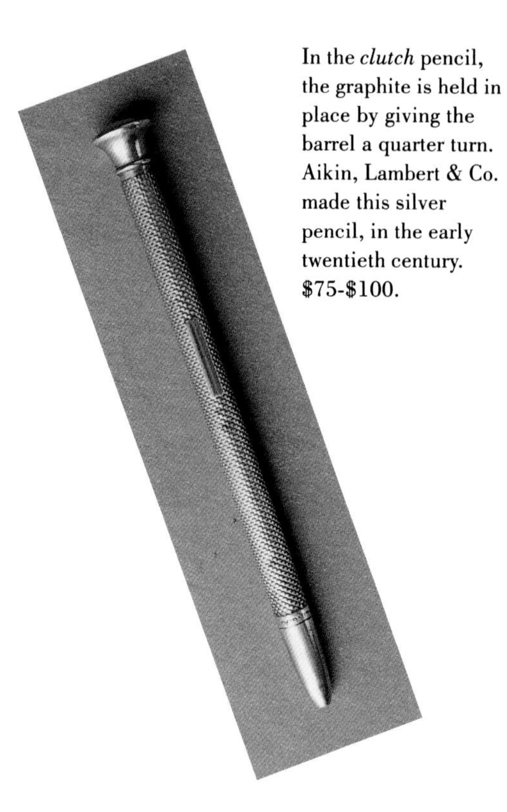

In the *clutch* pencil, the graphite is held in place by giving the barrel a quarter turn. Aikin, Lambert & Co. made this silver pencil, in the early twentieth century. $75-$100.

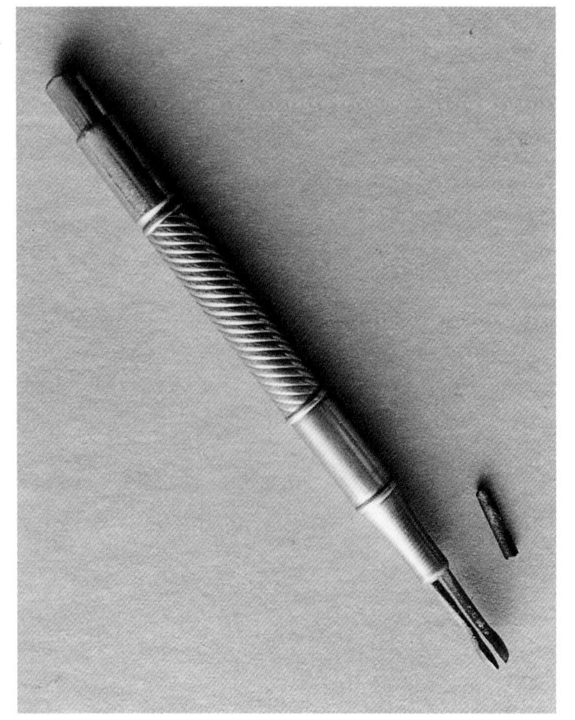

Same as above, showing the prongs that hold the graphite.

This pencil is similar to the Mabie *Magazine* pencil, which operated very much like the clutch pencils, but with more turns to the barrel.

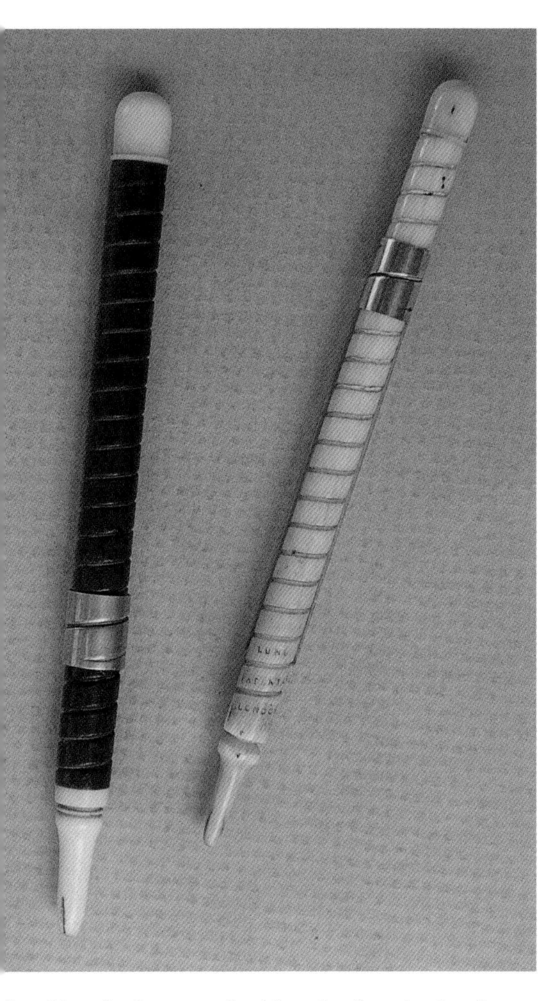

Lund bought the patent for this style of mechanism from Gabriel Riddle, who worked with Sampson Mordan. The collar pushes the lead down the chamber as it is turned. $125-$150.

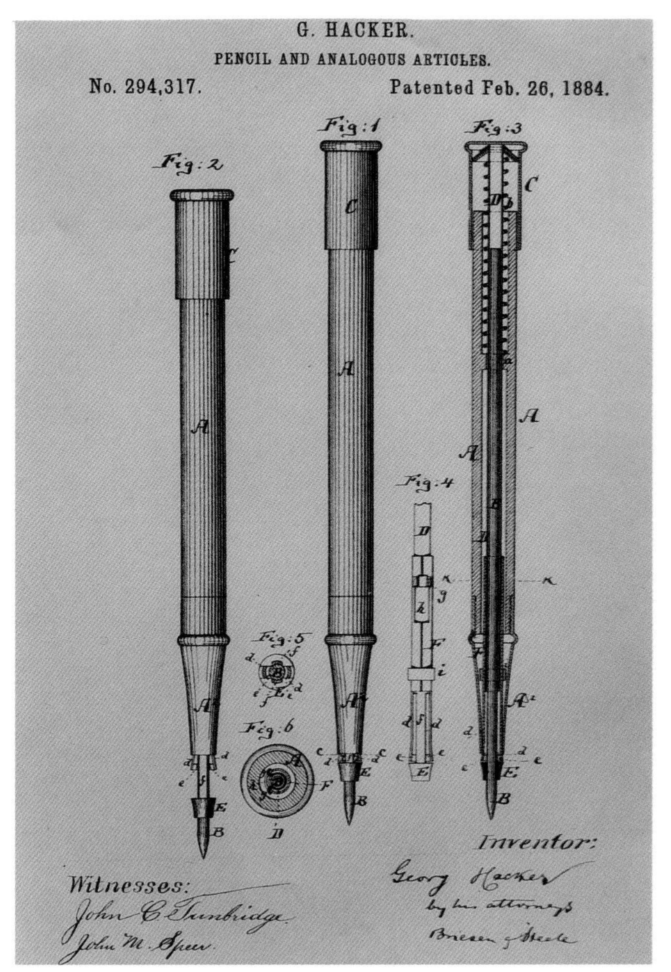

Patent for a pencil mechanism from 1884.

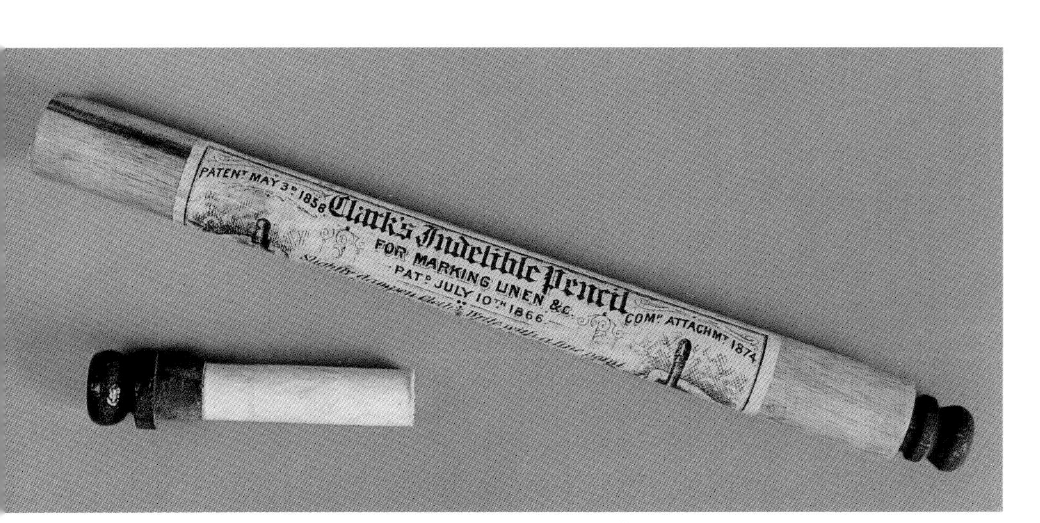

Clark's Indelible Pencil was attached to one wooden knob and fitted inside this wooden cylinder. $35-$50.

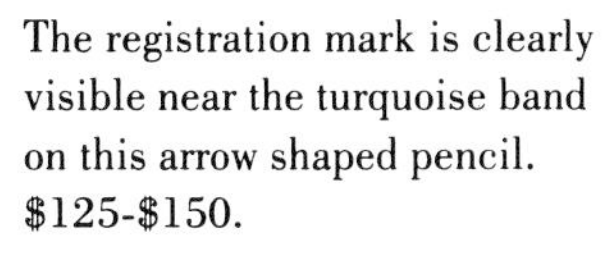

The registration mark is clearly visible near the turquoise band on this arrow shaped pencil. $125-$150.

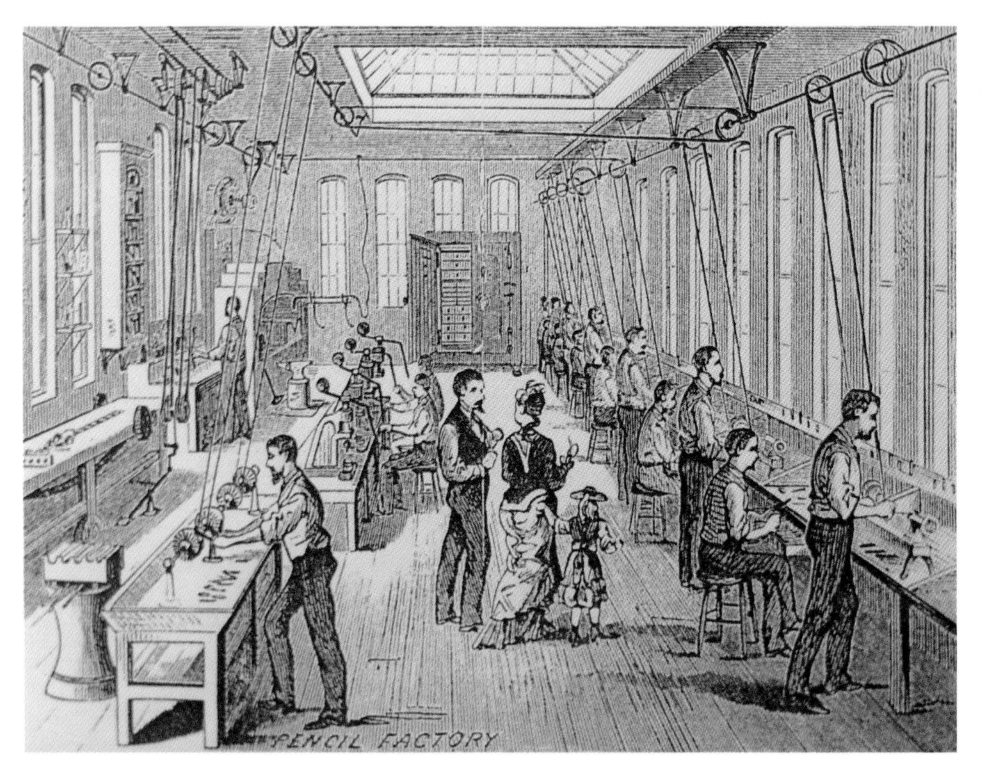

Illustration of a pencil factory from *Scientific American*, circa 1870.

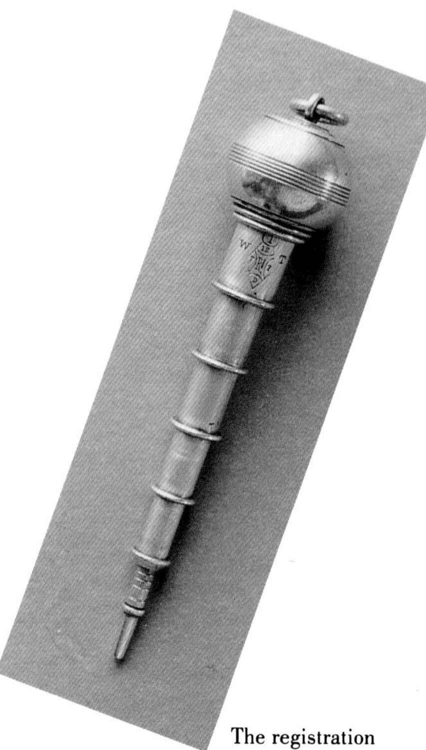

The registration mark on a S. Mordan & Co. figural pencil (see Chapter 5).

This pencil is hallmarked.

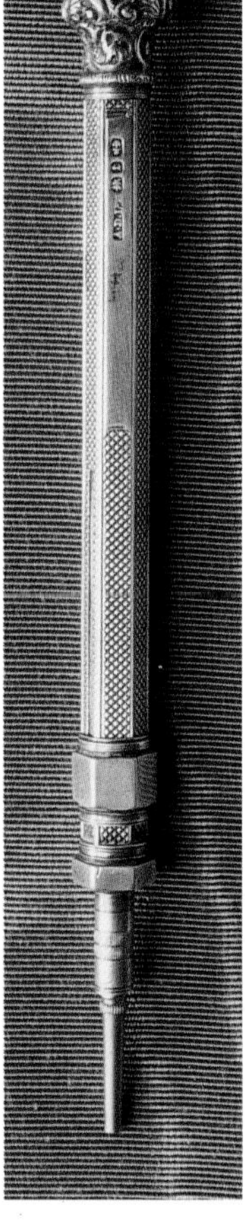

Pencil showing hallmark.

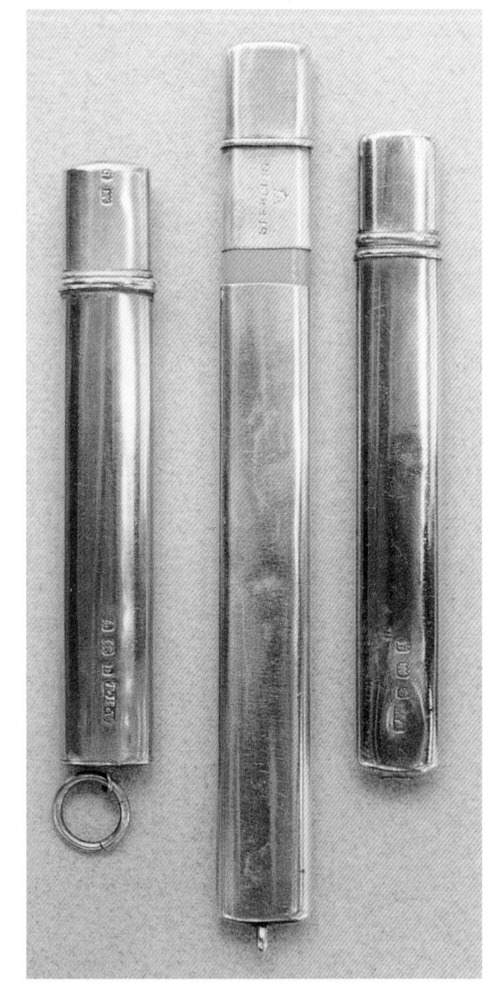

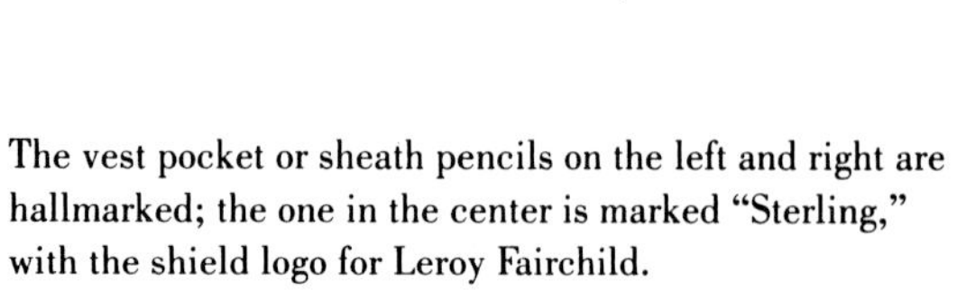

The vest pocket or sheath pencils on the left and right are hallmarked; the one in the center is marked "Sterling," with the shield logo for Leroy Fairchild.

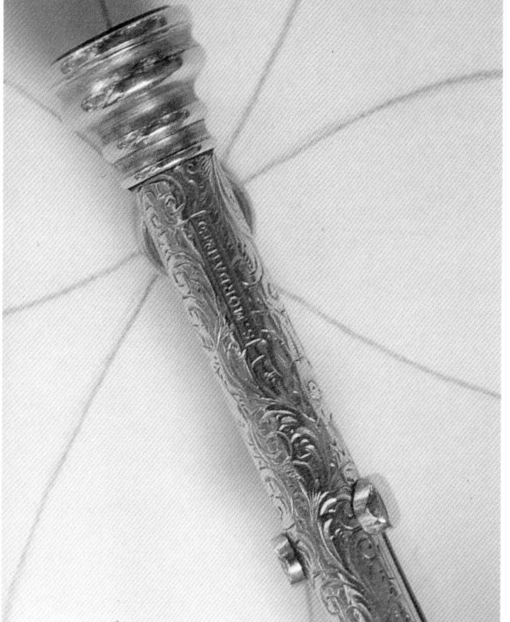

Mark for S. Mordan & Co.

Hallmarked pencil.

American pencil with "Fairchild" mark.

This unidentified mark consists of an ellipse with a line through it (see left), the word "Sterling" and an "X."

Elliptical mark with a horizontal line through it, possibly made by Hicks but not certainly.

Pencil made by Aikin, Lambert & Co., marked "Solid Gold."

Notes on Chapter Four

[1] Genevieve Cummins and Nerylla Taunton. *Chatelaines, Utility to Glorious Extravagance*. Antique Collectors' Club, 1996.

[2] Neil Davis, "Early One Mordan," *Journal of the Writing Equipment Society*, no. 35, 1992.

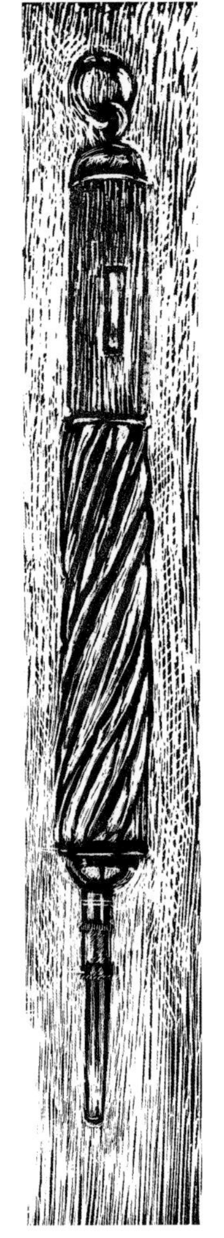

Chapter Five

Figural Pencils

"Why did the Victorians make so many pencils in the shape of so many things other than a pencil?" This is one of the questions most often asked by our contemporaries about figural pencils. Many people wonder why pencils of the Victorian era are so ornamental, figural, or both. Firstly, it should be apparent to anyone interested in antiques that our Victorian ancestors liked ornamentation. They were intrigued by designs from nature and often sought to make utilitarian objects more interesting by embellishing them with vines, flowers, birds, etc. Rich sources of gold and silver were discovered in the United States, and processes were developed for the manufacture of less expensive, plated jewelry. A middle class was beginning to emerge, partially as a result of the Industrial Revolution. Sumptuary laws that had not allowed anyone of the lower classes to wear jewelry had been softened or eliminated. But this era was also a time of strict rules for proper etiquette, including those requiring individuals to adhere to propriety when wearing jewelry. Certain types of jewelry were considered inappropriate for day, men were encouraged not to wear gaudy stones, and women were warned that extravagance could lead to ruin.

Juxtaposed against these restrictions, however, were many other factors. Because silver and gold jewelry could now be made from native silver or gold instead of melted down coin, and because these materials became considerably more accessible in the United States, greater quantities of affordable jewelry were made. Because there were more people who wanted to show off their newly earned wealth either by buying jewelry for themselves or as gifts, more jewelry was purchased. The establishment and development of the gold-plating industry in Providence, Rhode Island, made it possible for ever increasing numbers of husbands to buy gold-plated chains and other trinkets for themselves and their wives. The use of steam engines and other forms of mechanization further enhanced the ability of manufacturers to create more wares to ply the market with. By the 1880s companies were offering silverware and even silver pencils as premiums. Purchase enough soap, and you could own this sterling silver pencil made by Gorham.[1] Instead of relying on the travelling salesmen to reach a small market, companies began relying on catalogues to sell their products to an increasingly large market.

Gorham "Sterling Silver as Premiums" ad, showing a pencil protector. *The Cosmopolitan Press*, 1897.

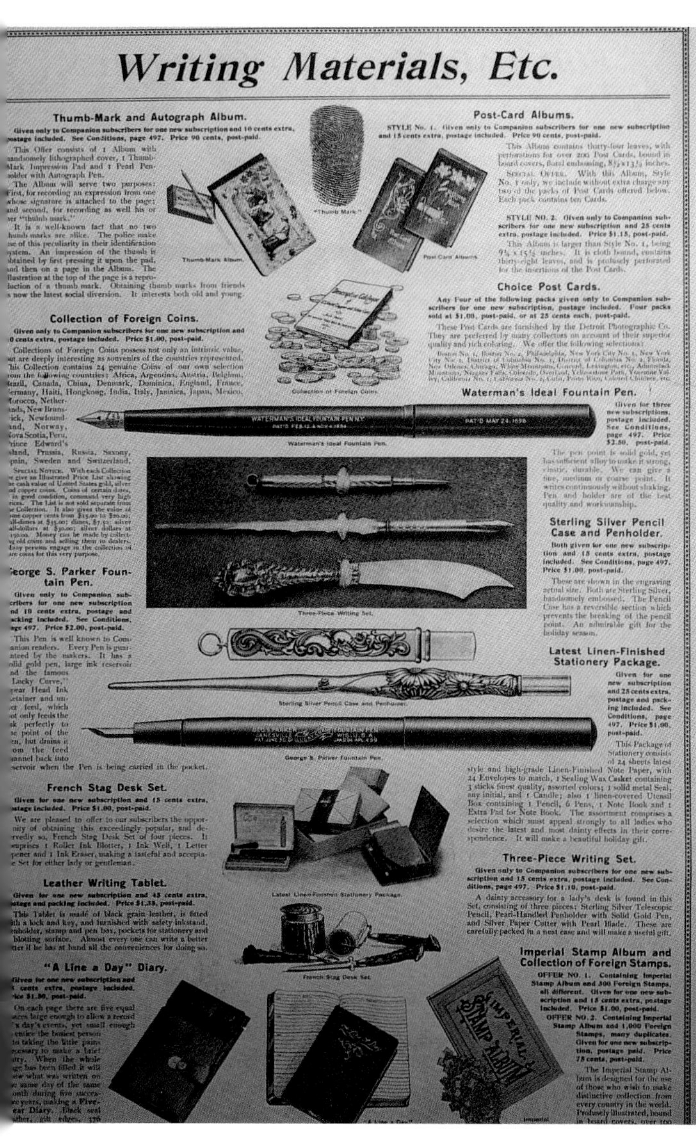

Writing Materials, Etc.

Thumb-Mark and Autograph Album.

Given only to Companion subscribers for one new subscription and 10 cents extra, postage included. See Conditions, page 497. Price 90 cents, post-paid.

This Offer consists of 1 Album with andsomely lithographed cover, 1 Thumb-Mark Impression Pad and 1 Pearl Pen-olde with Autograph Pen.

The Album will serve two purposes: First, for recording an expression from one whose signature is attached to the page; and second, for recording as well his or er "thumb mark."

It is a well-known fact that no two humb marks are alike. The police make se of this peculiarity in their identification ystem. An impression of the thumb is btained by first pressing it upon the pad, and then on a page in the Album. The llustration at the top of the page is a repro-duction of a thumb mark. Obtaining thumb marks from friends s now the latest social diversion. It interests both old and young.

"Thumb Mark."

Thumb-Mark Album.

Post-Card Albums.

STYLE No. 1. Given only to Companion subscribers for one new subscription and 15 cents extra, postage included. Price 90 cents, post-paid.

This Album contains thirty-four leaves, with perforations for over 200 Post Cards, bound in board covers, floral embossing, 8½ x 13½ inches.

SPECIAL OFFER. With this Album, Style No. 1 only, we include without extra charge any two of the packs of Post Cards offered below. Each pack contains ten Cards.

STYLE NO. 2. Given only to Companion subscribers for one new subscription and 25 cents extra, postage included. Price $1.15, post-paid.

This Album is larger than Style No. 1, being 9¼ x 15½ inches. It is cloth bound, contains thirty-eight leaves, and is profusely perforated for the insertion of the Post Cards.

Post Card Albums.

Collection of Foreign Coins.

Given only to Companion subscribers for one new subscription and 0 cents extra, postage included. Price $1.00, post-paid.

Collections of Foreign Coins possess not only an intrinsic value, ut are deeply interesting as souvenirs of the countries represented. his Collection contains 24 genuine Coins of our own selection om the following countries: Africa, Argentina, Austria, Belgium, razil, Canada, China, Denmark, Dominica, England, France, ermany, Haiti, Hongkong, India, Italy, Jamaica, Japan, Mexico, orocco, Nether-ands, New Bruns-ick, Newfound-and, Norway, ova Scotia, Peru, rince Edward's sland, Prussia, Russia, Saxony, pain, Sweden and Switzerland.

SPECIAL NOTICE. With each Collection e give an Illustrated Price List showing e cash value of United States gold, silver d copper coins. Coins of certain dates, in good condition, command very high rices. The List is not sold separate from e Collection. It also gives the value of me copper cents from $15.00 to $20.00; lf-dimes at $35.00; dimes, $7.50; silver lf-dollars at $30.00; silver dollars at 1900.00. Money can be made by collect-g old coins and selling them to dealers. any persons engage in the collection of are coins for this very purpose.

Collection of Foreign Coins.

Choice Post Cards.

Any Four of the following packs given only to Companion subscribers for one new subscription, postage included. Four packs sold at $1.00, post-paid, or at 25 cents each, post-paid.

These Post Cards are furnished by the Detroit Photographic Co. They are preferred by many collectors on account of their superior quality and rich coloring. We offer the following selections:

Boston No. 1, Boston No. 2, Philadelphia, New York City No. 1, New York City No. 2, District of Columbia No. 1, District of Columbia No. 2, Florida, New Orleans, Chicago, White Mountains, Concord, Lexington, etc., Adirondack Mountains, Niagara Falls, Colorado, Overland, Yellowstone Park, Yosemite Valley, California No. 1, California No. 2, Cuba, Porto Rico, Colored Children, etc.

Waterman's Ideal Fountain Pen.

WATERMAN'S IDEAL FOUNTAIN PEN N.Y.
PAT'D FEB.12 & NOV.4 1884
PAT'D MAY 24. 1898

Waterman's Ideal Fountain Pen.

Given for three new subscriptions, postage included. See Conditions, page 497. Price $2.50, post-paid.

The pen point is solid gold, yet has sufficient alloy to make it strong, elastic, durable. We can give a fine, medium or coarse point. It writes continuously without shaking. Pen and holder are of the best quality and workmanship.

Sterling Silver Pencil Case and Penholder.

Both given for one new subscription and 15 cents extra, postage included. See Conditions, page 497. Price $1.00, post-paid.

These are shown in the engraving actual size. Both are Sterling Silver, handsomely embossed. The Pencil Case has a reversible section which prevents the breaking of the pencil point. An admirable gift for the holiday season.

Three-Piece Writing Set.

eorge S. Parker Fountain Pen.

Given only to Companion subscribers for one new subscription nd 10 cents extra, postage and acking included. See Conditions, ge 497. Price $2.00, post-paid.

This Pen is well known to Companion readers. Every Pen is guaranteed by the makers. It has a olid gold pen, large ink reservoir nd the famous Lucky Curve," pear Head Ink etainer and un-er feed, which ot only feeds the k perfectly to e point of the en, but drains it om the feed annel back into servoir when the Pen is being carried in the pocket.

Sterling Silver Pencil Case and Penholder.

George S. Parker Fountain Pen.

Latest Linen-Finished Stationery Package.

Given for one new subscription and 25 cents extra, postage and packing included. See Conditions, page 497. Price $1.00, post-paid.

This Package of Stationery consists of 24 sheets latest style and high-grade Linen-Finished Note Paper, with 24 Envelopes to match, 1 Sealing Wax Casket containing 3 sticks finest quality, assorted colors; 1 solid metal Seal, any initial, and 1 Candle; also 1 linen-covered Utensil Box containing 1 Pencil, 6 Pens, 1 Note Book and 1 Extra Pad for Note Book. The assortment comprises a selection which must appeal strongly to all ladies who desire the latest and most dainty effects in their correspondence. It will make a beautiful holiday gift.

French Stag Desk Set.

Given for one new subscription and 15 cents extra, stage included. Price $1.00, post-paid.

We are pleased to offer to our subscribers the oppor-nity of obtaining this exceedingly popular, and de-rvedly so, French Stag Desk Set of four pieces. It mprises 1 Roller Ink Blotter, 1 Ink Well, 1 Letter pener and 1 Ink Eraser, making a tasteful and accepta-e Set for either lady or gentleman.

Latest Linen-Finished Stationery Package.

Three-Piece Writing Set.

Given only to Companion subscribers for one new subscription and 15 cents extra, postage included. See Conditions, page 497. Price $1.10, post-paid.

A dainty accessory for a lady's desk is found in this Set, consisting of three pieces: Sterling Silver Telescopic Pencil, Pearl-Handled Penholder with Solid Gold Pen, and Silver Paper Cutter with Pearl Blade. These are carefully packed in a neat case and will make a useful gift.

Leather Writing Tablet.

Given for one new subscription and 45 cents extra, stage and packing included. Price $1.35, post-paid.

This Tablet is made of black grain leather, is fitted ith a lock and key, and furnished with safety inkstand, enholder, stamp and pen box, pockets for stationery and blotting surface. Almost every one can write a better tter if he has at hand all the conveniences for doing so.

French Stag Desk Set.

"A Line a Day" Diary.

Given for one new subscription and cents extra, postage included. ice $1.30, post-paid.

On each page there are five equal aces large enough to allow a record a day's events, yet small enough entice the busiest person to taking the little pains cessary to make a brief try. When the whole ge has been filled it will ow what was written on e same day of the same onth during five succes-e years, making a Five-ear Diary. Black seal ather, gilt edges, 376

Imperial Stamp Album and Collection of Foreign Stamps.

OFFER NO. 1. Containing Imperial Stamp Album and 300 Foreign Stamps, all different. Given for one new subscription and 15 cents extra, postage included. Price $1.00, post-paid.

OFFER NO. 2. Containing Imperial Stamp Album and 1,000 Foreign Stamps, many duplicates. Given for one new subscription, postage paid. Price 75 cents, post-paid.

The Imperial Stamp Album is designed for the use of those who wish to make distinctive collection from every country in the world. Profusely illustrated, bound in board covers, over 100

IMPERIAL STAMP ALBUM

"A Line a Day."

Imperial

Pencils as premiums for ordering *Youth's Home Companion*, October 1905.

The solutions to the quandary created by the fact that lots of affordable jewelry was readily available while at the same time proper etiquette required one to be judicious in the actual wearing of the jewelry were varied. Men discovered that, while wearing very much jewelry was considered improper, they could justify wearing a gold watch with a gold chain. Attached to the gold chain could be fobs representing fraternal organizations, seals for wax, or even a figural pencil. (Despite a common misperception, many of the small figural pencils were originally made for men. They are sometimes described in trade catalogues as watch-chain charms. Some are still found with their original fob chain still attached.) There were patents for watch key, pencil combinations, and watch-chain bars that had pencils (as well as toothpicks) in them. Women, who were traditionally encouraged to wear more jewelry, followed the fashions carefully. If diamonds were considered too flashy to wear in daytime, they became a part of the evening costume. Following a tradition already established by their inclusion in chatelaines, pencils (which were actually functional) could be worn at any time. They were offered in catalogues as charm pencils to be worn on a chain or a bracelet. In the Blue Book of 1878, Tiffany's offered "shopping bracelet[s], with pencil attached, gold or silver, some richly jeweled."[2] Ingeniously, both men and women found that they could

Seal with a dove carrying a letter, set into the finial of a white metal pencil. Unlike most finials, this one does not unscrew.

wear a pencil in the shape of a horse-head or one studded with stones or embellished with multi-colored gold and still be considered appropriately dressed.

During the Victorian era (and into the early twentieth century) there was a heightened interest in what we might now call sentimentality. Remembrances and tokens of love and friendship abounded. Pencils made wonderful gifts, as they not only spoke of the respect accorded to the receiver (owning a pencil implied one could write) but they could also be endowed with additional meaning. One maker of mechanical pencils specialized in a pattern of stamping where the die spelled out such mottoes as "Friendship" or "A Token." Many pencils had seals set in them, carved with images such as a dove carrying a letter in its beak. Pencils were inscribed with names and dates (a tradition that continues today) and with such cryptic messages as "Rauheneck, half past six." Phrases like "Wine, Women and Song" were spelled out in pictures on vest-pocket pencils. A huge red stone, mounted in the terminal of a pen-pencil combination made by John Rauch, reflects the name of the owner: Wm. E. Bloodgood.

Jewelry in the mid- to late-nineteenth century was often highly symbolic, and most customers understood what the images or stone represented. Tiffany's even published a book on birthstones. The most common of the metaphoric usage of stones, however, would be the use of jet to signify mourning. When Prince Albert died in 1861, Queen Victoria began to wear mourning jewelry made of this black substance. The masses emulated her by purchasing jewelry made of jet if they could afford it, or of black glass if they couldn't. The huge numbers of casualties in the Civil War sadly gave Americans ample cause to wear mourning, creating an even greater demand for black accessories.

Other materials, including black hard rubber and gunmetal, were also used to create the subdued effect the Victorians sought. (True to Victorian form, this effect was counter-balanced by the flashing of light off faceted jet and by the glitter of gold mountings.) Pencil-case makers, naturally, followed this trend and made use of these black materials in their products (it should not be assumed, however, that all or even most of the black hard-rubber pencils were used exclusively for mourning).

Another illustration of Victorian sentimentality was the proliferation of jewelry made of or incorporating human hair. From watch chains fabricated of hair to lockets containing a cherished lock, the universal message of *memento mori* was embraced by the people of this era. Again, pencil-case makers were certainly aware of popular trends. An example of this is a combination pen-pencil with a tiny hinged lid in its terminal, which when lifted reveals a lock of hair curled in a compartment covered with a small disk of clear glass. (Oddly, although photography had become exceedingly popular and some jewelry was made with a small photograph set into a bezel, only one such Victorian pencil has been discovered to date.)

Young man pulling his jacket aside to show his watch chain.

Not only did jewelry and other accessories become increasingly available to greater numbers of people during the nineteenth century, but the ability to travel did as well. As remembrances of trips, pencils were inscribed with names of cities or historic sites. Another popular device used by the makers of souvenir pencils was the *Stanhope*. The Stanhope (also called "Peeper") is created by drilling a small hole in a bone or ivory staff and by then fitting a miniature crystal through the opening. A miniscule image is attached to one end of the crystal, so that when the viewer holds the staff up to light and looks through the peephole, he or she sees the picture. Stanhopes were named after Charles Stanhope, the third Earl of Stanhope (1753-1816), who invented a lens made from a small glass rod. Essentially, images attached to the flat end of the rod are magnified by the convex surface of the other end. This concept

Photograph of man wearing a watch chain with a charm, circa 1880.

was further developed by others, who discovered that tiny photographic images could be created by utilizing a camera with multiple lenses (one was able to create as many as 450 2mm images) and that these pictures could be attached to the lens invented by Stanhope. Rene Dagron recognized the commercial potential of this process in 1859, and Queen Victoria popularized it by wearing jewelry set with microphotographs of family members in the 1860s.

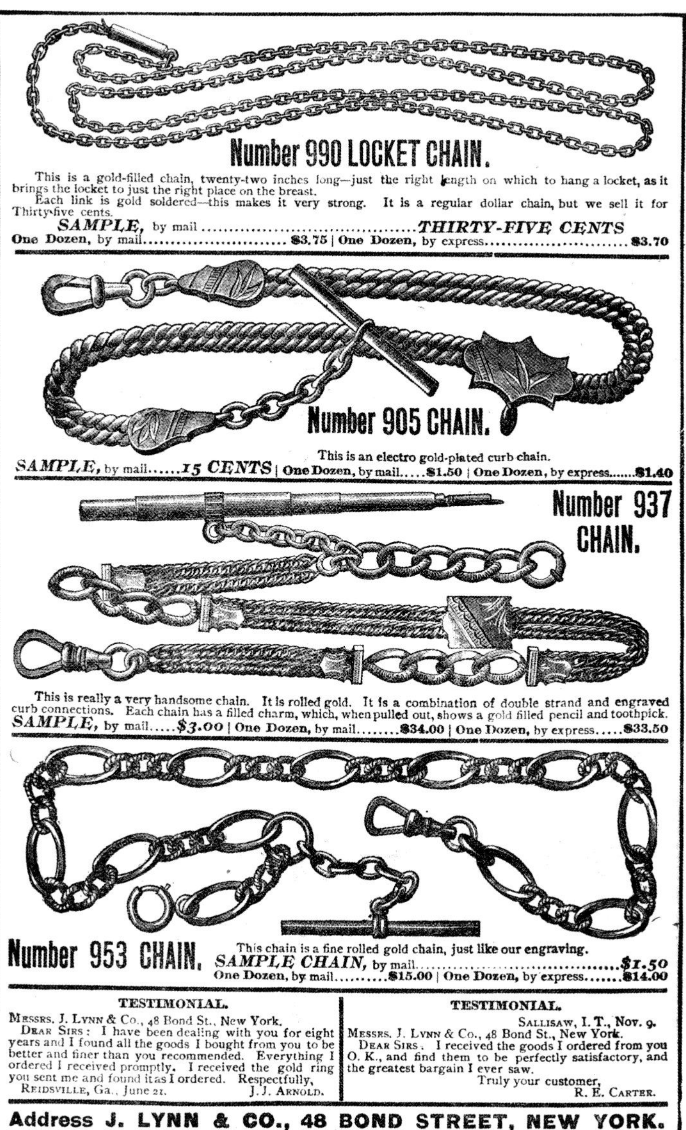
Number 990 LOCKET CHAIN.

This is a gold-filled chain, twenty-two inches long—just the right length on which to hang a locket, as it brings the locket to just the right place on the breast.

Each link is gold soldered—this makes it very strong. It is a regular dollar chain, but we sell it for Thirty-five cents.

SAMPLE, by mail .... *THIRTY-FIVE CENTS*
One Dozen, by mail .... $3.75 | One Dozen, by express $3.70

Number 905 CHAIN.

This is an electro gold-plated curb chain.

SAMPLE, by mail..... *15 CENTS* | One Dozen, by mail.....$1.50 | One Dozen, by express.......$1.40

Number 937 CHAIN.

This is really a very handsome chain. It is rolled gold. It is a combination of double strand and engraved curb connections. Each chain has a filled charm, which, when pulled out, shows a gold filled pencil and toothpick.

SAMPLE, by mail.....*$3.00* | One Dozen, by mail........$34.00 | One Dozen, by express.....$33.50

Number 953 CHAIN. This chain is a fine rolled gold chain, just like our engraving.

SAMPLE CHAIN, by mail........ *$1.50*
One Dozen, by mail.........$15.00 | One Dozen, by express.......$14.00

TESTIMONIAL.

Messrs. J. Lynn & Co., 48 Bond St., New York.
Dear Sirs: I have been dealing with you for eight years and I found all the goods I bought from you to be better and finer than you recommended. Everything I ordered I received promptly. I received the gold ring you sent me and found it as I ordered. Respectfully,
Reidsville, Ga., June 21. J. J. Arnold.

TESTIMONIAL.

Sallisaw, I. T., Nov. 9.
Messrs. J. Lynn & Co., 48 Bond St., New York.
Dear Sirs: I received the goods I ordered from you O. K., and find them to be perfectly satisfactory, and the greatest bargain I ever saw.
Truly your customer,
R. E. Carter.

Address J. LYNN & CO., 48 BOND STREET, NEW YORK.

Page from a nineteenth century catalog showing watch chains with "a filled charm, which, when pulled out, shows a gold-filled pencil and toothpick." *Courtesy, The Winterthur Library: Printed Book and Periodical Collection.*

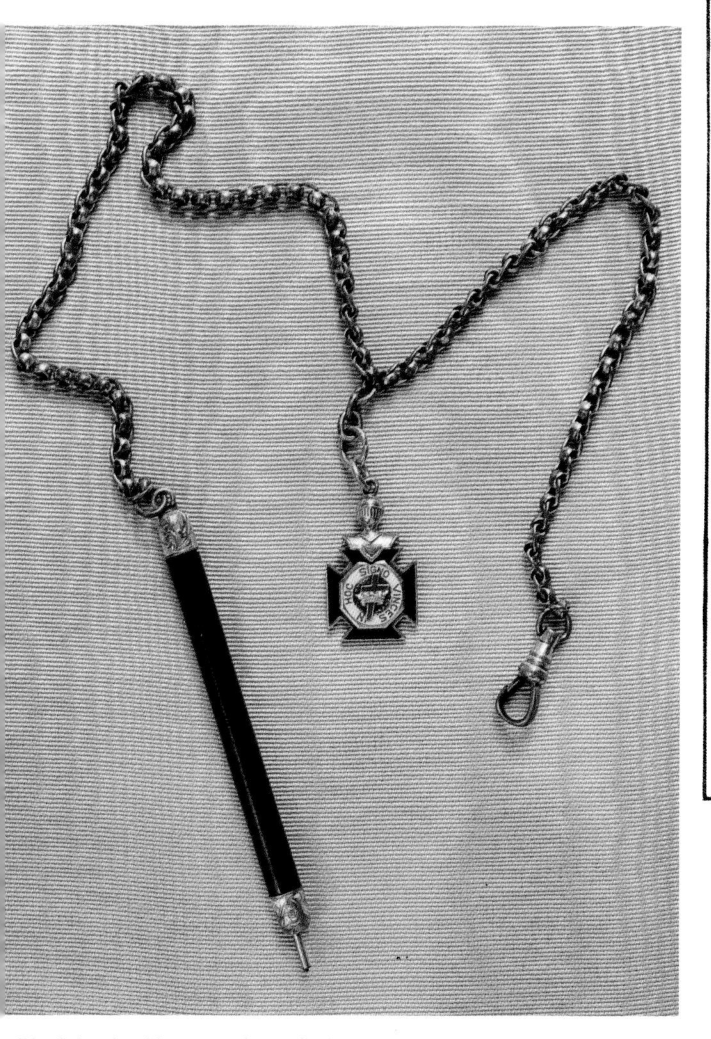

Black hard-rubber pencil attached to watch chain, with a charm. $35-$75.

Most of the pencils made with "Peepers" in them were made between 1880 and 1920. They are often made of base metal, bone, or wood, and are sometimes figural. The Stanhopes in pencils usually depict a tourist site, although they occasionally reveal a nude or partially clad woman.[3]

Gold-filled watch chain bars with pencils inside. $40-60.

Watch chain bars open, showing pencils and gold toothpick.

Gold-filled watch chain bars, one open showing pencil. $40-60.

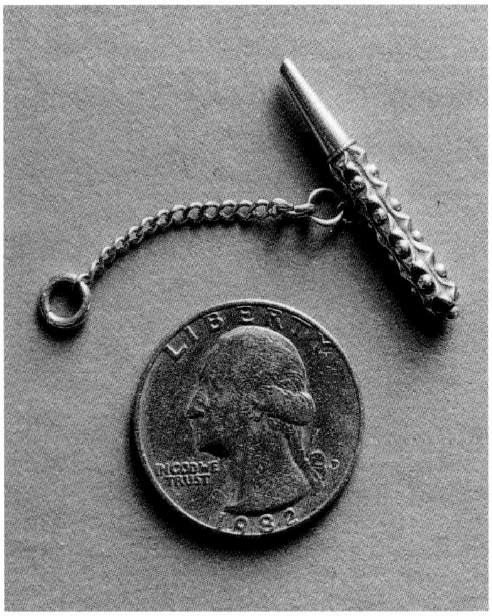

Gold-filled charm pencil for a watch chain, 1-1/4". $40-60.

Sterling silver magic pencil in the shape of a horse head, with red stone eyes. Marked Edward Todd. $500-$800.

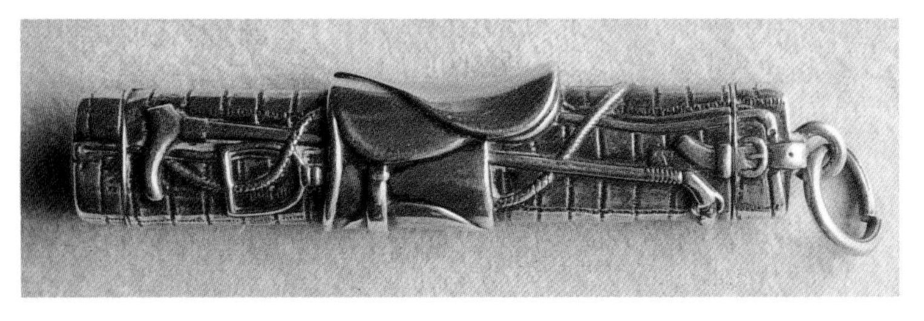

Sterling silver magic pencil with riding motif. Marked Gorham. $400-$600.

Photo of a woman wearing a sautoir (long chain) used for watches, pencils, etc.

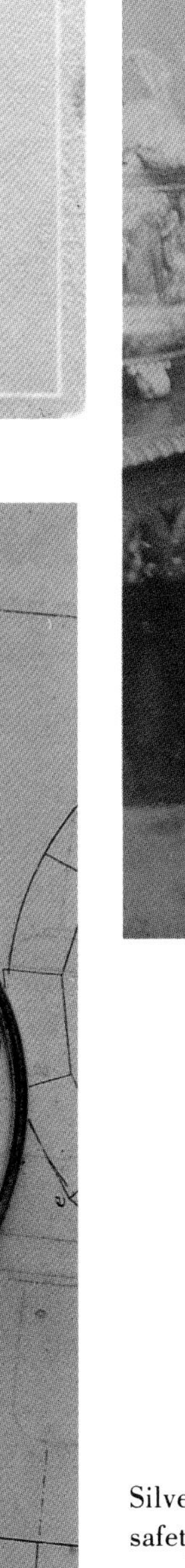

Nineteenth-century photograph of a woman wearing a long chain tucked into her pocket, holding a watch or pencil, etc.

Silver bracelet with a pencil attached with a safety chain. Unmarked. $150-$175.

Silver four-colored pencil, inscribed "Rauheneck, half past six." $150-$225.

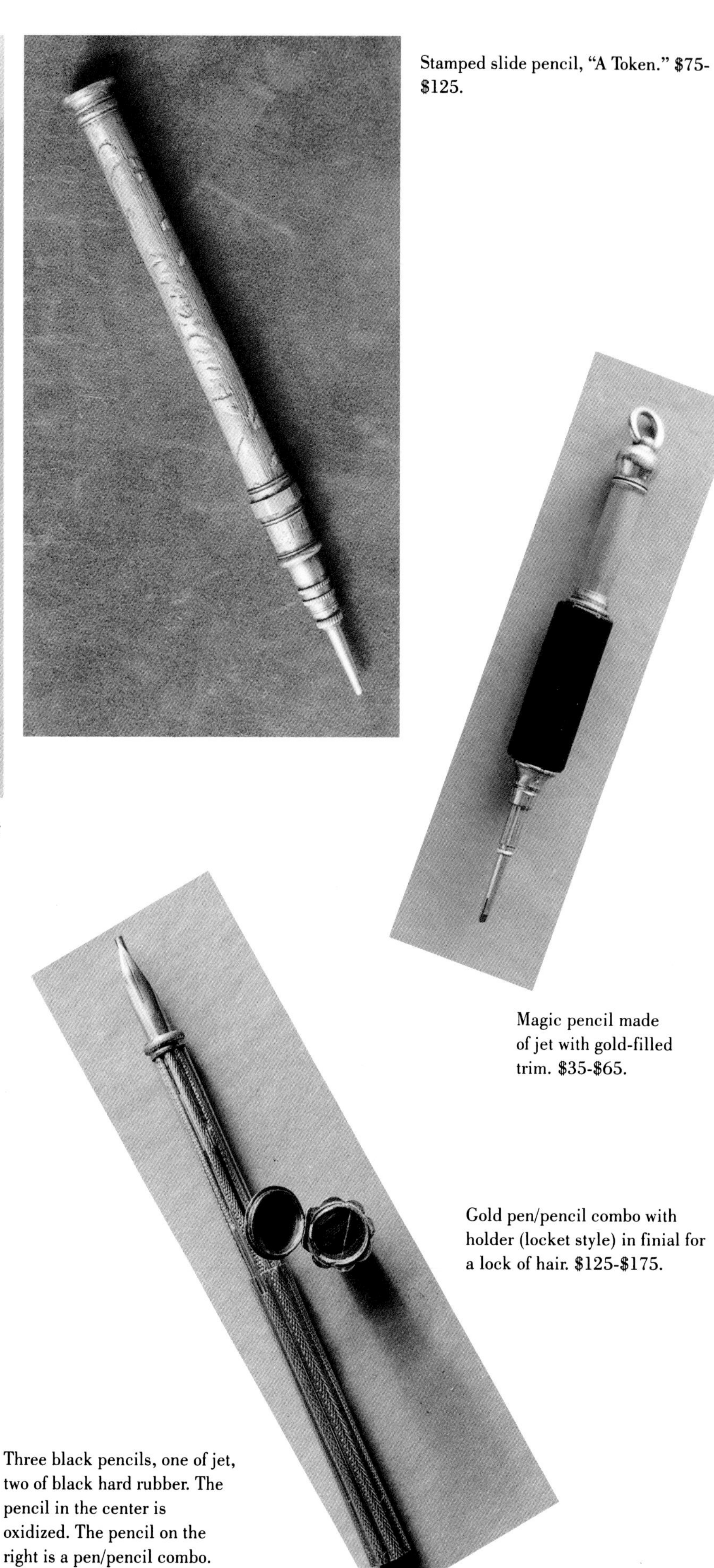

Stamped slide pencil, "A Token." $75-$125.

Magic pencil made of jet with gold-filled trim. $35-$65.

Gold pen/pencil combo with holder (locket style) in finial for a lock of hair. $125-$175.

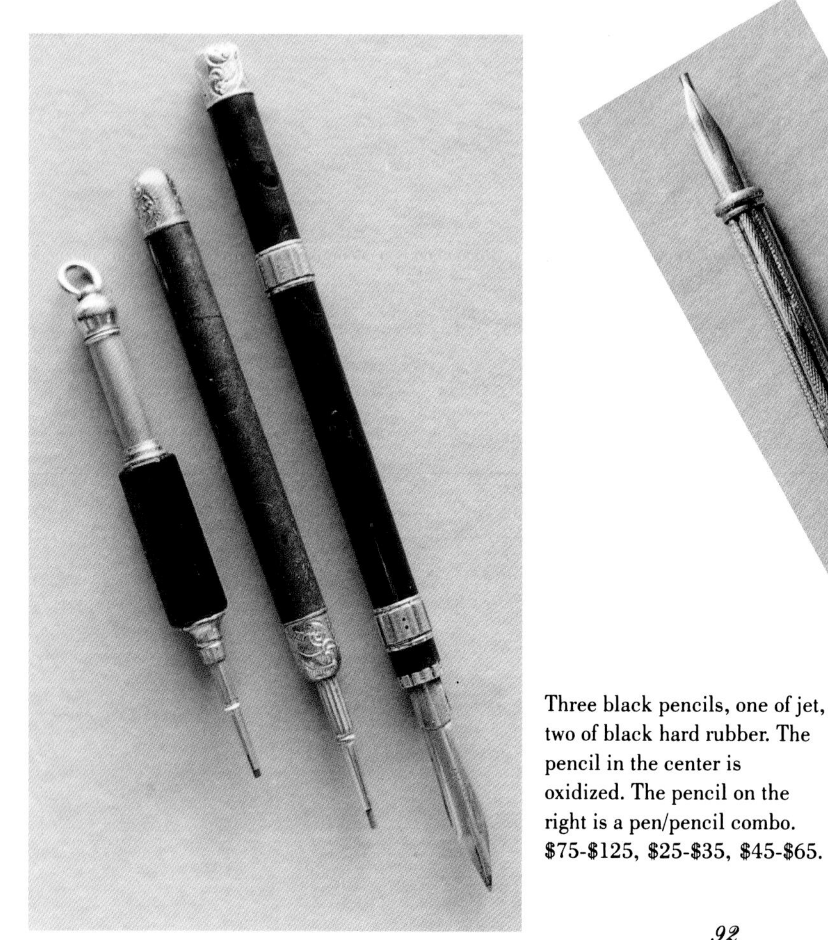

Three black pencils, one of jet, two of black hard rubber. The pencil in the center is oxidized. The pencil on the right is a pen/pencil combo. $75-$125, $25-$35, $45-$65.

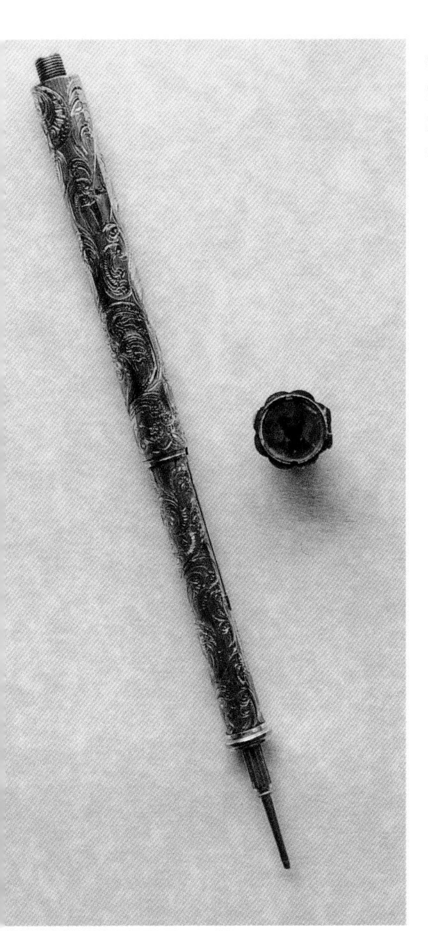

Gold pencil with holder in the finial for a small photograph (missing cover). $75-125.

Detail.

Detail of gold-filled pencil with large red stone. Inscribed "Wm. Bloodgood." Made by Rauch. $125-$150.

Silver sheath pencil, souvenir of Stuttgart. $35-$55.

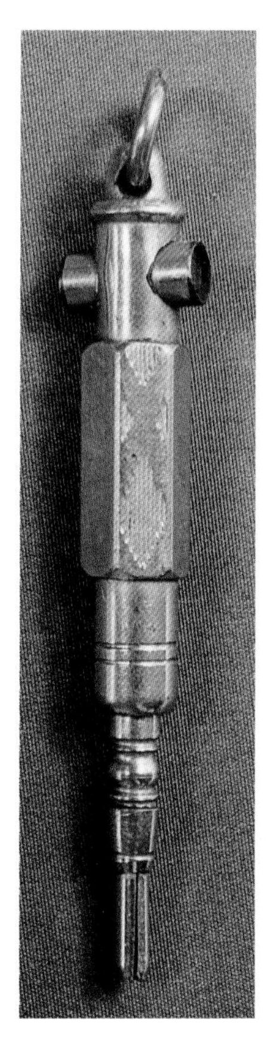

White metal pencil with Stanhope. $50-$75.

Figural pencil in the shape of an axe, with a Stanhope. $50-$75.

Gold porte-crayon commemorating Queen Victoria's Golden Jubilee (1837-1887). Marked Gowland Bros., with Mordan arrow, and M. T. & Co. $250-$350.

Another view of the same pencil, showing the crown.

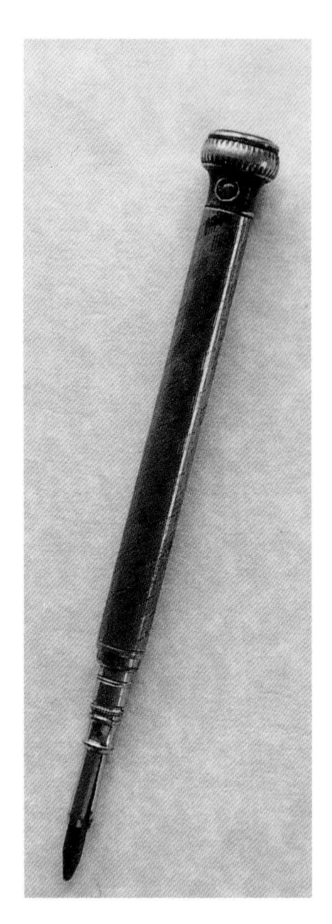

Brass engine-turned pencil with Stanhope depicting Queen Victoria and her Golden Jubilee. $50-$75.

Souvenir pencils commemorating momentous events were also sold, at both the high and low ends of the market. These pieces offer an excellent example of nineteenth-century ingenuity. Two examples recognizing Queen Victoria's Golden Jubilee (celebrating her fifty years on the throne) are shown here. One is made of gold and has an elaborate cast slider consisting of a crown on one side and a lamb in front of rays of light on the other, with "JUBILEE 1887 M.T. Co" engraved on the band running under those images. Although Queen Victoria chose to appear in public on her Jubilee wearing a hat rather than a crown, she is pictured wearing a crown of exactly this shape in a formal portrait photograph taken in 1887. This pencil is marked "Gowland Bros." and has an arrow pointing away from the pencil. Although there is speculation that the arrow represents Sampson Mordan & Co.,[4] the marking M. T. Co could indicate Mabie Todd & Company. A perfect illustration of how these pencils were made to appeal to both the high and low ends of the market, the other example of a commemorative Jubilee item is this simple pencil made of engine-turned brass. Here, a microphotograph of "H. M. Queen

Photograph of Queen Victoria wearing her crown (she chose to wear a hat in public during her Jubilee, rather than her crown).

Victoria Jubilee 1837-1887" is shown in a Stanhope, surrounded by pictures of Windsor Castle, Osbourne House, Balmoral Castle, and Westminster Abbey.

Gold snake pencil. Pencil telescopes out of the snake's mouth. $400-$600.

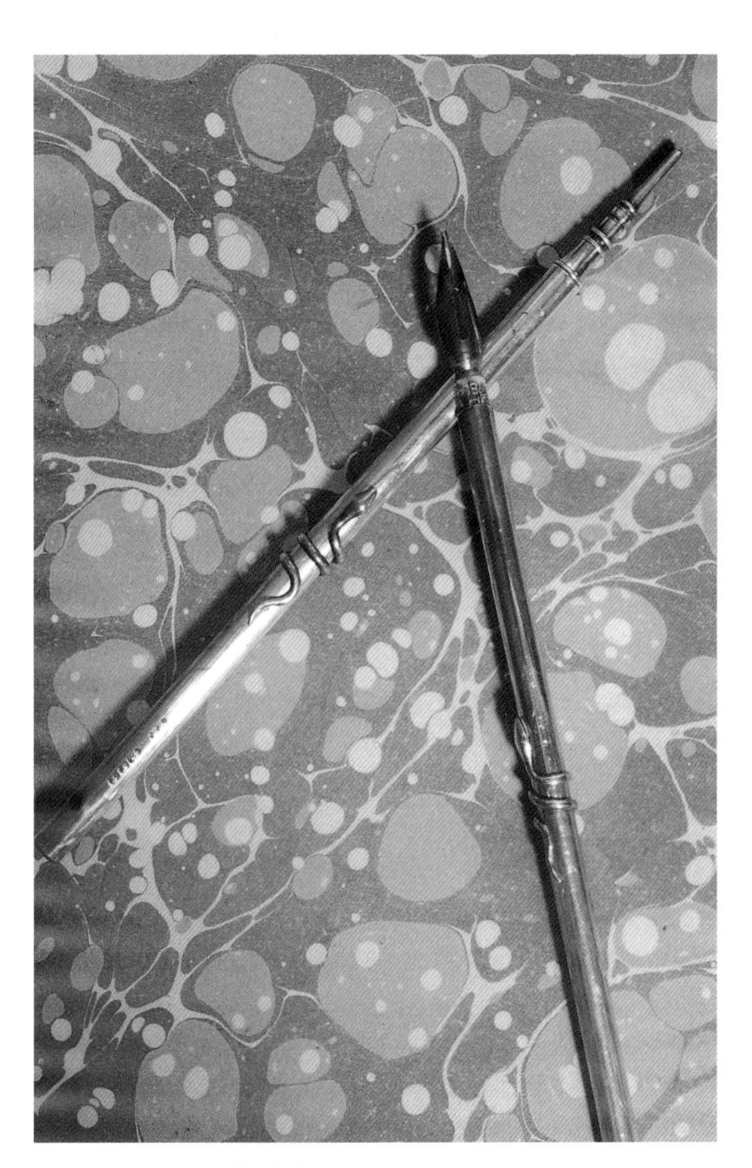

Desk set hallmarked "W. H.", with a snake coiling around both the pen and the pencil. Set $250-$350.

Sterling silver magic pencil, with a snake curling around cattails. Marked Fairchild-Johnson. $600-$900.

Sterling silver sheath pencil, with a snake appearing to weave in and out of the case. Marked Edward Todd. $175-$225.

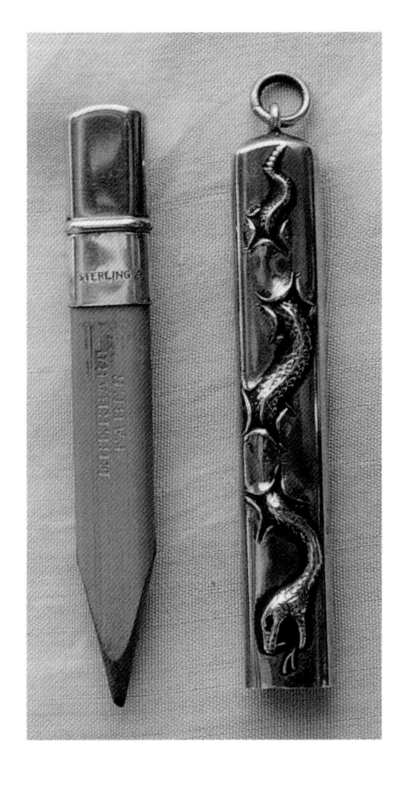

Sterling silver telescoping pencil, with a chased image of a man and a snake. Unmarked. $175-$225.

Stylistically, the nineteenth century offered great variation. The Victorians were highly eclectic, and this is reflected in their decorative arts. The metal work produced during this era exhibited this diversity: from interest in the historic (Gothic or Medieval, Neoclassical, etc.), geographic or cultural (patterns inspired by those created in Japan, for example), to naturalism, the nineteenth-century designer had much to choose from. Many themes used in nineteenth-century jewelry show up in the pencils as well. Designs incorporating snakes, for example, were a favorite of Queen Victoria, who chose to wear a snake bracelet to her first session of Parliament in 1837 and whose engagement ring was in the shape of a coiled snake (most likely symbolizing undying love). Victoria once again created a fashion that was followed throughout her reign and was reinforced by masterpieces created by jewelers like Lalique and Fouquet. Snakes had gone in and out of favor as a motif in jewelry for centuries, but they were unquestionably popular in the 1800s. Snakes encircled fingers as rings and arms as bracelets. They show up on brooches, necklaces, scarf rings, and match safes. As for writing instruments, some of the most sought after pieces are those embellished with snakes. The form is sensual and symbolic. A sometimes repellent subject fabricated of gold or silver and enhanced with diamonds, rubies, garnets, or turquoise, the snake retains its mystique as it encircles or weaves in and out of pens and pencils created in the nineteenth century. In a sense temptation becomes domesticated, tamed.[5]

Archeological discoveries and revivals of classical imagery inspired another popular motif. Owen Jones's book, *Grammar of Ornament* (1857), motivated many artisans to use historic designs and forms. With the completion of the Suez Canal in 1870, and the popularity of Verdi's *Aida* (1871), Egyptian themes became prevalent in the decorative arts. The Egyptian theme was evident in jewelry and silverware created between 1865 and 1875.[6] Pocketknives, letter openers, and staffs for dip pens were made in the shape of mummies. Tabletop sized obelisks were made in stone, crystal, etc., and were commonly used as household ornaments. Monuments shaped like obelisks stood in town squares, as well as in cemeteries. Magic pencils in the shapes of Egyptian mummies and obelisks decorated with hieroglyphics started appearing by the 1880s.

Until recently thought to have been made in the 1920s (after the discovery of Tutankamen's tomb), these beautiful pencils were actually made much earlier, perhaps in some cases as souvenirs for those who were able to travel to Egypt. Objects collected during one's grand tour were highly prized, and may have been given as gifts upon one's return (much like we bring home souvenirs for our friends and families today).[7] Because mummy or pharaoh pencils are among the most common of the figural pencils still available at antique markets, we can surmise that they were originally made in relatively large quantities. Although most of these pencils do not have a maker's mark, some are marked with 800 (meaning eight hundred parts silver per thousand). Mordan is known to have made pencils in the shape of an obelisk in the 1880s.

Figural pencils in the shape of a cross were another Victorian favorite. These pieces were made of gold (or were gold filled), silver, or black hard-rubber, and the pencil invariably slid out from the bottom of the vertical shaft. Keys were another motif that captured the imagination of pencil-case makers. One of the original shapes used for figural pencils (and one offering numerous variations) was the gun. Pencils disguised as miniature pistols and rifles were made by the 1840s. The pencil was pushed out of the guns' barrel. A rifle-shaped pencil may also have a toothpick. "Guns" that were actually compendiums consisting of pencils, pens, and inkwells were made into the twentieth century.

Miniature bottles were made to be telescoping pencils or were made to hold tiny pencils. Additional objects that served as subject matter for the artisans who designed figural pencils were: brooms, cats, dogs, frogs, rabbits, pigs, axes, hands, pipes, eggs, owls, oars, rowboats, matchsticks, trowels, cigars, ears of corn, cannons, nails, screws, hammers, butter knives, arrows, buggy whips, globes, canes, golf clubs, tennis rackets, umbrellas, mallets, tops, nuts, and acorns, etc.

Enamelled telescoping pencil in the shape of a pharaoh. $125-$150.

Back view.

Double-faced pharaoh. $135-$175.

Magic pencil in the shape of an obelisk. Gold wash on silver with blue enamel. $225-$300.

Enamelled telescoping pencil in the shape of a pharaoh. $125-$150.

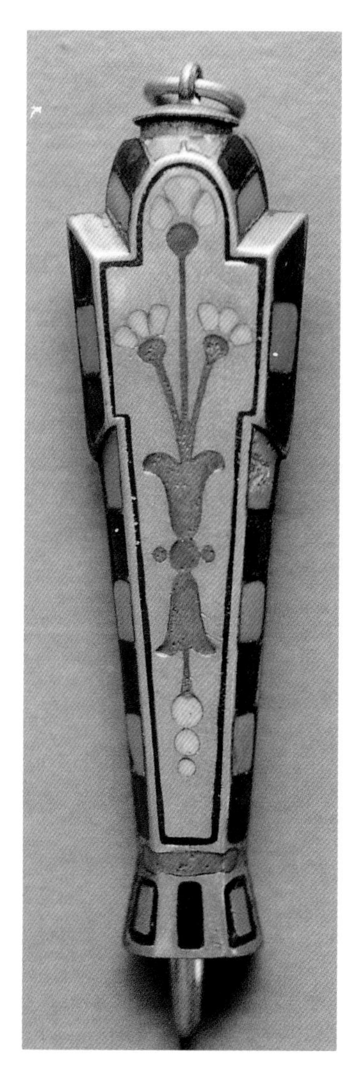

Back view.

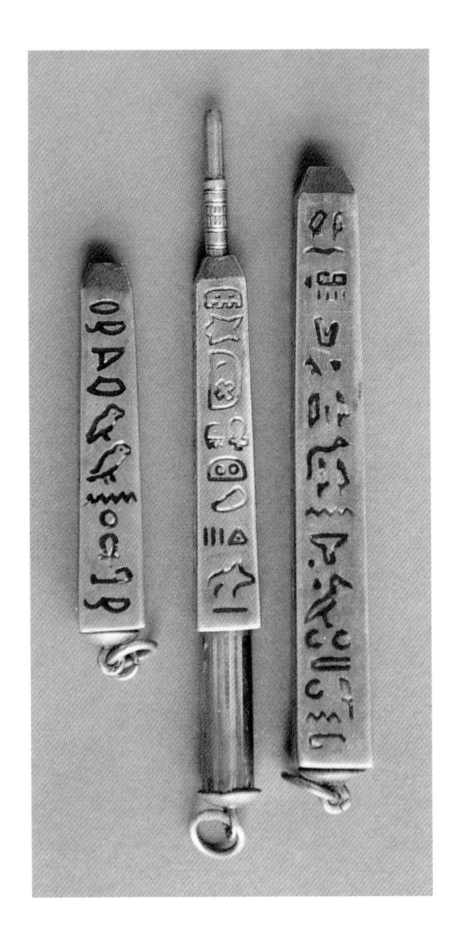

Three magic pencils shaped like obelisks. Large pencil, $225-$300. Smaller pencils $175-$200.

Different sized pencils, letter-opener, and a pocketknife shaped like Egyptian pharaohs or mummies.

Sterling silver and enamel telescoping pencil, with fine detail. $225-$275.

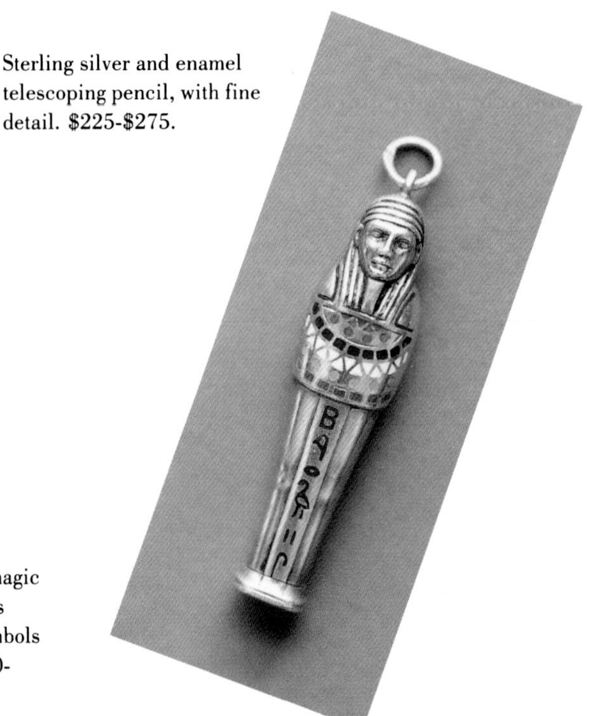

Silver sheath pencil and gold magic pencil in the shape of mummies holding the crook and flail, symbols representing the pharaoh. $150-$225, and $225-$250.

Two silver pencils in nearly identical shapes, one telescoping and one propelling. Rather large, at 3-1/2". $250-$300.

Gold pencil in the shape of a cross, with turquoise mounted slider. $150-$200.

Magic pencil, chased black hard rubber cross with gold-filled trim. $150-$175.

Unusually shaped, Egyptian motif magic pencil. Gold and enamel. $250-$350.

Silver slider pencil in the shape of a cross. $150-$175.

Gold-filled slider pencil in the shape of a cross. $150-$175.

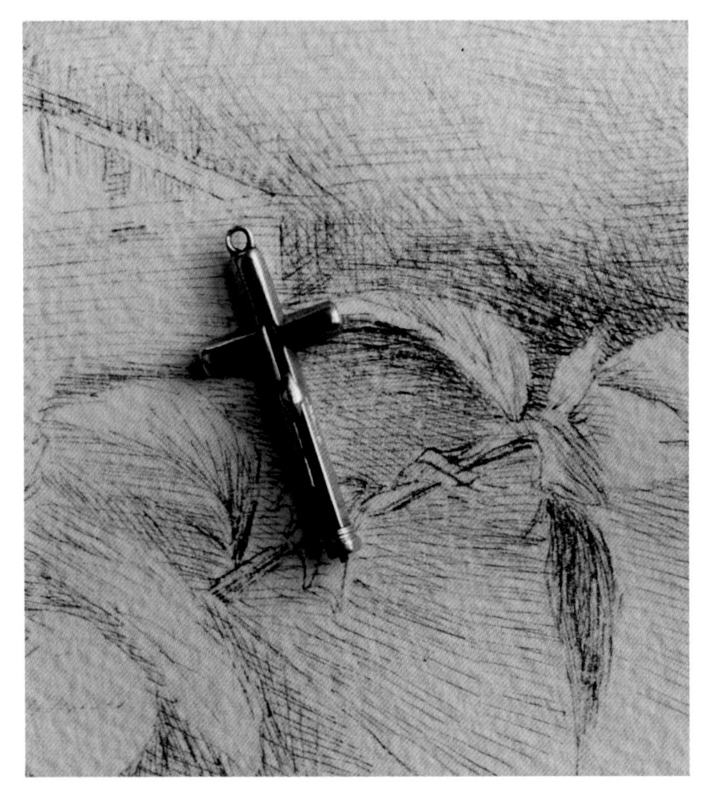

Another pencil in the shape of a cross, $135-$175.

Silver pencil in the shape of a pistol. Marked S. Mordan, July 6, 1840. $400-$600.

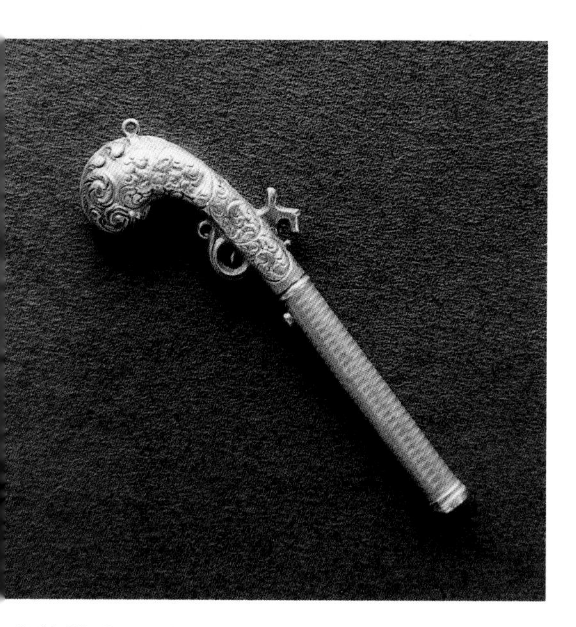

Gold filled pencil in the shape of a pistol. Pencil slides out by pushing a small pin on the bottom of the barrel. Unmarked. $350-$500.

White metal gun with a wooden handle. The pencil "shoots" out of the barrel when the trigger is pulled. 5-3/4" long when the pencil is extended. $175-$225.

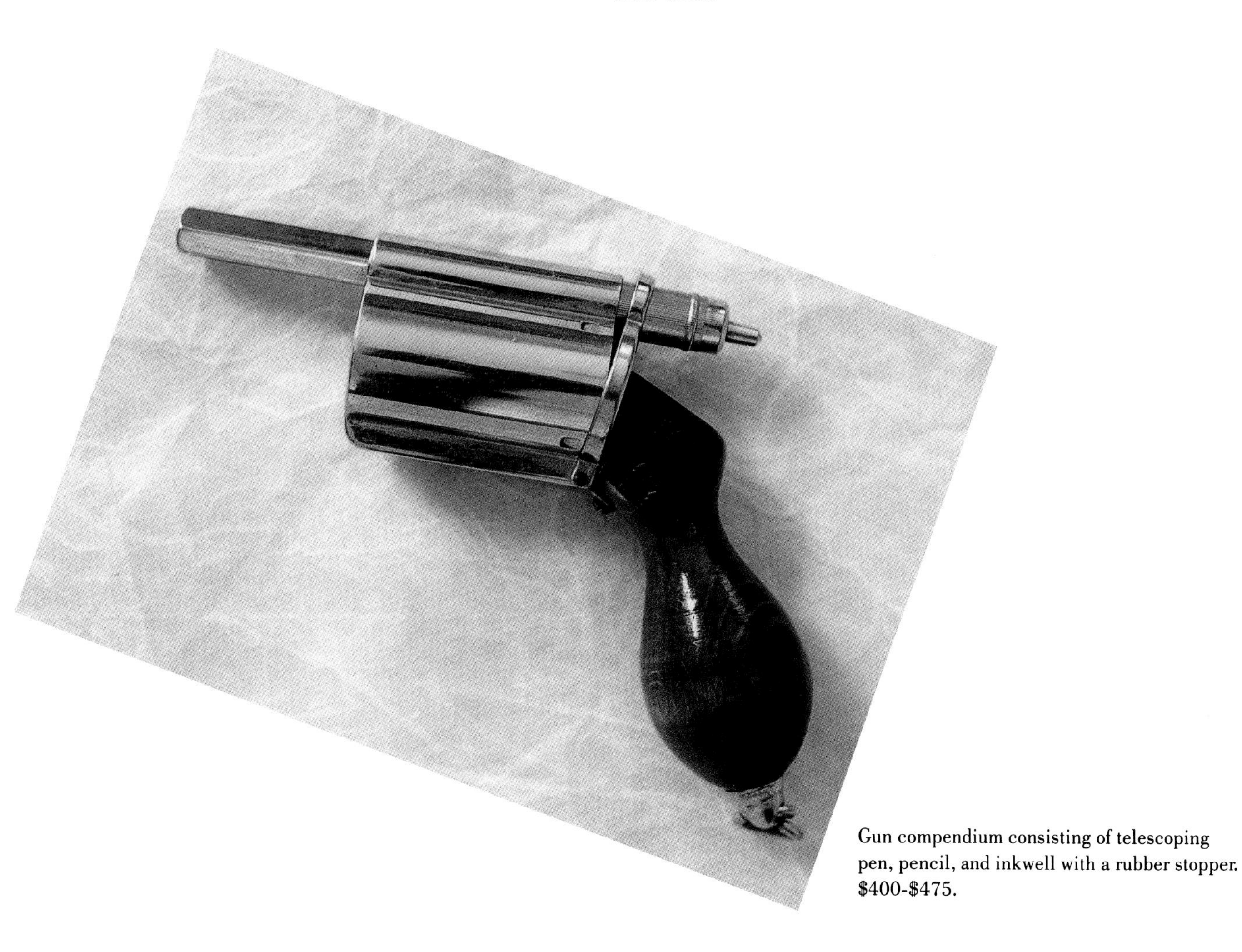

Gun compendium consisting of telescoping pen, pencil, and inkwell with a rubber stopper. $400-$475.

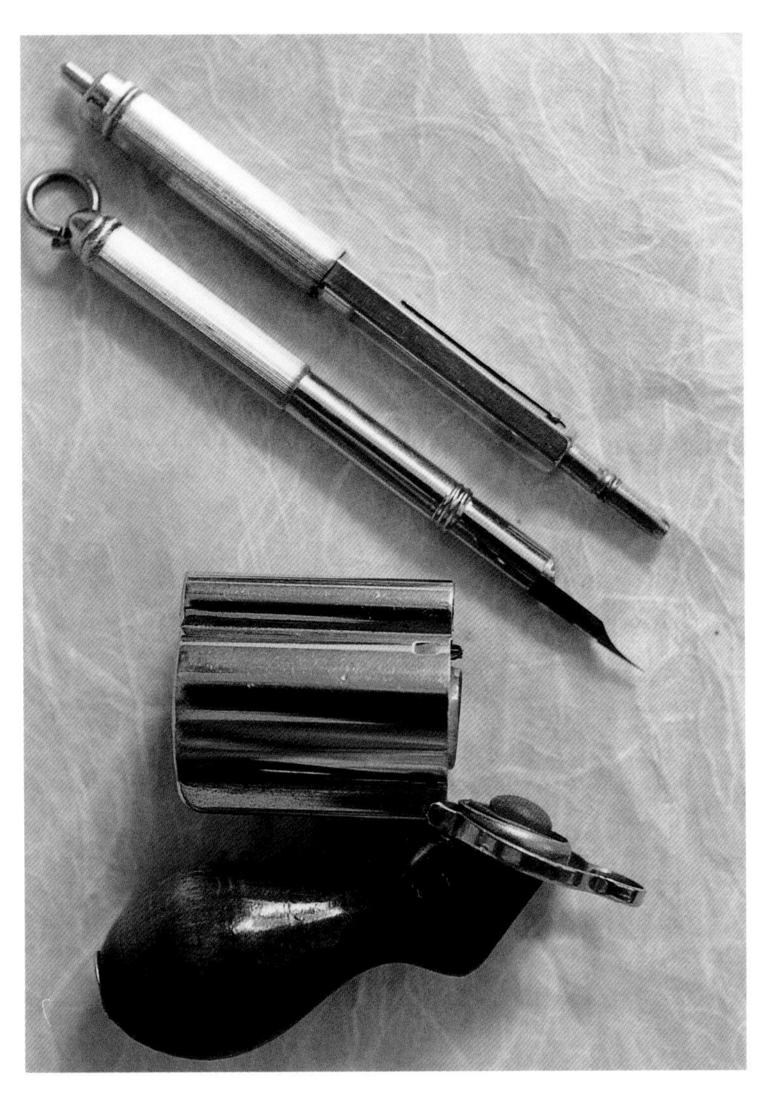

Same as above but open to show detail.

Champagne bottle in bucket. $175-$225.

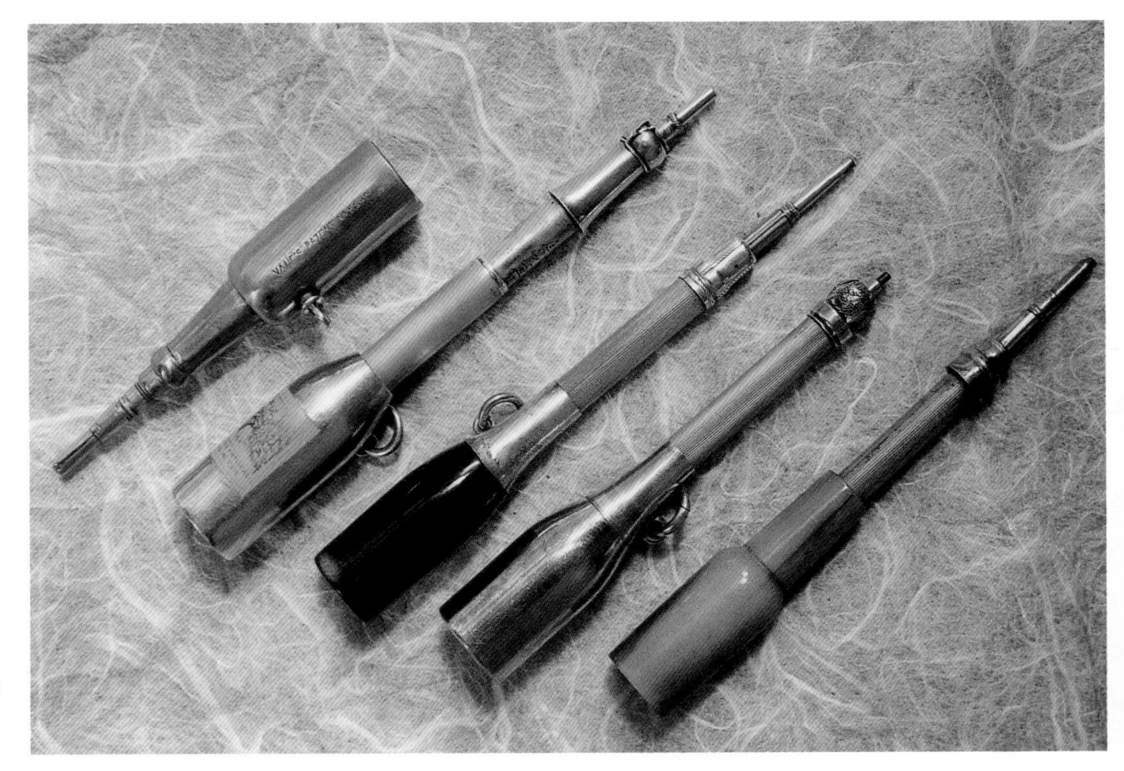

Pencils in the shape of bottles. From left to right; Vale's patent gravity mechanism, $150-$175. Silver and enamel with registration mark, $300-$400. Celluloid and metal, $75-$125. Gold and silver, $300-$400. Horn and metal $75-$125.

Silver and enamel bottle (with Apollinaris on its label) shaped pencil on advertisement circa 1880.

Wooden pencils in miniature bottles. $40-$60.

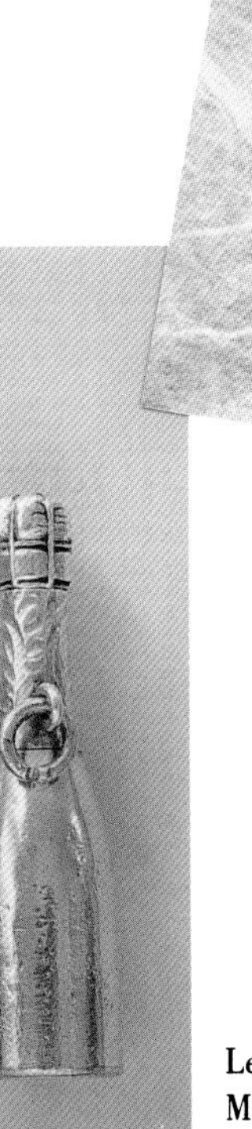

Left to Right: Bottle Pencils. Hard rubber pencil, Monopole champagne in brass and nickel plate, magic pencil by Leroy Fairchild. *Photo Courtesy of Chris Odgers.*

Silver and gold urn marked Edward Todd. $400-$600.

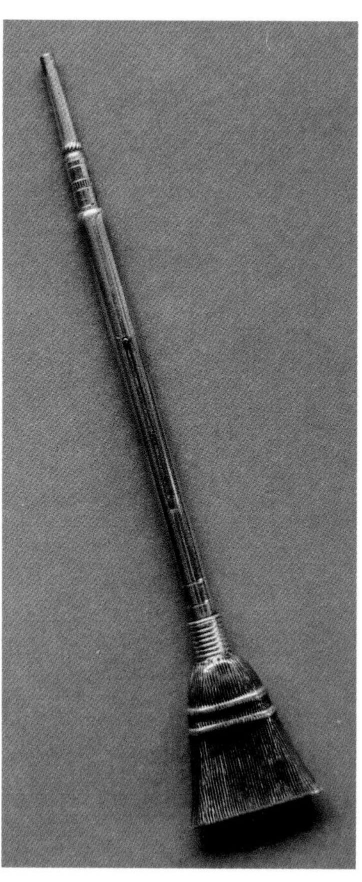

Sterling silver broom. $350-$500.

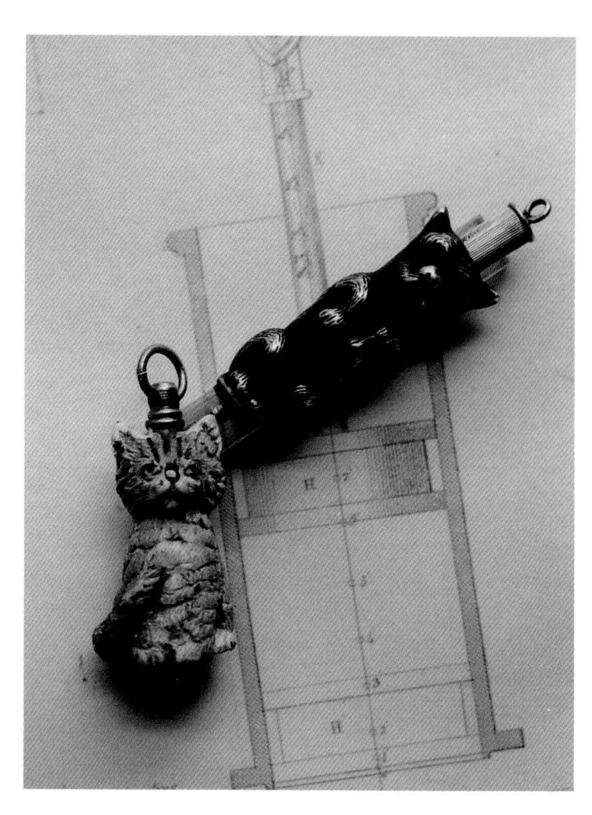

Composition cat shaped pencil, $125-$150. Silver and gold dog shaped pencil marked Aikin, Lambert, $350-$425.

Composition bulldog magic pencil. $175-$225.

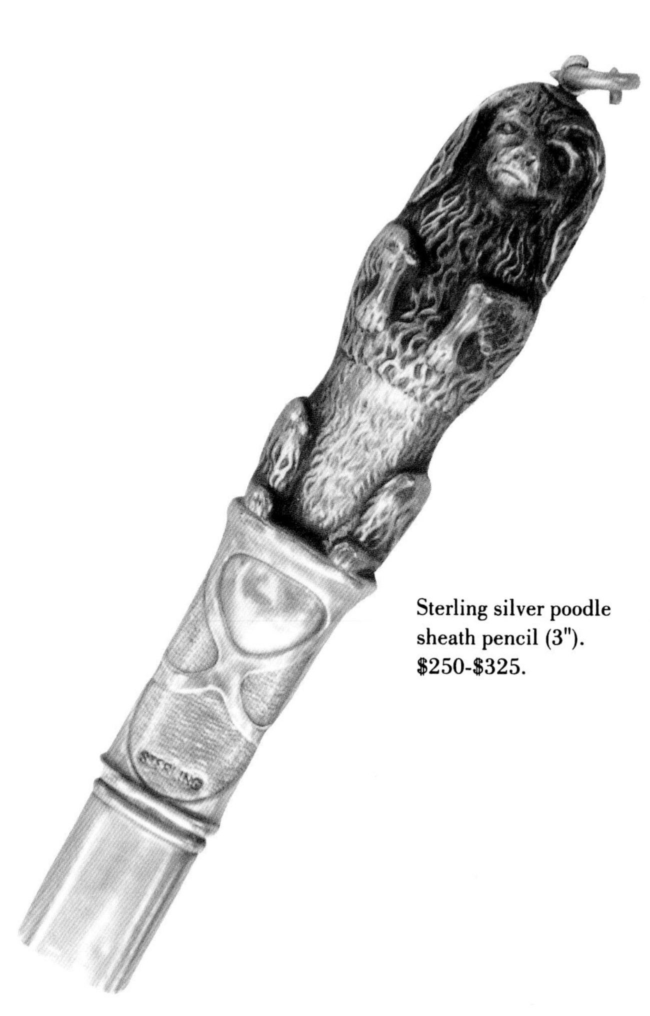

Sterling silver poodle sheath pencil (3"). $250-$325.

Magic pencil in the shape of a dog's head made of tiger-eye. Eyes missing. $400-$800.

Finial shaped like a ferocious animal. Marked Rausch. $225-$325.

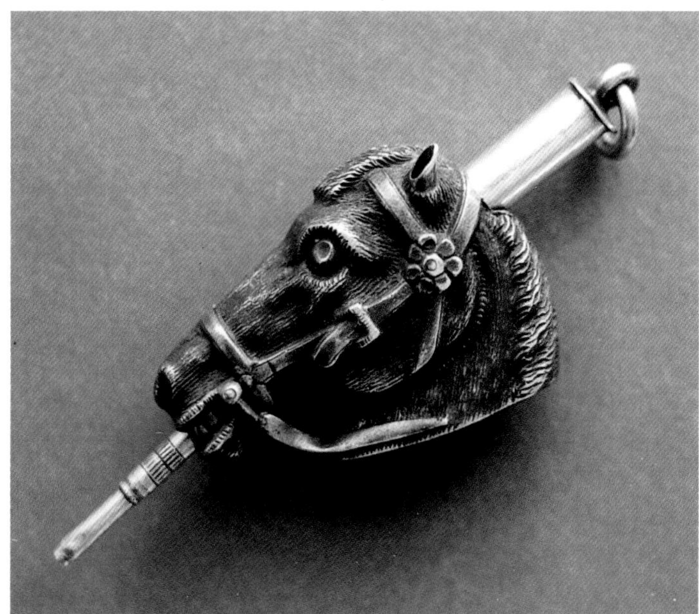

Finely detailed silver pencil in the shape of a horse's head. Unmarked, $450-$650.

Magic pencil with lizard and frog made of silver and rose gold. Unmarked. $175-$225.

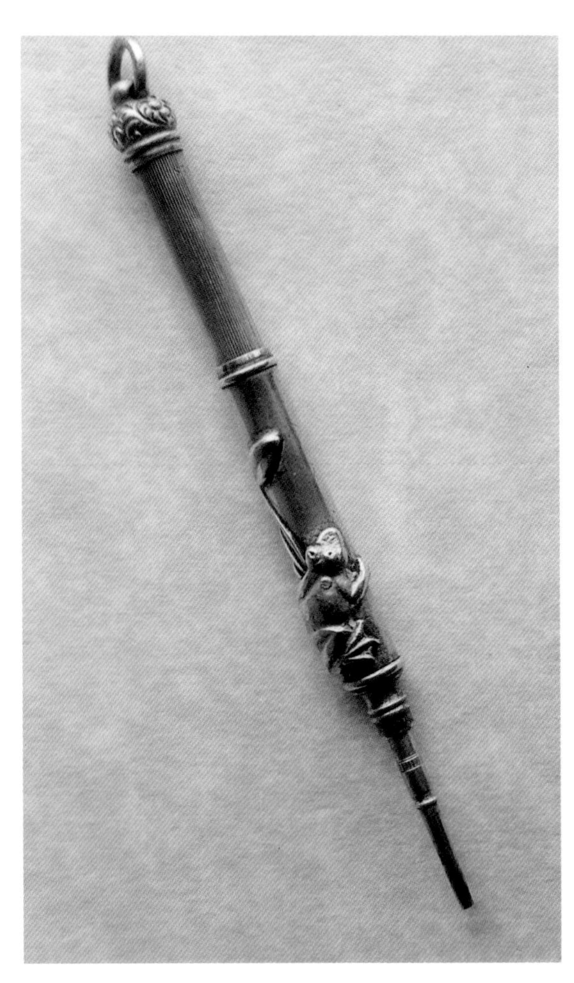

Silver magic pencil decorated with a frog. $175-$225.

Composition frog, with metal feet. Unmarked. $225-$300.

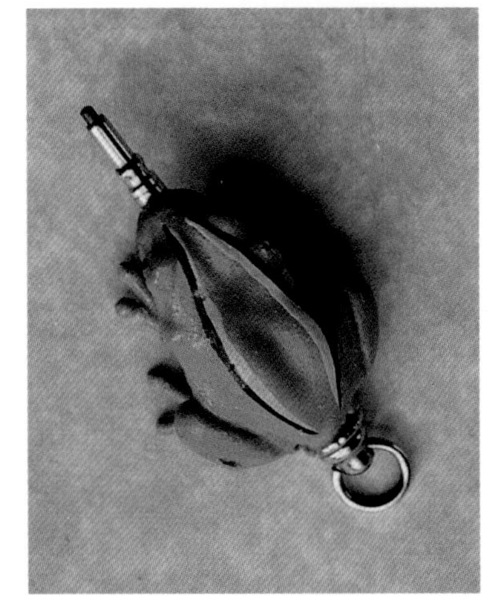

Composition and metal frog-shaped pencil, in remarkable condition. $325-$425.

Large silver magic pencil, with scene of a man standing near the ocean, with dolphins jumping over the waves, and a woman and an anchor, also by the ocean. $325-$450 (in working condition).

Left: silver sheath pencil with four leaf clovers and a good luck boar, $175-$225.
Right: silver sheath or vest pocket pencil with die stamped symbols of good luck, $125-$150.

Sterling silver vest-pocket pencil with a bas-relief lobster, marked "B" with a vertical arrow. $175-$225.

S. MORDAN & Co., Ltd.

PENCIL CASES, FANCY DESIGNS.

916. OWL PENCIL.

944. PIG PENCIL.

4641. PENCIL AND DESK SEAL.

877. CRICKET BAT PENCIL AND PAPER KNIFE. Pencil and Paper Knife. 4½ inches long. Olive Wood.

926. MATCH PENCIL.

1551. BRAMAH KEY PENCIL. ⅗ Size. With Reading Glass.

929. BAMBOO PENCIL. With Reserve for Leads.

917. TENNIS RACKET PENCIL. With Gilt Wirework. Made in 3 larger sizes.

887. CHAMPAGNE BOTTLE PENCIL.

8875. Any Brand Enamelled, or Bottle Copied to order.

897. BEER BOTTLE PENCIL.

For Trade Prices of above, see Price List page 6.

21 D

Page of figural pencils from a reproduction catalog of S. Mordan & Co., Ltd. (provided by Jim Marshall) originally published in 1898. Mordan not only offered a wide range of figural pencils, but also advertised that they were "always glad to consider proposals for the introduction of new inventions in any of the above or similar classes of goods."

Bakelite pencil shaped like a cigar, "Reina Victoria." Probably made by A.T. Cross. $75-$85.

S. Mordan pencil in the shape of a pig (the stones for the eyes are missing), registration mark. 1.25" closed. $350-$600.

Same as above left, but opened to show pencil.

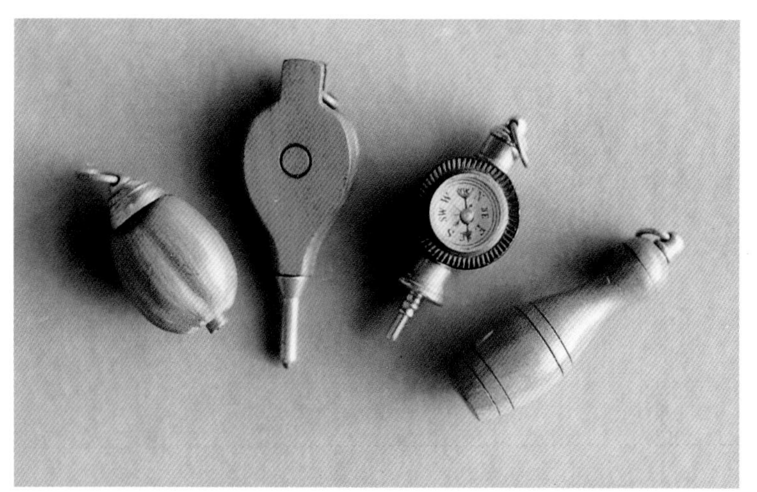

Early twentieth century novelty pencils, made of wood, nickel silver, or hard rubber: nut ($50-$75), bellows ($50-$75), compass ($50-$75), bowling pin ($50-$75).

Magic pencil finely crafted to look like a tiny ear of corn. Silver, no maker's mark, 1.25" closed. $400-$600.

Sterling silver pencil in the shape of a hand-rolled cigar (cheroot), twist mechanism. 3.5", $275-$350. Photographed on a leather cigar pouch, initialed L. E. W. (Louis Edson Waterman?)

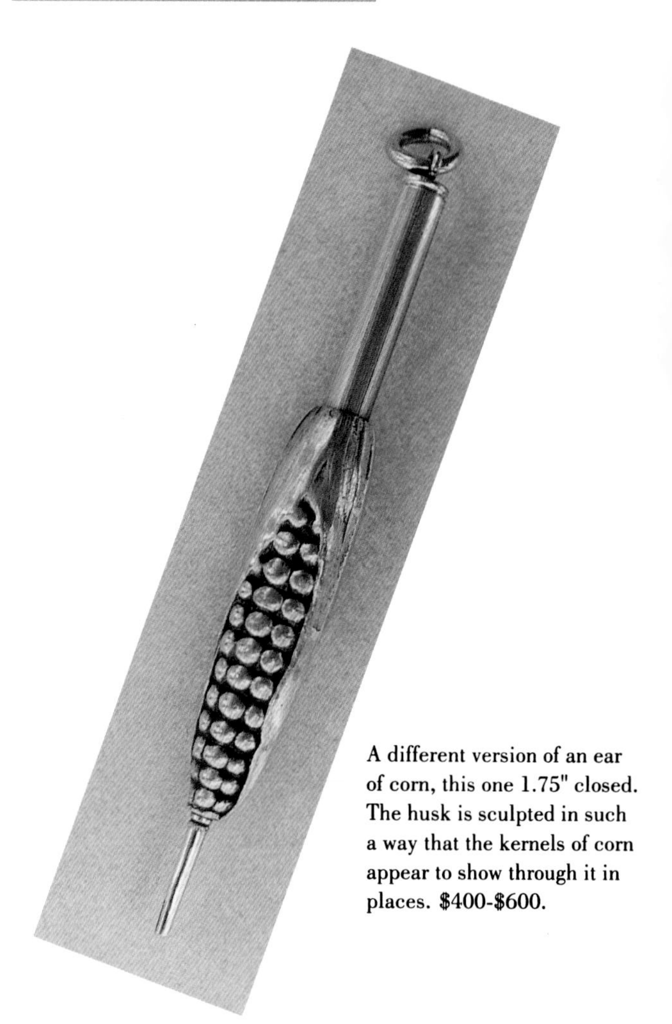

A different version of an ear of corn, this one 1.75" closed. The husk is sculpted in such a way that the kernels of corn appear to show through it in places. $400-$600.

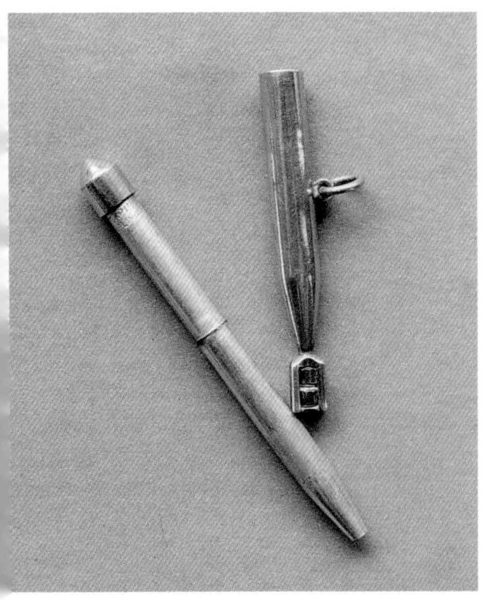

Sterling and rose gold telescoping pencil made to fit into a sheath shaped like a torpedo. Marked Edward Todd, patent applied. Pencil opened is 2.5" long. $250-$300.

Same as above, with the pencil in the sheath. Approx. 2.25" closed.

Two gold-filled screw pencils shaped like cannons, one marked W. S. Hicks. 2" closed, $175-$225.

Pencil in the shape of a cannon, made of green and pink gold, presented in 1885. $350-$425. *Photo Courtesy of Chris Odgers.*

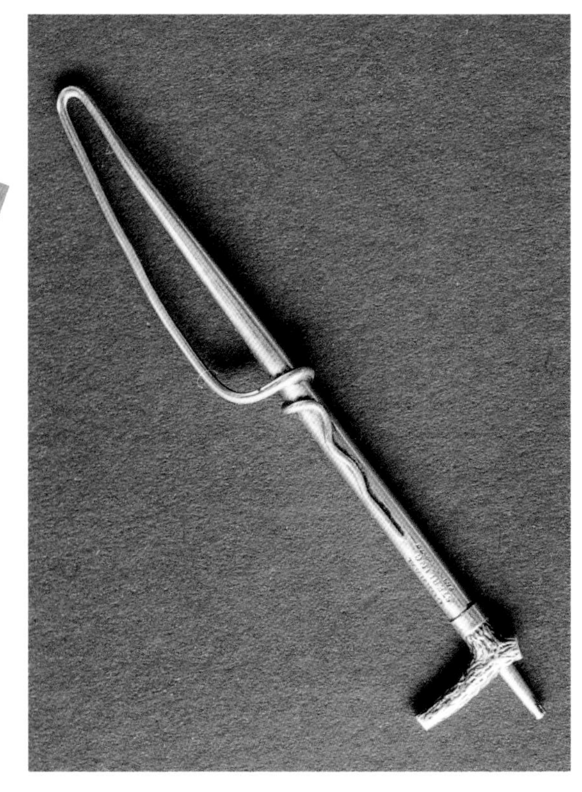

Marked Gorham, sterling silver slide pencil and collar-button hook, made to resemble a buggy whip. Patented October 8, 1878. 3.25", $425-$600.

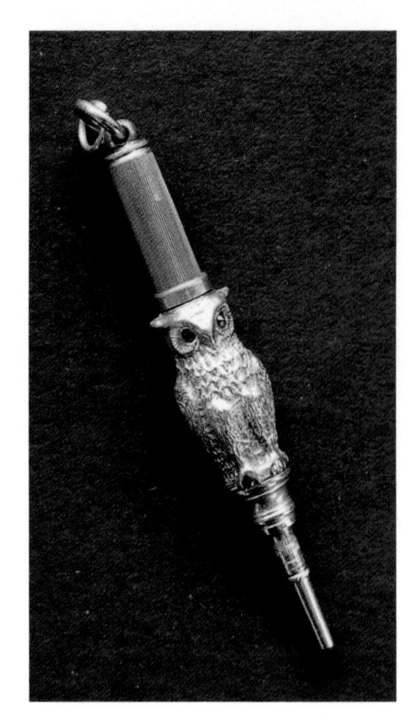

Magic pencil patented March 21, 1871, gold-filled owl with red stone eyes, made by W. S. Hicks. 1.25" closed, $375-$500.

Same as above, but in silver.

Silver colored owl, drop pencil. 1.5" closed, $175-$225.

Silver pencil in the shape of an owl, photographed on a Pendleton lithograph.

Novelty pencil shaped like a miniature deer foot. 1.5" closed, $75-$100.

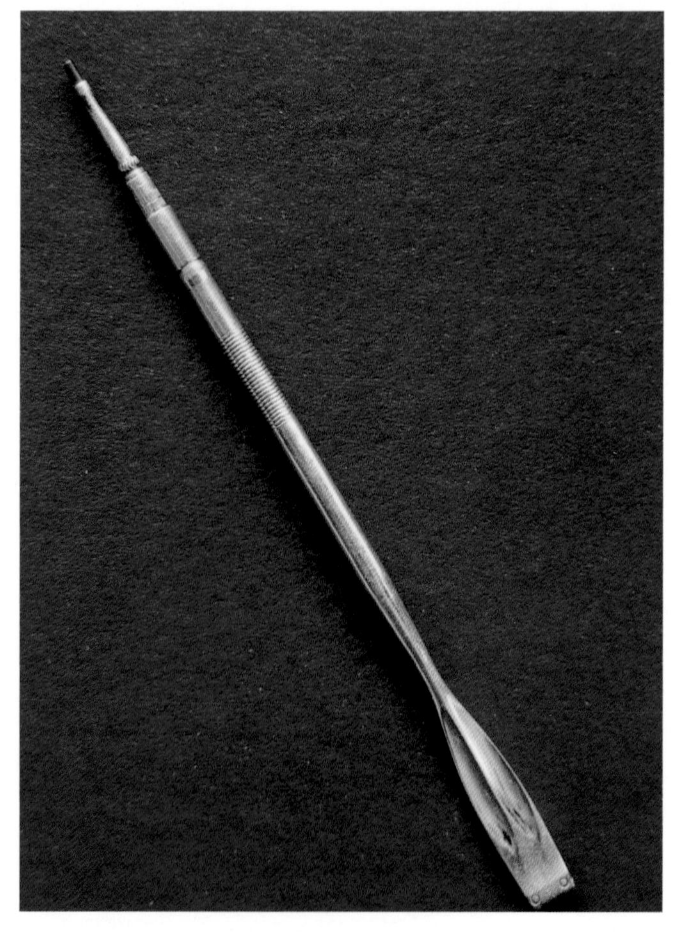

Silver oar slide pencil with minuscule bolts on the gold tip. Nearly 3.5" closed, $375-$425.

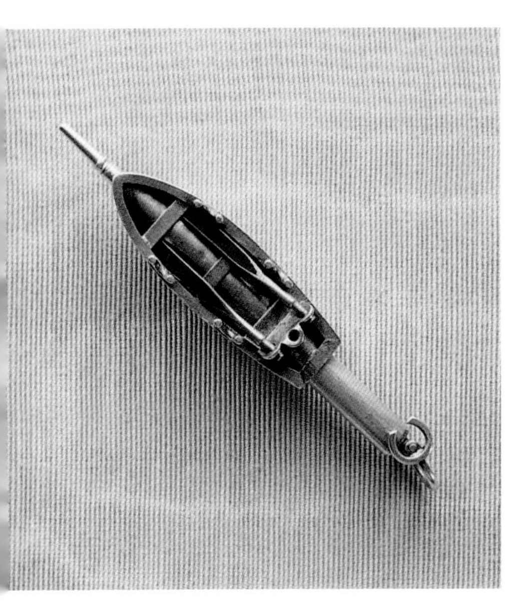

Silver rowboat with blue enameled oars, magic pencil. Remarkable detail with lapstrake hull and tiny oarlocks, perhaps made to commemorate a college team (Cambridge or Oxford). 2", closed, $500-$800.

Sterling silver spinning top pencil, made by S. Mordan & Co. 1.5" closed.

Same as left. Pencil extends six levels (five shown here), and has a registration mark indicating that it was made Dec. 7, 1873. 3.5" open, $600-$700.

Hand-engraved, 925 silver, twist mechanism pencil fabricated in the shape of a butter knife. Marked H. G. & S. Ltd.

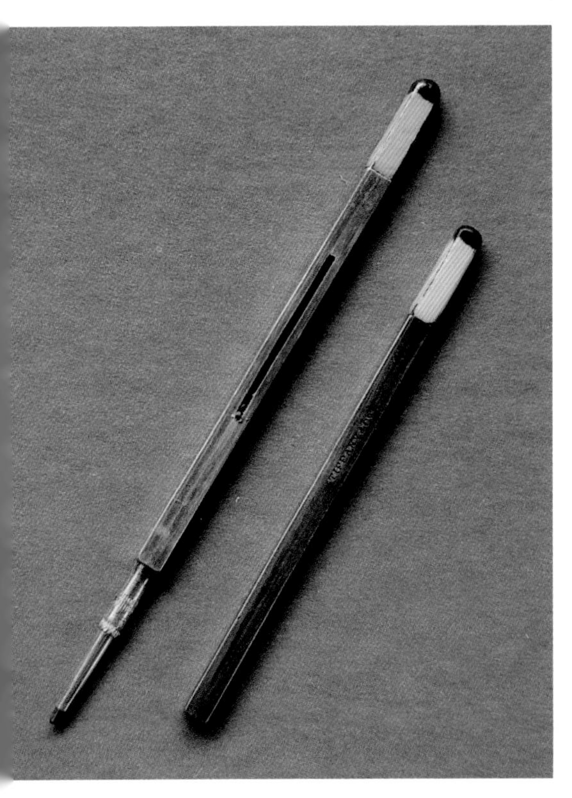

Tiffany sterling silver match stick slide pencils, with blue and yellow enameled tips, also made by Mordan and others. 2.5", $150-$175.

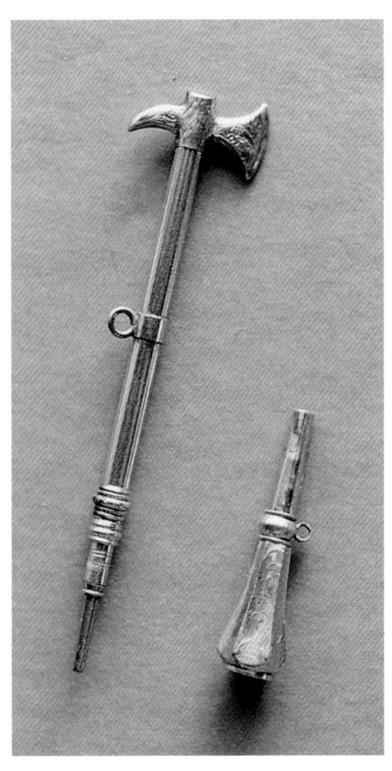

Gold and gold-filled pin-slide pencil in the shape of an axe, with a watch key attachment. Originally made as a watch fob. 3.5" total length, $275-$325.

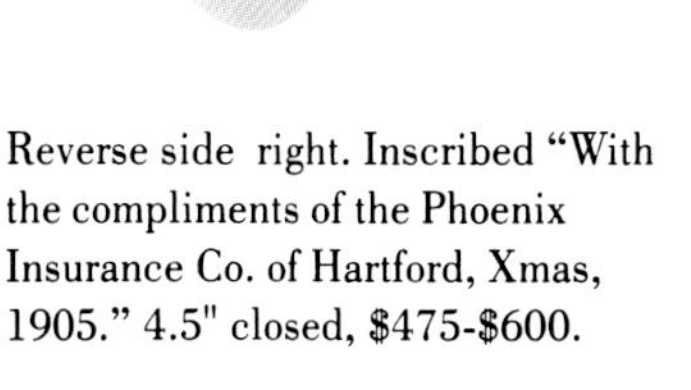

Reverse side right. Inscribed "With the compliments of the Phoenix Insurance Co. of Hartford, Xmas, 1905." 4.5" closed, $475-$600.

Hand-engraved magic pencil made by S. Mordan & Co., with a registration mark indicating it was made in 1882. The blade of the butter knife is engraved on both sides. 5.5" opened, $625-$800.

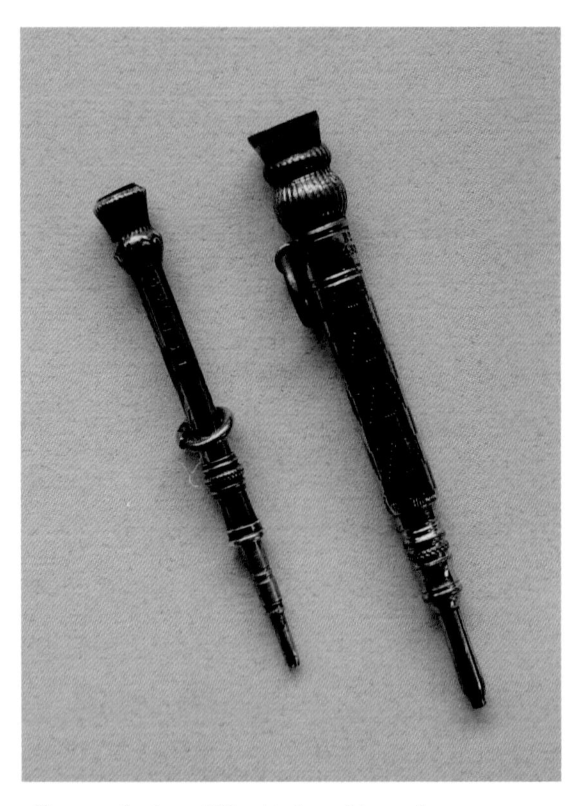

Two pencils shaped like thistles, with purple stone or glass set in the terminals. The larger pencil is made of non-precious materials ($50-$75), the smaller one is made of silver and marked T. H. V., hallmarked Birmingham, 1900 ($150-$175).

Highly detailed, magic pencil shaped like a woman in a kimono holding a fan on one side (see next photograph for the other side), marked Edward Todd & Co., with original watch fob chain still attached. 1.25", $500-$700.

Reverse side of above, with a male figure in a kimono using a barely visible toothpick.

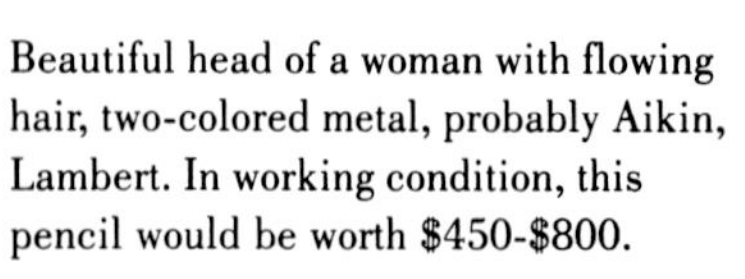

Beautiful head of a woman with flowing hair, two-colored metal, probably Aikin, Lambert. In working condition, this pencil would be worth $450-$800.

Sterling silver pencil shaped like a nail, mid-1800s. 3.5", $175-$225.

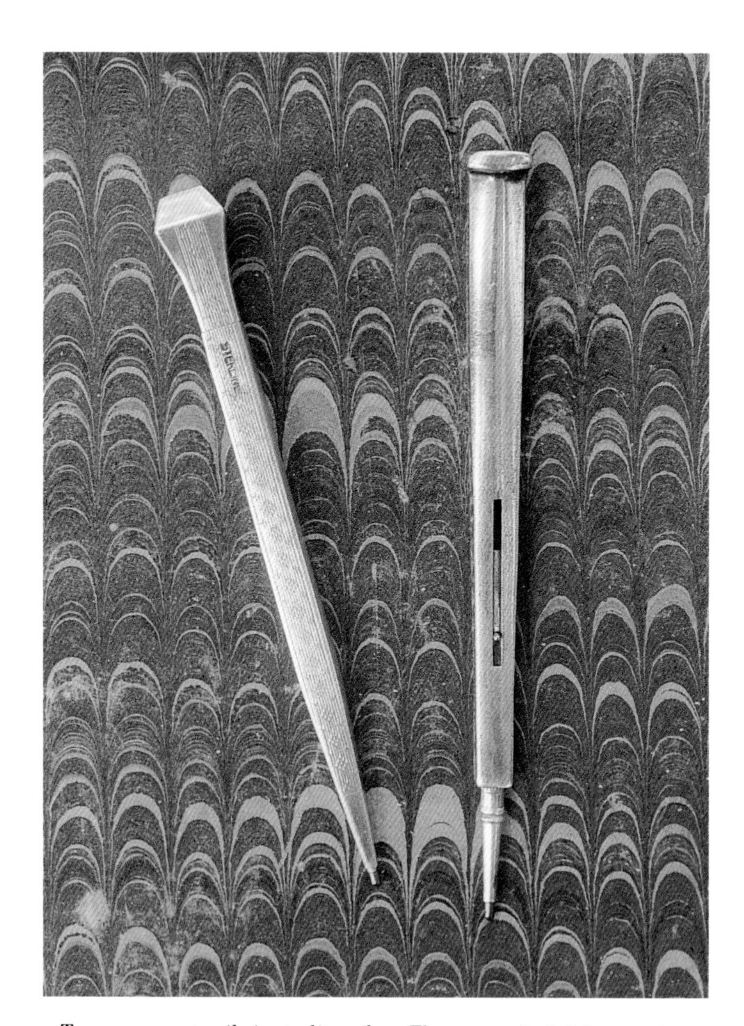

Two square-cut nails in sterling silver. The one on the left has a twist mechanism and is engine-turned, marked sterling ($175-$200), the other has a pin-slide and is marked "Sterling E." Possibly made by Eagle Pencil Co. Eagle used the designation E. on the pens and pencils they sold that were made of silver. $125-$150.

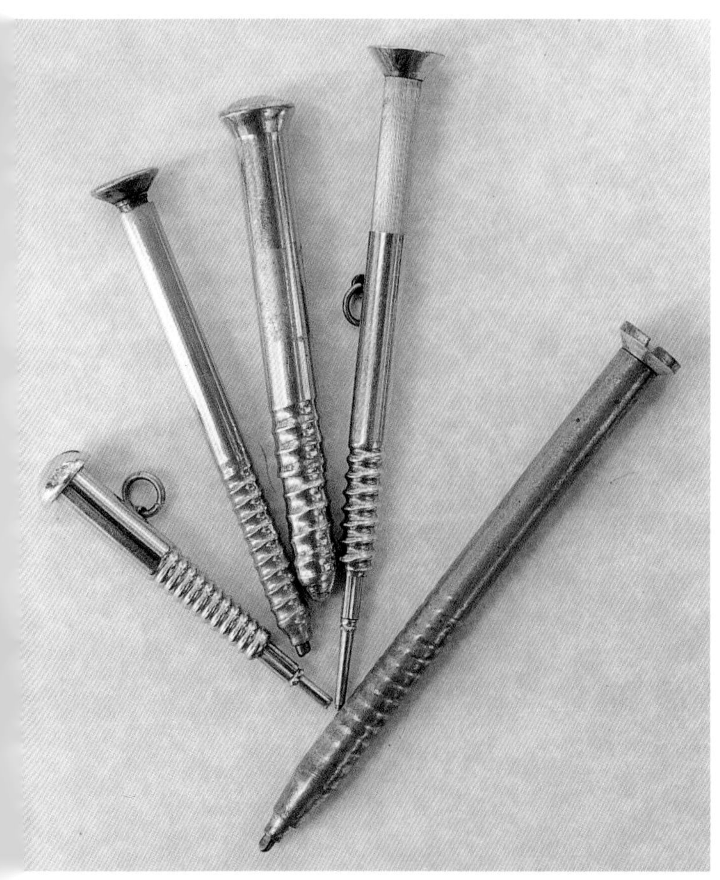

Left: Various pencils in the guise of screws, two with screw mechanisms, two are magic pencils and one has a cedar pencil that unscrews from the shaft. Left to right: $125-$175, $75-$100, $50-$75, $150-$175, $35-$50.

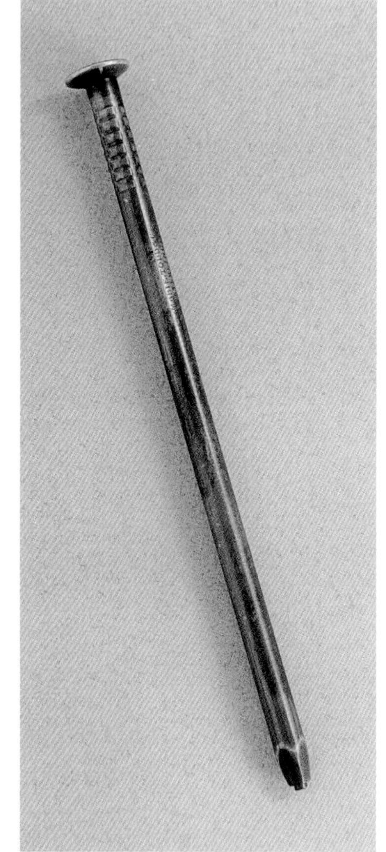

Right: Sterling silver, nail shaped pencil made by Mabie, Todd & Co. $175-$225. *Photo Courtesy of Chris Odgers.*

Sterling silver pin-slide, square nail shaped toothpick in its original box, made by Mabie, Todd & Co., $150-$175. Note that the same makers made pencils and metal toothpicks in the same form.

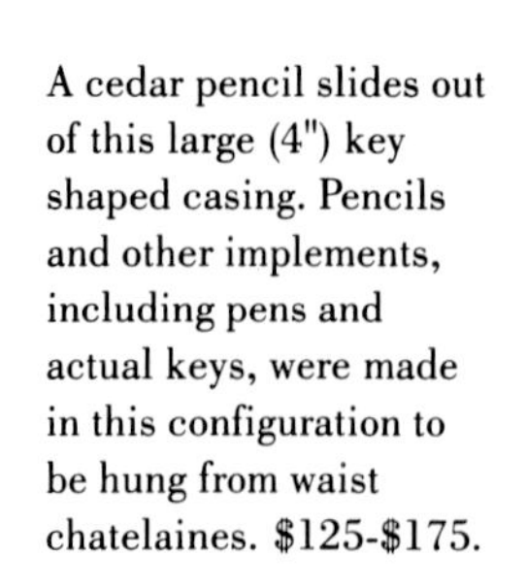

A cedar pencil slides out of this large (4") key shaped casing. Pencils and other implements, including pens and actual keys, were made in this configuration to be hung from waist chatelaines. $125-$175.

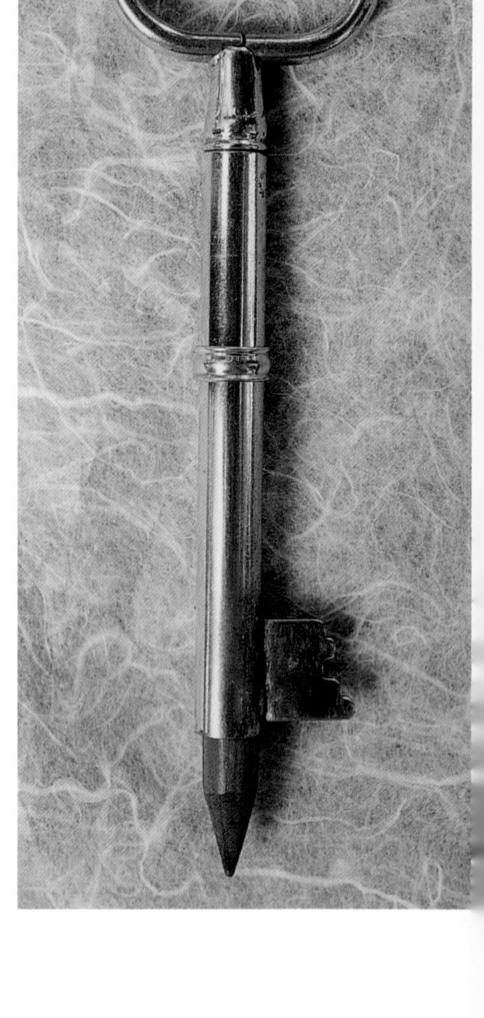

Pencil made to look like a nail by A. T. Cross, advertising Pittsburgh Steel Co. $75-$125.

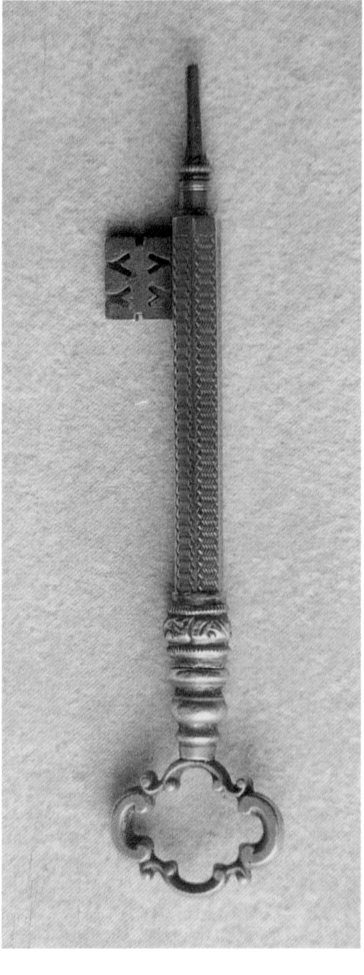

Silver key pencil, with engine turning on the barrel. Unmarked, $325-$400.

Gold-filled magic pencil in the shape of a simple key. $325-$400.

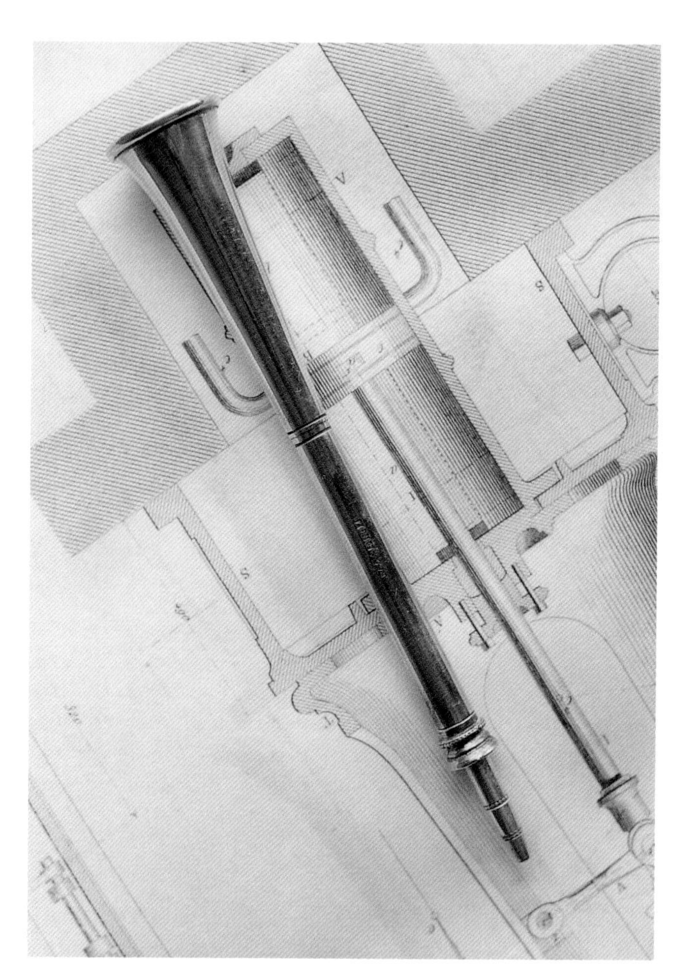

Silver coach horn pencil, hallmarked. $325-$500.

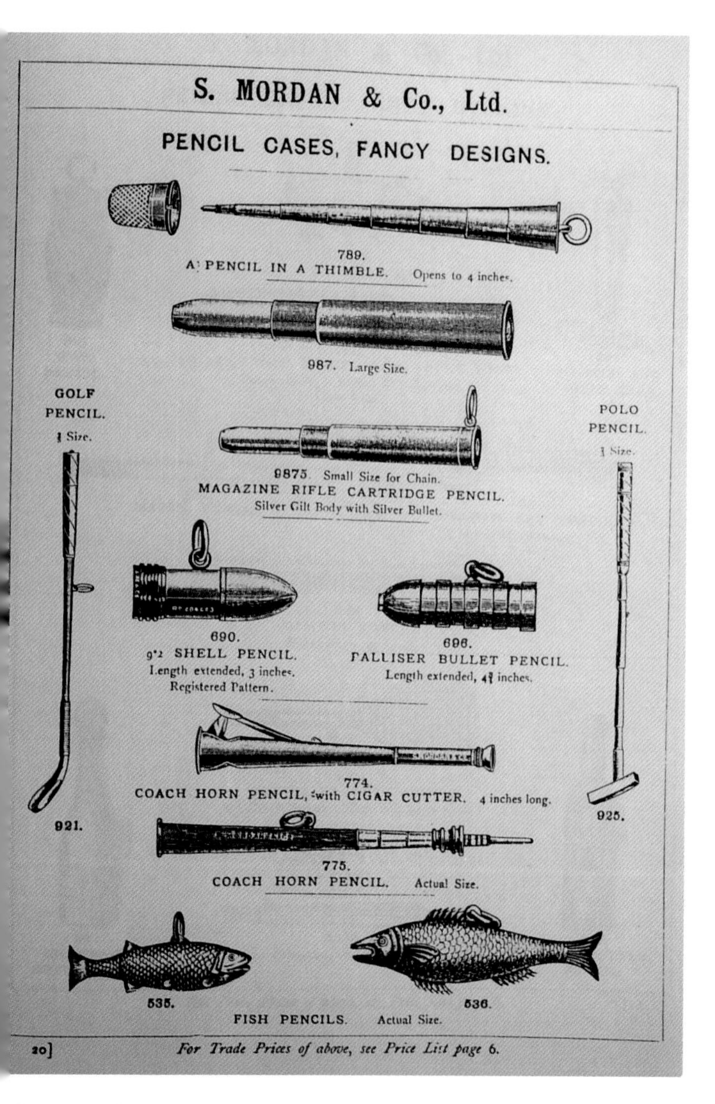

S. MORDAN & Co., Ltd.

PENCIL CASES, FANCY DESIGNS.

789.
A PENCIL IN A THIMBLE. Opens to 4 inches.

987. Large Size.

GOLF PENCIL. ¾ Size.

POLO PENCIL. ¾ Size.

9875. Small Size for Chain.
MAGAZINE RIFLE CARTRIDGE PENCIL.
Silver Gilt Body with Silver Bullet.

690.
9·2 SHELL PENCIL.
Length extended, 3 inches.
Registered Pattern.

696.
PALLISER BULLET PENCIL.
Length extended, 4⅞ inches.

774.
COACH HORN PENCIL, with CIGAR CUTTER. 4 inches long.

921.

925.

775.
COACH HORN PENCIL. Actual Size.

535.

536.

FISH PENCILS. Actual Size.

20] *For Trade Prices of above, see Price List page 6.*

Page from S. Mordan Catalogue, 1898, showing figural pencils including the coach horn.

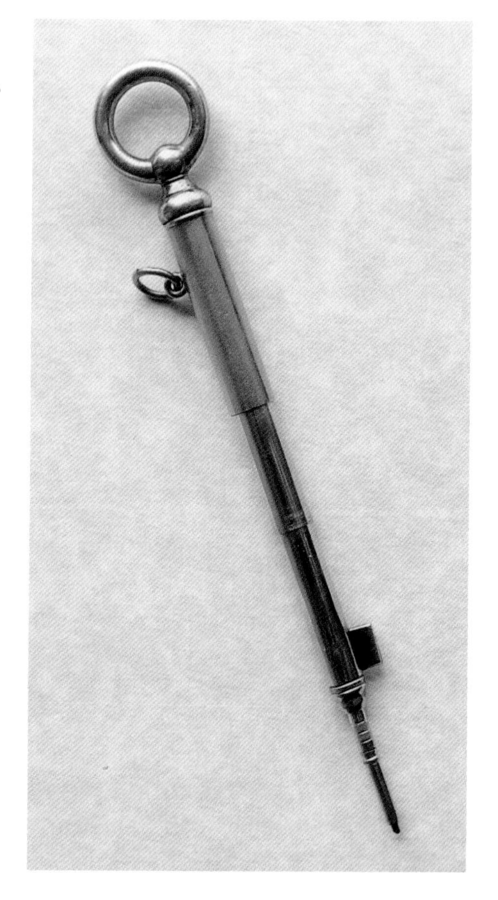

Same as above left, but open. Notice the combination telescoping-slide mechanisms.

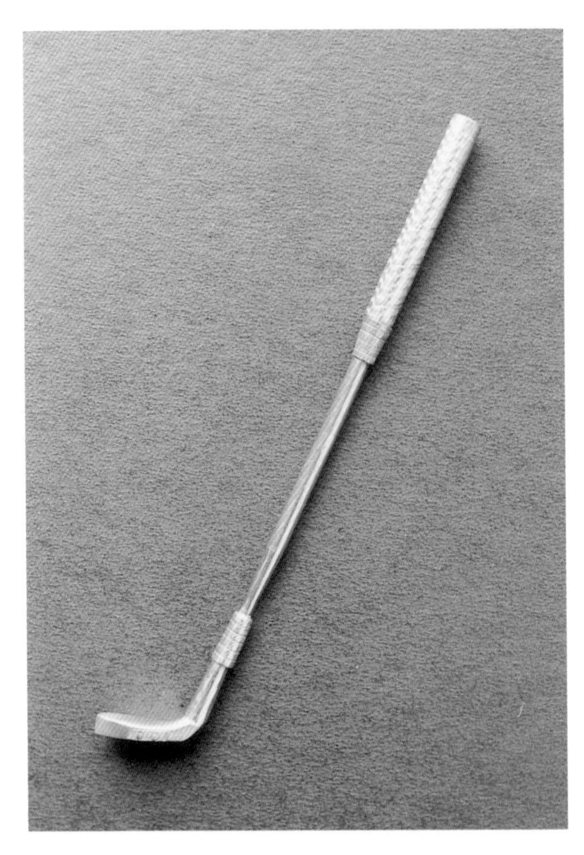

Silver golf club pencil. $325-$425.

Silver figure of a man playing the bagpipes. Unmarked, 3.25", in working condition worth $650-$800 or more.

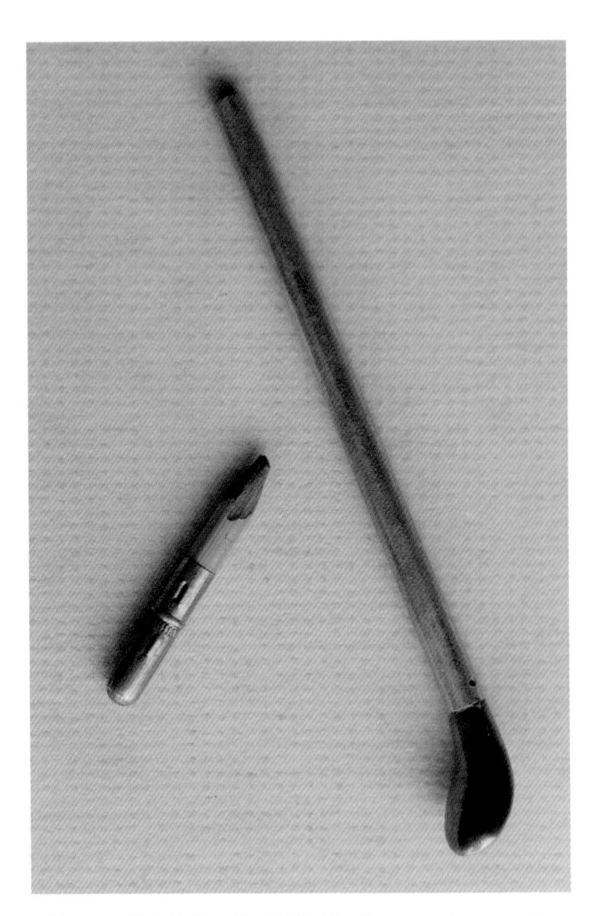

Brass and nickel, golf club holder for a wooden pencil (which fits into the end of the shaft). $35-$50.

Reverse side of above. The amount of detail on this piece is remarkable, and it is an extremely unusual piece.

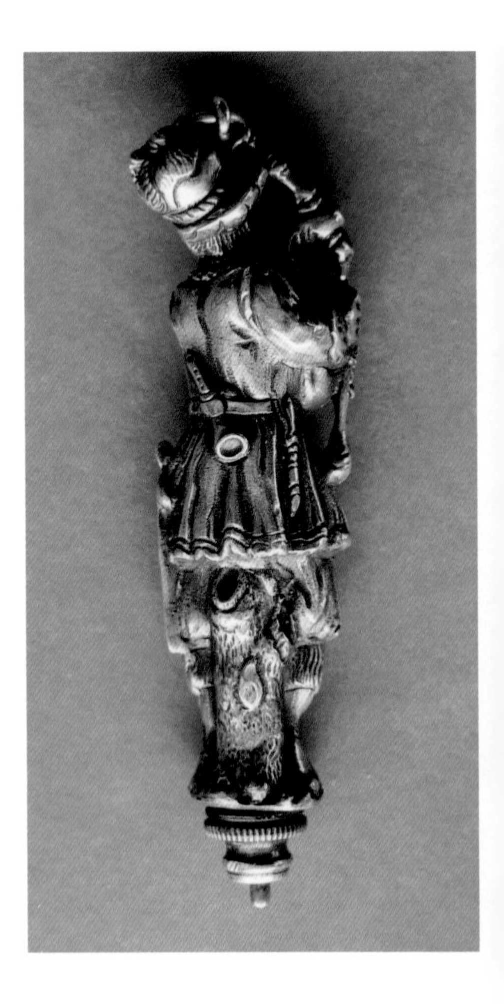

Tablet decorated with a print of a well-dressed man holding a top hat, and cane which is actually a wood and celluloid pencil. The tablet is 8.5" long by 2.5" wide, the pencil is 4.25" long. Circa 1870, $50-$75.

Globe pencil, papier-mâché, 1", $50-$75.

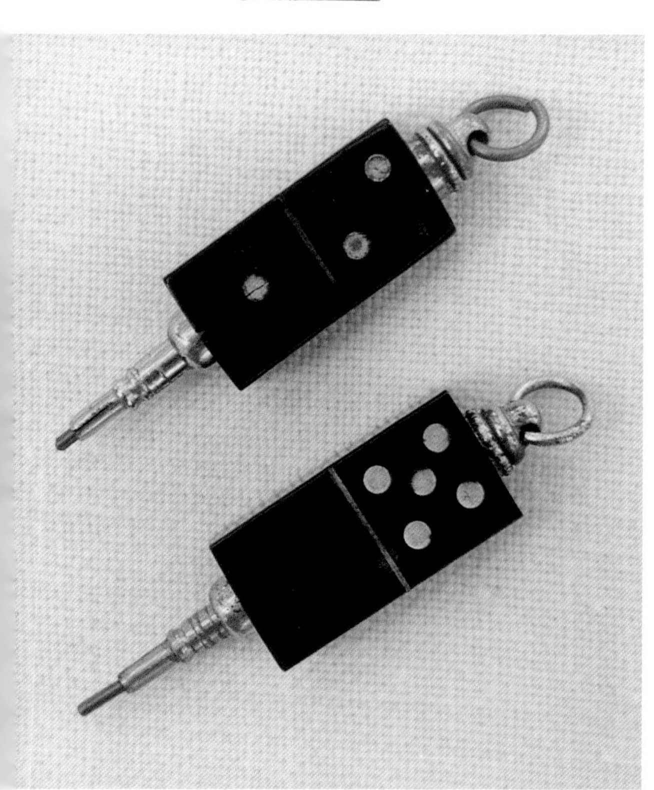

Ebony domino pencils. 1.75" opened, $75-$125.

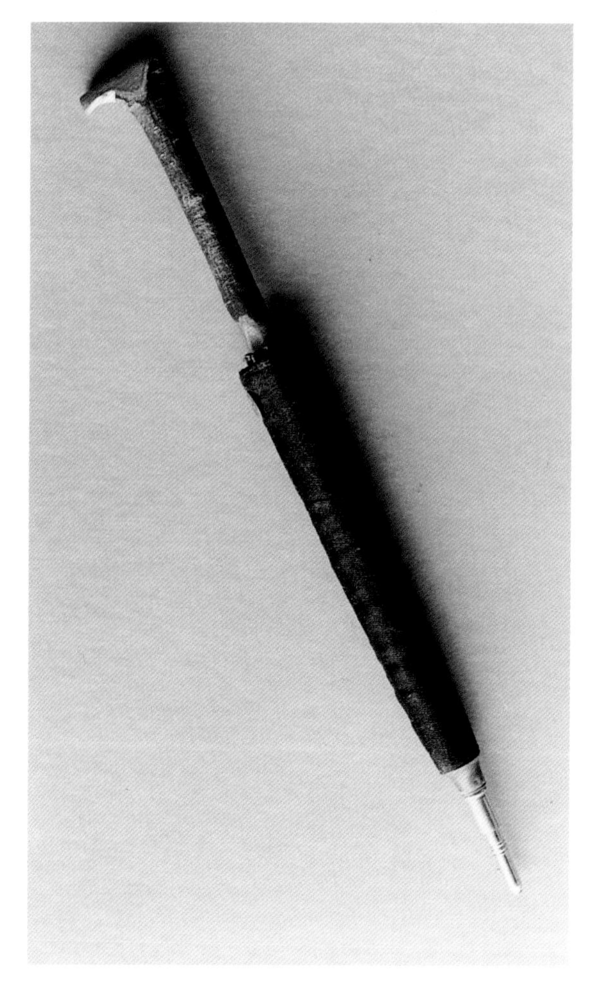

Rustic wood, fabric and metal "umbrella" which actually holds a dip pen inside the umbrella, and pencil in the tip. 9", $60-$75.

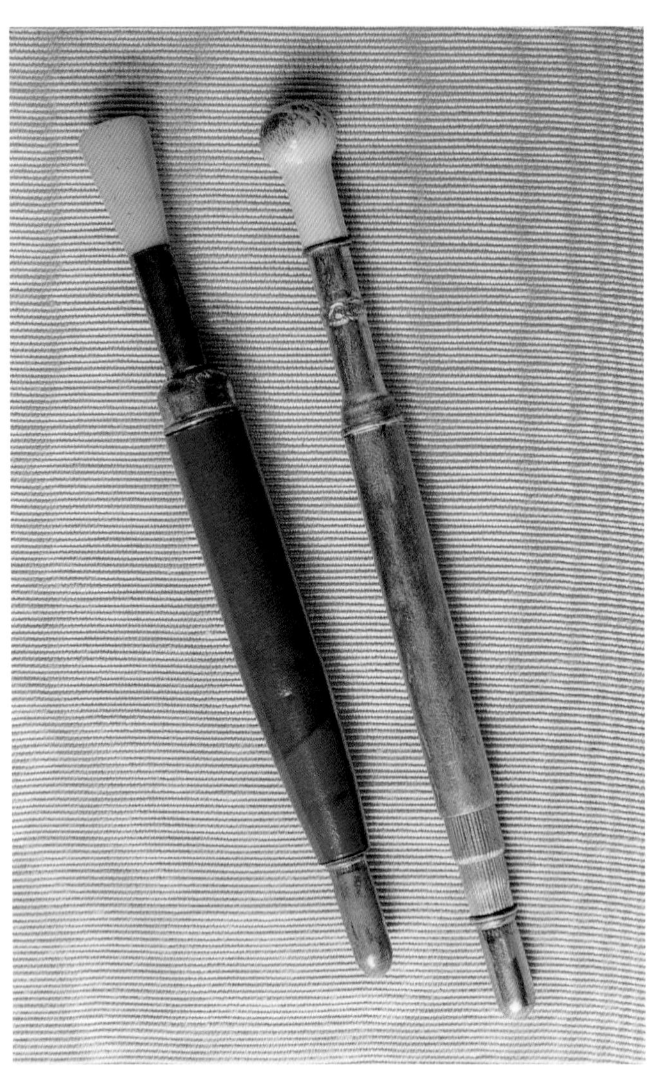

Two celluloid and brass pencil holders made to look like an umbrella and a cane. Early twentieth century, $35-$50.

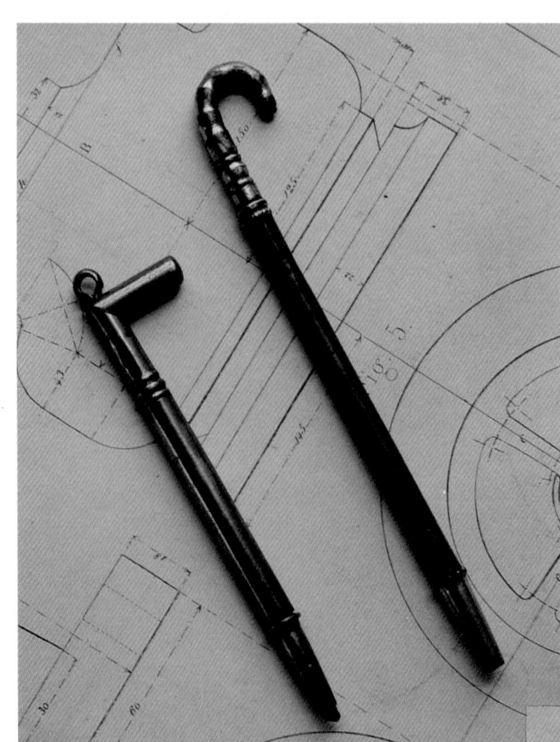

Nickel and brass pencils imitating canes. 3"-3.5", $45-$60.

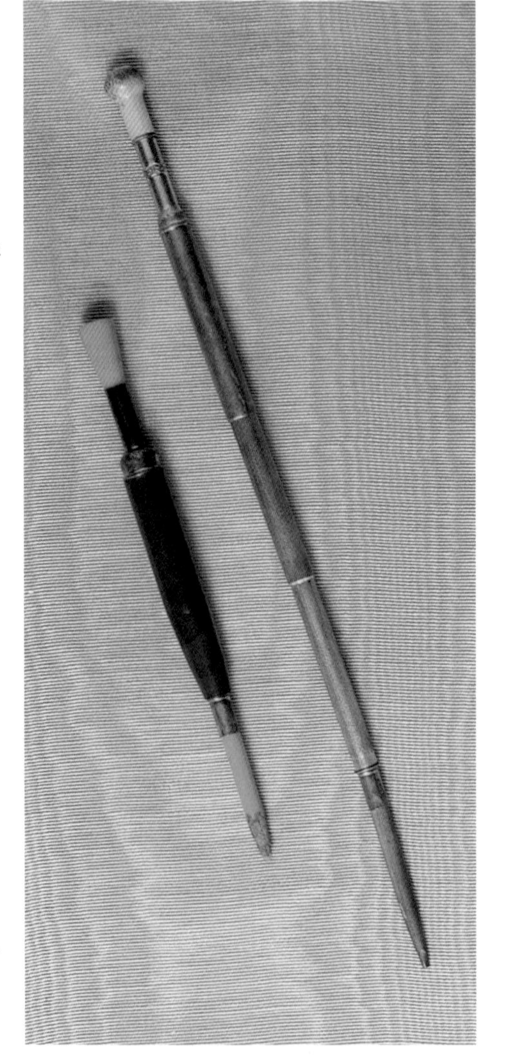

Same as above, but open to show wooden pencils. The cane telescopes out more than seven inches.

Gold-filled pencil in an umbrella-shaped sheath, with a turquoise stone set in the terminal. 3.25", $175-$250.

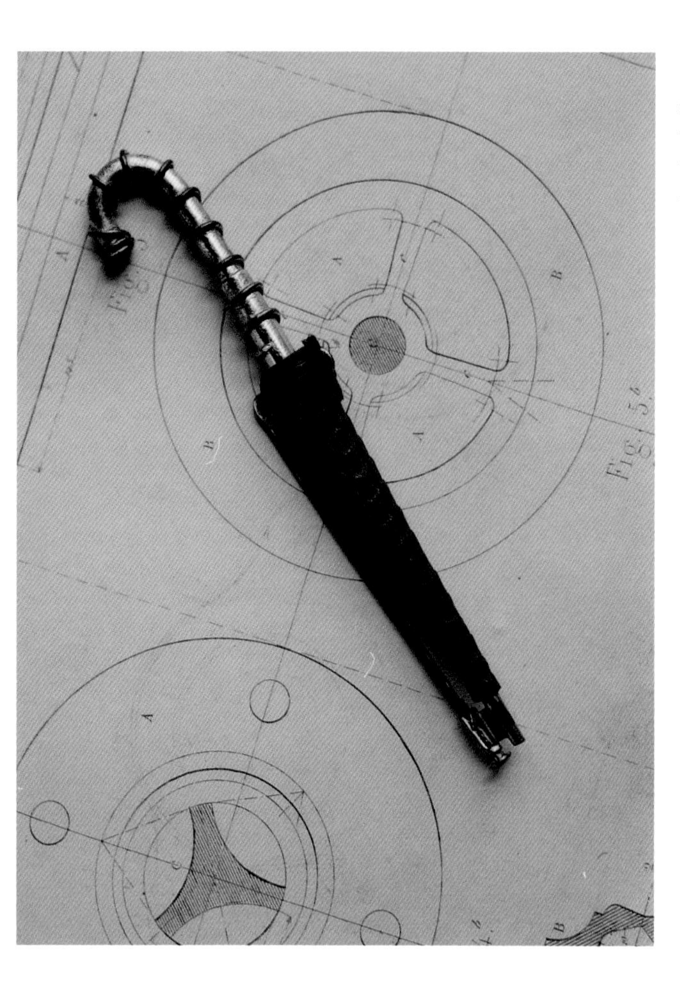

Brass and snake skin pencil in the form of an umbrella. This is one of the few pencils originally made to be worn as a pin. Probably French, 3.25", $125-$175.

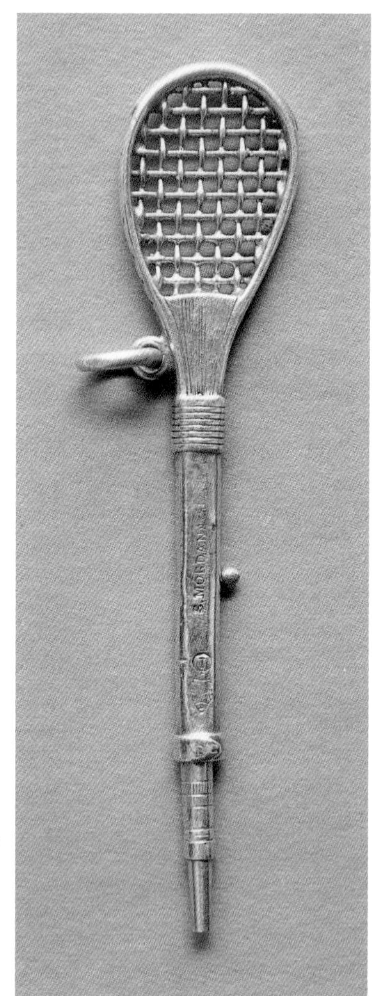

Silver S. Mordan & Co. pin-slide pencil, shaped like a tennis racket. Registration mark, 1878. 2" closed, 2.5" open, $350-$500.

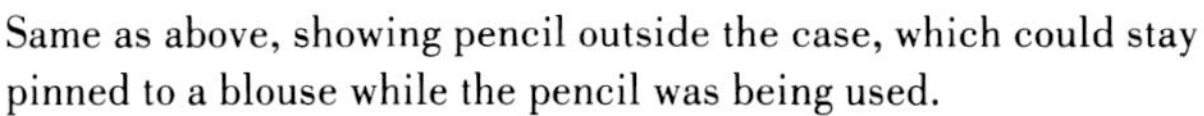

Same as above, showing pencil outside the case, which could stay pinned to a blouse while the pencil was being used.

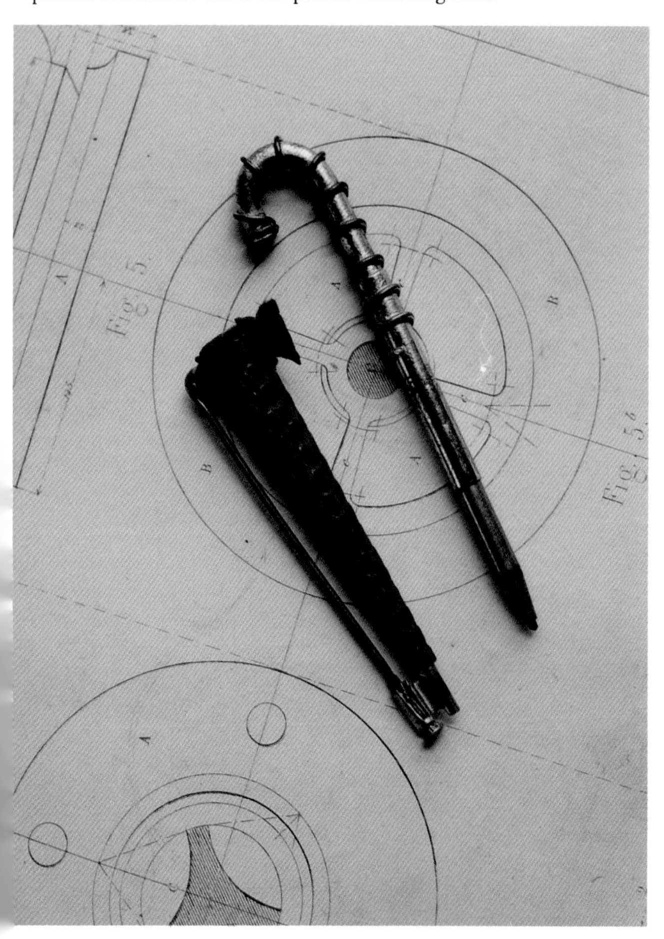

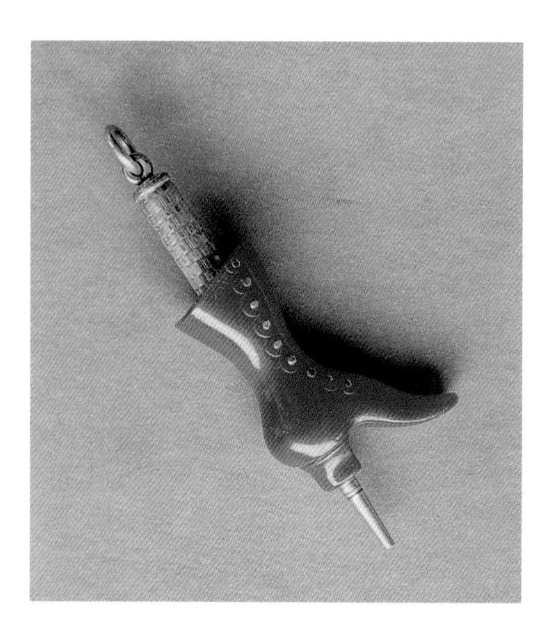

Coral-colored celluloid, high-topped shoe figural pencil. Mark indecipherable, 1.5" closed, 2.5" open, $225-$275. Sometimes celluloid will look very much like the material it's made to imitate (tortoiseshell or coral, for example), but a visible seam will often make correct identification possible.

Mallet pencil made of nickel-plated brass. 2.5", $45-$60.

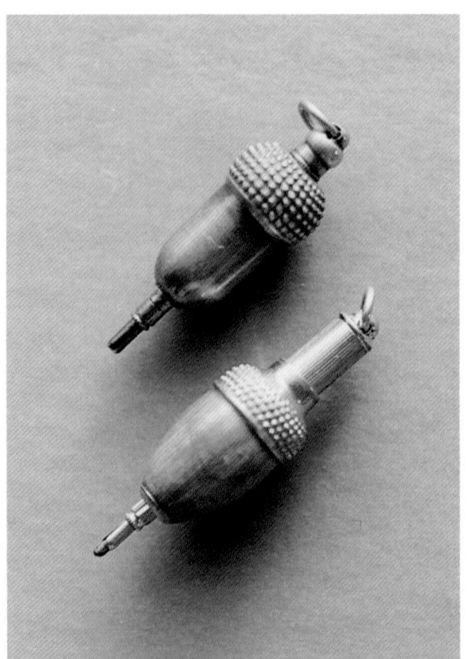

Two magic pencils, emulating acorns. Both 1.25" closed. Gold-colored metal pencil, $150-$175. Wood and metal pencil, $75-$125.

Same as above, open to show wooden pencil.

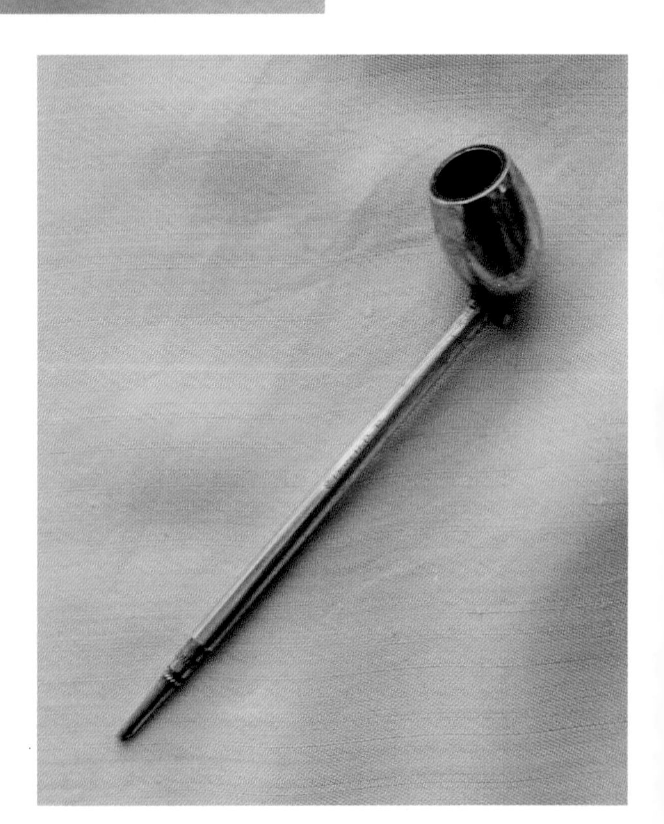

Sterling silver pipe, pin-slide pencil, 3.25" open. $250-$300.

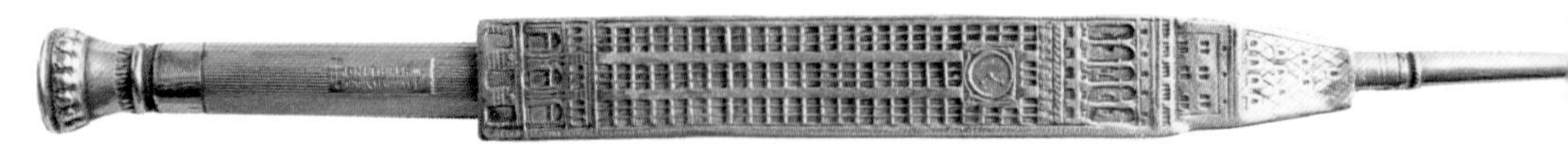

Metropolitan Life Building (New York City) magic pencil, marked Tiffany and Aikin, Lambert. The pull end may be a replacement, as the original version of this pencil had a square pedestal base. 3.5", with correct parts, $450-$700.

w-104 Daniel Low & Co.—STERLING SILVER NOVELTIES—Salem, Mass.

PENHOLDERS, PENCILS, PROTECTORS, TOOTHPICKS. All are sterling silver. Illustrations actual size.

We wish to call attention to the very fine hand chasing on our pencils. *Takes long leads. Adjustable to several sizes, holds leads perfectly rigid.

*S 424 Pencil, engine turned 1.25 · S 399 Tooth Pick, hand chased 1.00 · S 414 Pencil .85 · S 2429 Pencil .65 · S 55 Pencil .50 · S 655 Tooth Pick Case, quill .60 · S 409 Pencil 1.00 · S 1428 Pencil 1.25 · S 407 Pencil 1.35 · S 405 Pencil, hand engraved 1.75 · S 408 Pencil, engine turned 1.35 · S 420 Detachable like 422 2.50 · S 56 Pencil 1.25 · S 280 Pencil 1.50 · S 643 Case holds wooden tooth picks .75

S 651 Penholder grey 1.35 · S 652 Penholder, heavy, 1.25 653 plain 1.25 · S 210 Fine chasing 1.00 · S 647 Penholder, openwork 1.00 · S 641 Chased .85 · S 642 Pen holder, etched .50 · S 402 Pencil Protector .50 · S 415 Peanut Pencil 1.35 · S 421 Pencil Protector heavy 1.35 · S 398 Pencil Protector grey .85 · S 403 Pencil Protector .90 · S 401 Pencil Protector calendar 1.35 · S 208 Case for Koh-i-noor Pencil .75 · S 410 Screw Pencil, chased .75

S 422 New small size detachable Pencil. The pencil when closed fits entirely within case 2.00

S 1447 Pencil Sharpener, hand chased, shown open 1.00

Catalogue page from Daniel Low & Co. (who also made the Salem Witch souvenir spoons and pencils) showing the sterling peanut pencil.

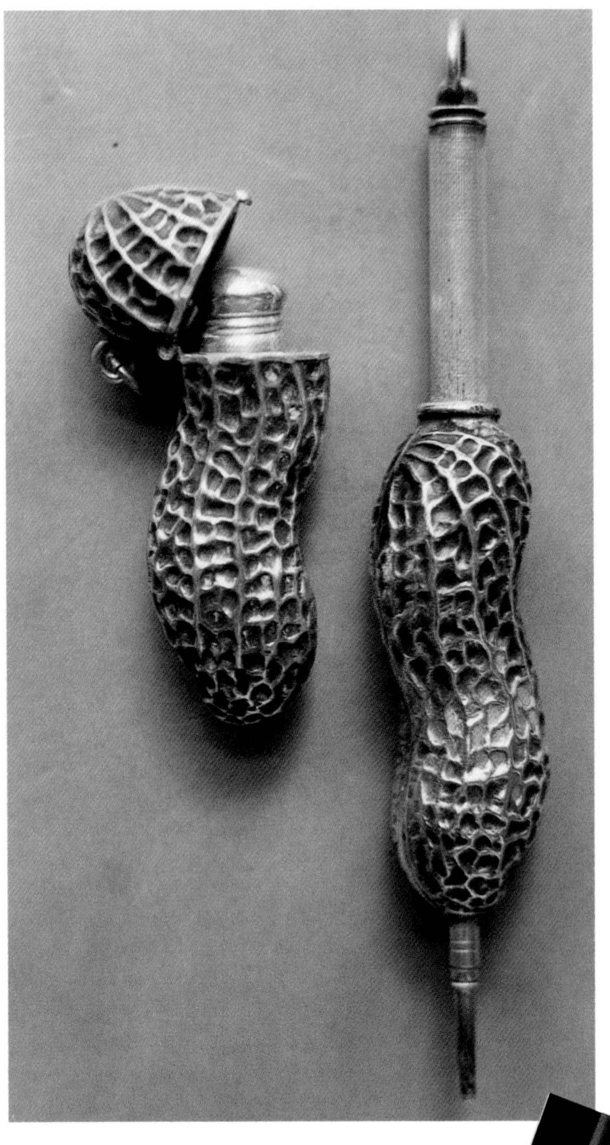

Sterling silver scent bottle in the shape of a peanut ($225-$250) next to a pencil made in the same style ($225-$250).

Reverse side of right.

Silver magic pencil, with a repoussé figure on each side. $125-$175.

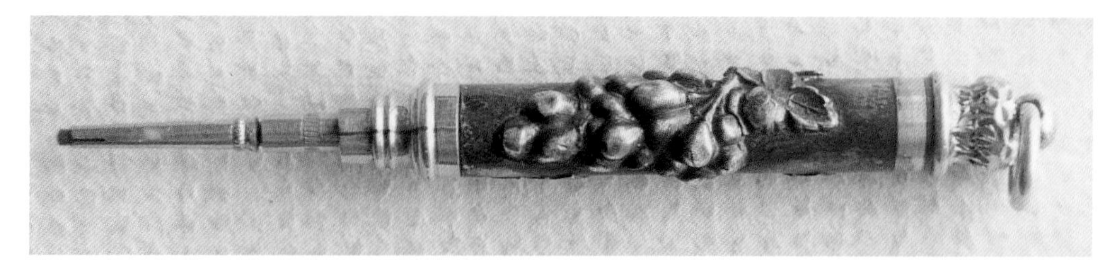

Gold and silver twist pencil with a bunch of grapes. 2", $125-$150.

Slide (with cabochon stones set in sliders) pen-pencil combination, gold metal with repoussé figure of a nude woman. $325-$350.

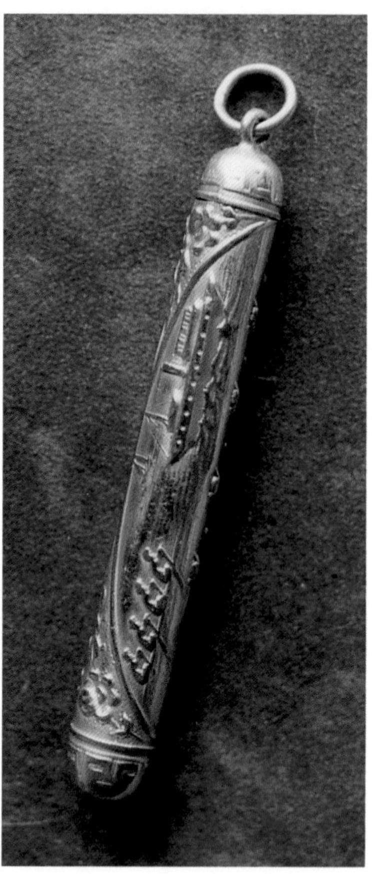

Gold magic pencil with a scene depicting four men rowing with ships in the background. $325-$350.

Gold-filled magic pencil with repoussé flowers and horseshoe, set with diamonds. $250-$275.

Gold-filled pen-pencil combination embellished with a portrait of a woman wearing a bonnet. $225-$250.

Notes on Chapter Five

[1] Gorham ad, *The Cosmopolitan*, May, 1897

[2] Tiffany's *Blue Book*, 1878

[3] See *Collecting Stanhopes*, by Douglas Jull.

[4] The Mordan arrow, according to Jim Marshall, has six feathers on one side and four on the other. The arrow was registered as a Mordan mark for solid gold, according to Culme.

[5] For an excellent description of this topic, see *Snake Charm.*

[6] See *Silver in America* for examples of Gorham silver designed by George Wilkinson.

[7] Some stories substantiate this. People remember receiving gifts from those who returned from such trips.

Chapter Six

Not Just a Pencil

From the beginning, pencil-cases were made to include other useful tools. The most typical would be the combination pens and pencils, pencils with seal terminals (for sealing wax) and holders for multi-colored pencils. However, these were by no means the only configurations offered during Victorian and Edwardian times. Perhaps one of the most complex combinations is the pencil-case made by John Sheldon in 1849. The pencil and pen would be pushed out from the barrel, as would a metal toothpick.[1] (Not uncommonly, toothpicks were offered in conjunction with pens and pencils. Those who could afford to use toothpaste may have been, in fact, the ones who ended up with the worst teeth. The toothpaste that was available in the nineteenth century was highly abrasive and frequently did more damage than good. Hence, wealthier patrons needed toothpicks to remove particles of food from their less-than-perfect teeth. Toothpicks were also made separately with the same mechanisms as pencils and were sold by the same manufacturers.)[2] The toothpick here doubles as a schilling gauge, used to measure a coin to ascertain whether any metal had been shaved from it or not. The terminal holds a waffle seal, and there's a compartment for extra graphite. Most unusual, however, is the diminutive postal scale that's housed within the barrel. A tiny clip holds the letter, which allows the sender to determine by the hand engraved numbers on the scale how much postage they'd have to pay.

Makers like Sampson Mordan created writing implements that included blades used for cutting quills, erasing, etc. Some combos also include buttonhooks. Pencils were incorporated into folding ivory or telescoping silver rulers. Whistles were combined with pencils and some had magnifiers attached. There were pencil/letter opener combinations, and a rare pencil (shaped identical to the chased snail pattern dip pen staffs) that also serves as a thermometer case has been discovered. Pencils were hidden inside the handles of canes and parasols.

Multi-colored pencils can be distinctively handsome. Many have full (or partial) bands with enameled insets indicating if the attached pencil would be black, blue, or red. Colored (as well as indelible) leads were introduced in the 1860s and quickly became popular. Some of these pencils have the slider bands placed one above the other on the barrel and some have the band broken into three equal parts, one segment for each color. Others have slider buttons that are set with appropriately colored stones (or enamel) that are used to push the pencil nozzle from the barrel for use.

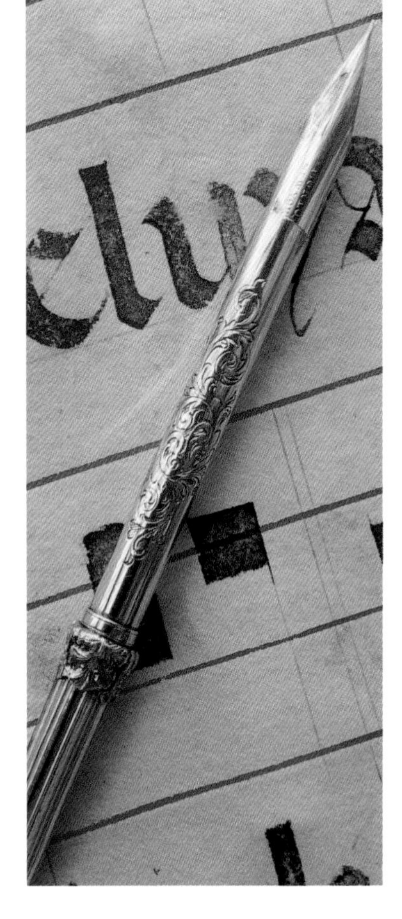

Combination pen-pencil in gold with hand-engraved ferrule and nib marked Morton's New York, 1854. Also marked "Registered, Decr. 11, 1847 P. & A." near the end of the reeded barrel. The end unscrews to hold extra graphite in a compartment separate from where the nib rests when the pen is inverted for storage, or for the pencil to be used. 5.75" as seen here, $250-$275.

Notes on Chapter Six

[1] See article by Peter Katz in the *Journal of the Writing Equipment Society*, no. 48, 1997, entitled "The Toothpick and the Propelling Pencil."

[2] A warning about the overuse of toothpicks was described in *Hall's Family Doctor* (1872, pages 338 and 339): "Every dentist knows that the more a tooth-pick is used, the more the yielding gum is pressed upward; and the larger the space between the teeth next to the gum, the larger will be the pieces of food that will lodge there, and the greater the necessity of the use of the tooth-pick for the remainder of the life, and to some extent the solid tooth itself may wear away...To be under the necessity of spending five or ten minutes after each meal in picking the teeth is a great and useless waste of time, and is essentially a loafing and indecent practice unless done in one's own private room."

This photograph shows the pen and pencil, and how the pencil mechanism is engaged with a slider pin (which also fits into a groove in the outer pencil casing so that the pencil can be withdrawn into the case when not in use).

Gold-filled collapsible pen-pencil, with engine turning. $75-$125.

Two collapsible pen-pencil combinations. The barrel on the silver pen-pencil is nearly 6" long when extended, and closes to 3-1/2". The silver piece is hand-engraved and is set with a citrine carved with the signet "Esther." $150-$175. The gold and rolled gold piece has an amethyst seal with a pot of flowers (?) and a nib marked "Wm. F. Utter, N. Y. 3." $150-$175.

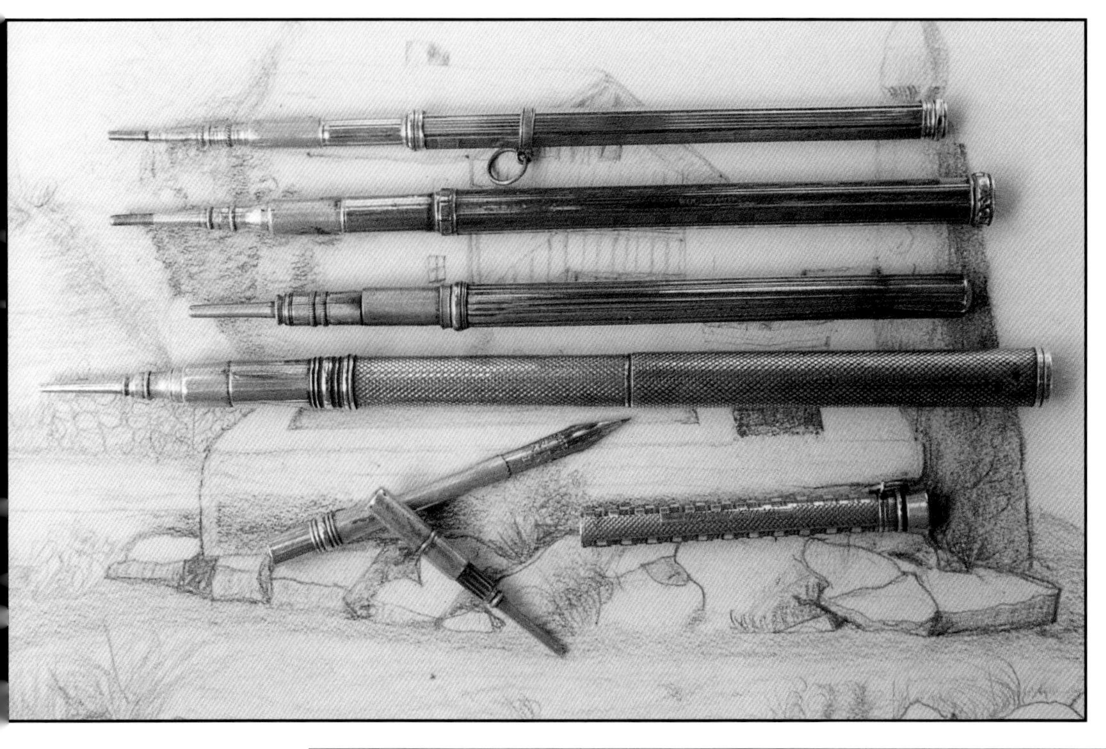

Five pen-pencil combinations from the mid-1800s, in silver. Top to bottom: hexagonal barrel with chatelaine ring, $75-$125; fluted barrel with engine turning and a compartment to hold extra graphite which concealed a tiny strip of paper with "P. P. Douglass Pen, C. W. Newton oposite [sic] Middletown Bank" hand-written in blue/black ink, $125-$150; reeded barrel reversible pencil, nickel silver, $50-$60; engine turned collapsible pen-pencil, $75-$125; smaller combo (2.75" closed, 4.25" with nib engaged), engine-turned, chatelaine ring, with nib marked "E. Wiley, Br'lyn, N. Y. 1."

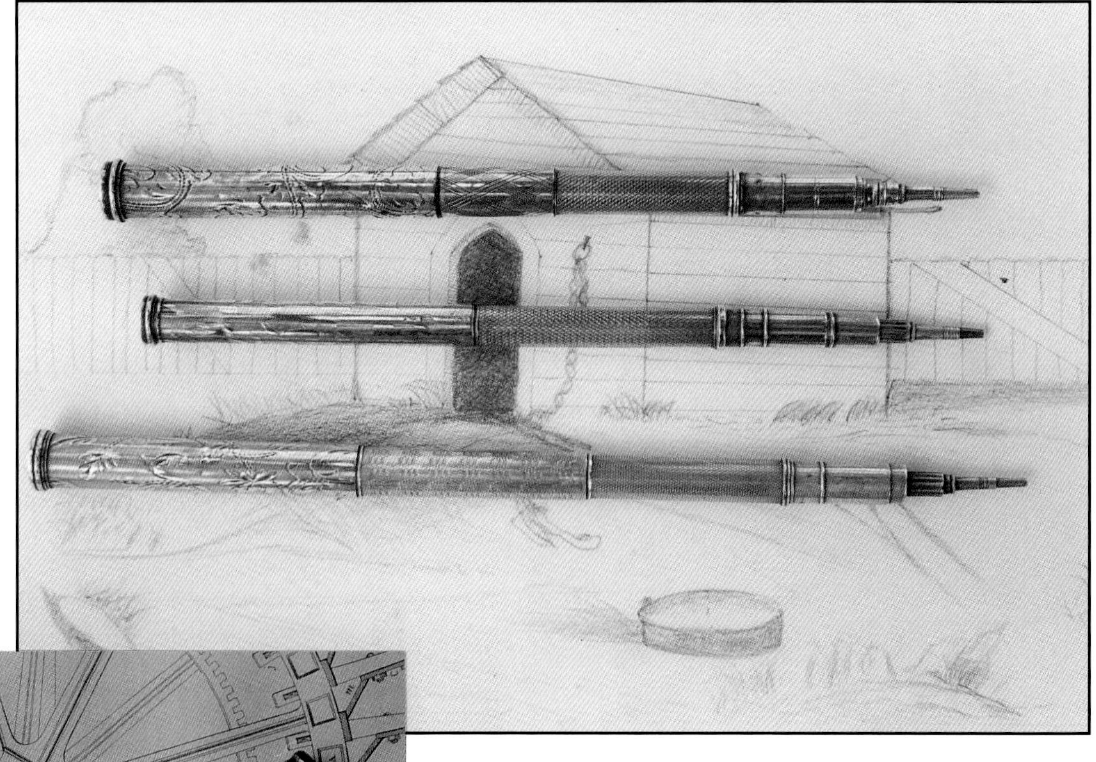

Three collapsible, reverse case combos, the largest one being 7" long. All three have different engine turned patterns on the various planes, $75-$125. Photographed on a pencil drawing of a doghouse, circa 1884.

Nickel-plated pen-pencil combo. $35-$50.

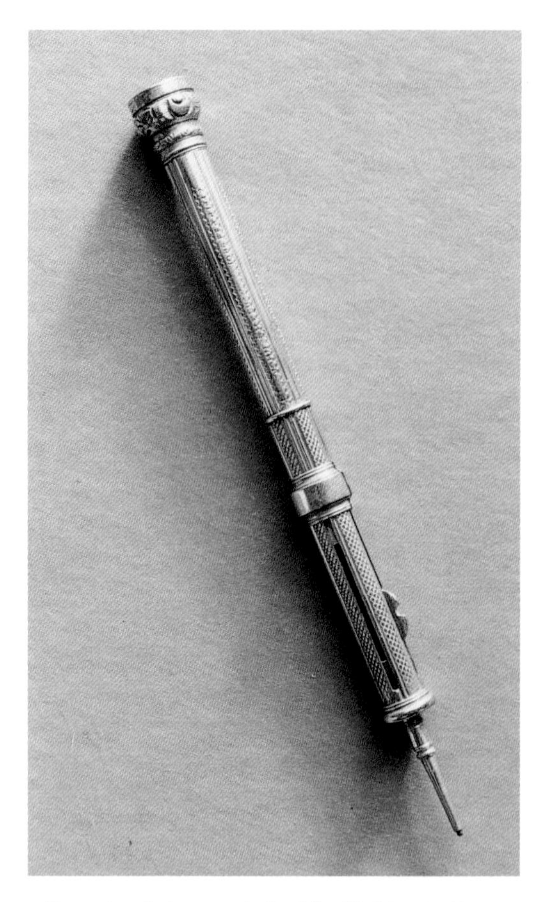

Exceptional piece made by John Sheldon in 1842. The pencil is shown extended in this photograph. $400-$600.

Gold pen-pencil with cast foliate sliders. $225-$300.

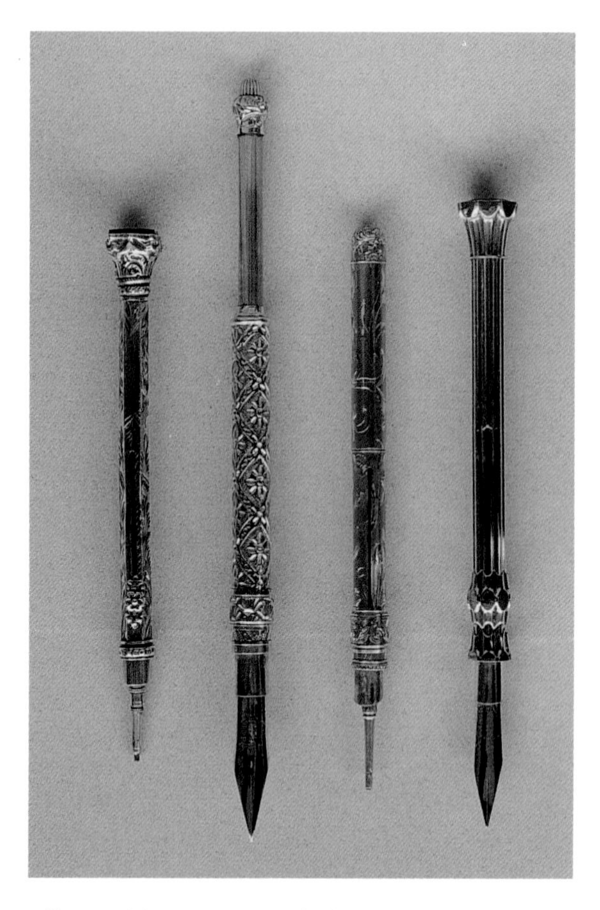

Four combination pen-pencils. Left to right: gold-filled with unusual terminal ($125-$150), gold-filled made by Leroy Fairchild ($125-$150), no makers mark, 14kt gold ($175-$225), silver made by W. M. Willmarth ($125-$150). *Photo Courtesy of Chris Odgers.*

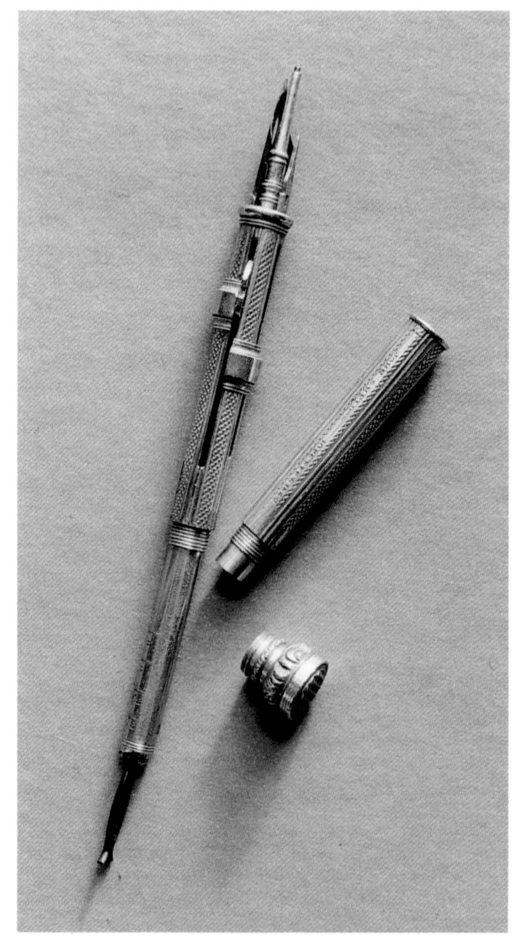

This view shows the Sheldon piece open, with its various components visible. Not only does this pencil-case also hold a pen and toothpick, but a schilling gauge (to ascertain whether any silver had been shaved off the edges of the coin) and a postal scale.

Left to right: rolled gold pen-pencil combo, 1854, with hexagonal sliders ($125-$175), rolled gold combo with slider pen, twist pencil mechanism ($125-$175), silver combo with penknife ($175-$225), gold telescoping pencil, barleycorn pattern ($150-$175).

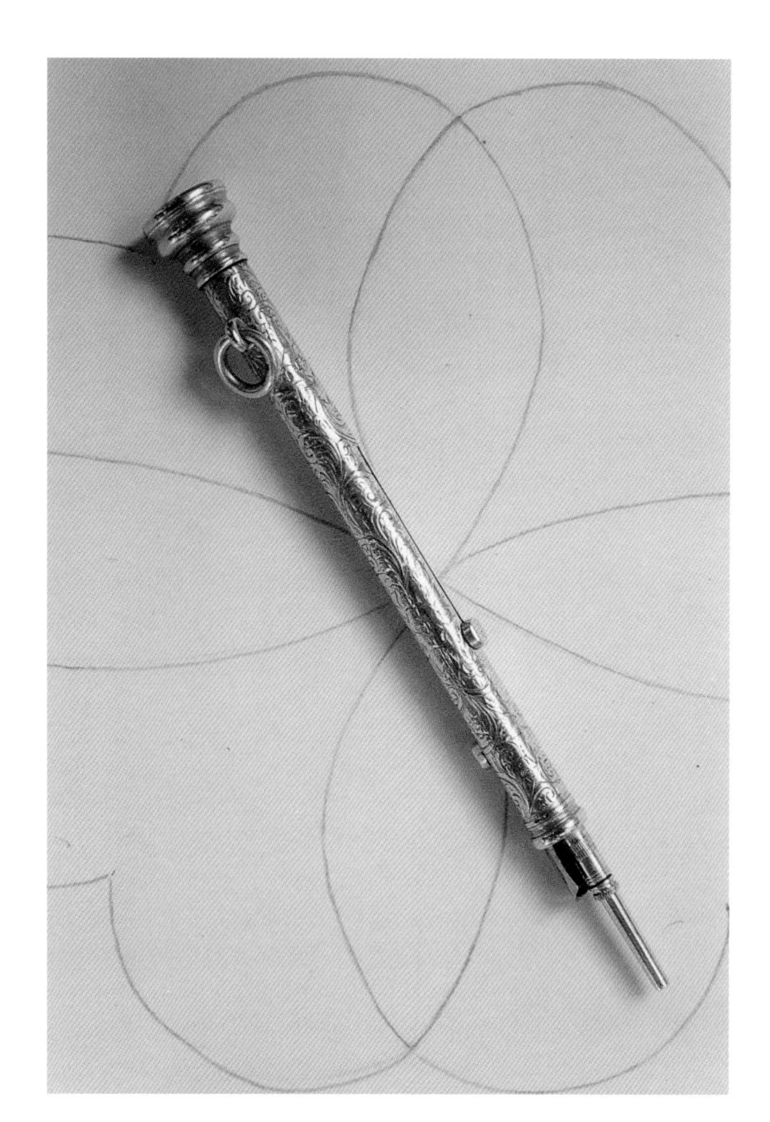

Gold combo, marked S. Mordan & Co., hand-engraved with bloodstone set in terminal. Gold button sliders, chatelaine ring, 4.5" open, $225-$325.

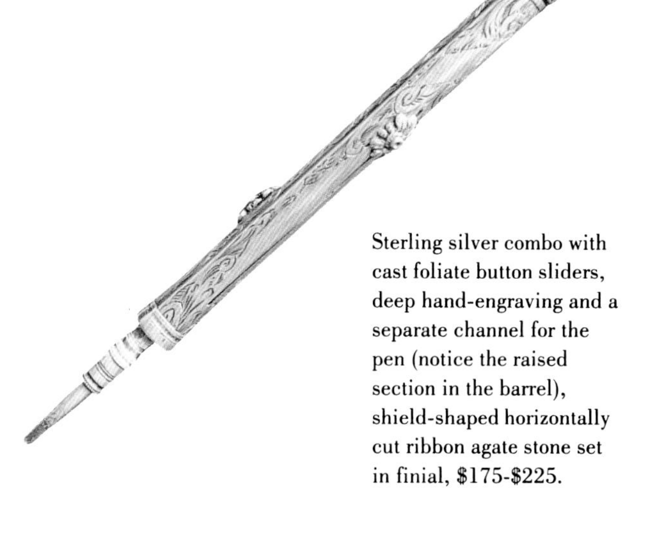

Sterling silver combo with cast foliate button sliders, deep hand-engraving and a separate channel for the pen (notice the raised section in the barrel), shield-shaped horizontally cut ribbon agate stone set in finial, $175-$225.

Rolled gold combo with slider rings, hand-engraved ($150-$175).

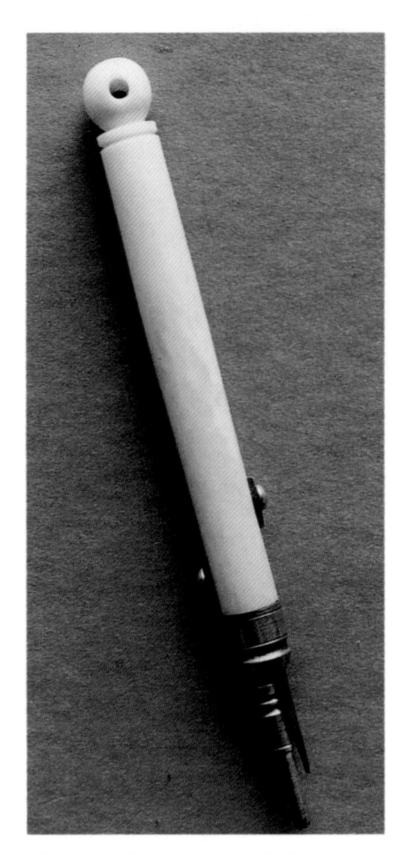

Bone combo with two nickel slider buttons, missing Stanhope, $50-$75.

Fluted rolled gold combo with two ring sliders and a turban shaped finial, nib marked A. F. Warren Ludden & Sollace, 4" unopened, 5" open, $125-$150.

French telescoping pen-pencil combo with a cast silver barrel and rolled gold trim, marked C. M. Depuse, $125-$175.

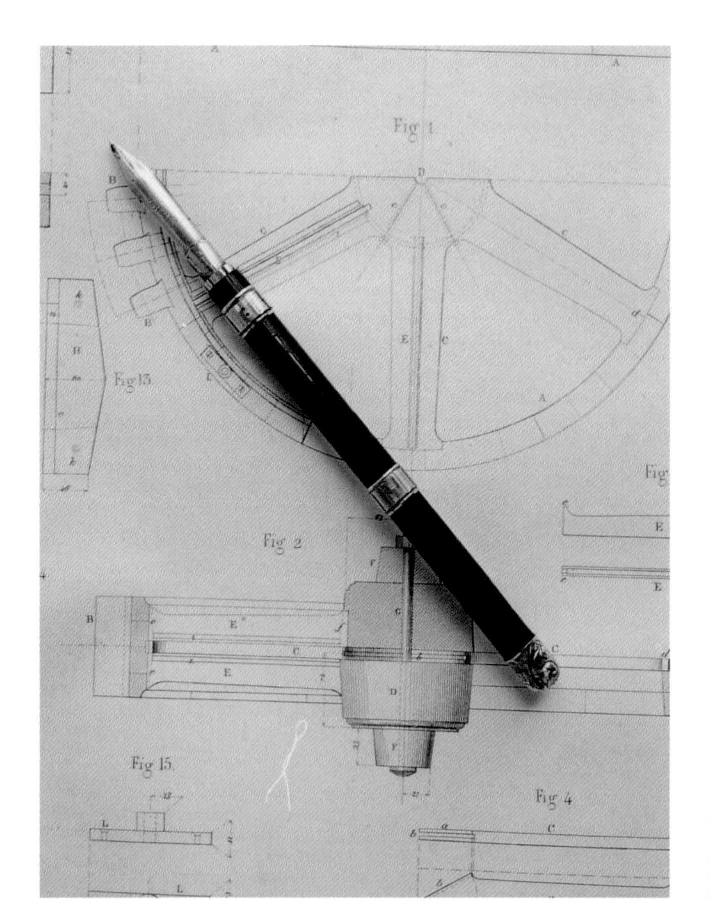

Unmarked, engine-turned, black hard-rubber combo with rolled gold trim, nib marked "The Brilliant," $100-$125.

Sterling silver case with telescoping pen and pencil. Notice that the finials are different from each other, so the owner can determine which implement he or she is selecting. 3.25" closed, pencil 5.5" extended, $150-$175. Photographed on a nineteenth century pencil drawing of a dog.

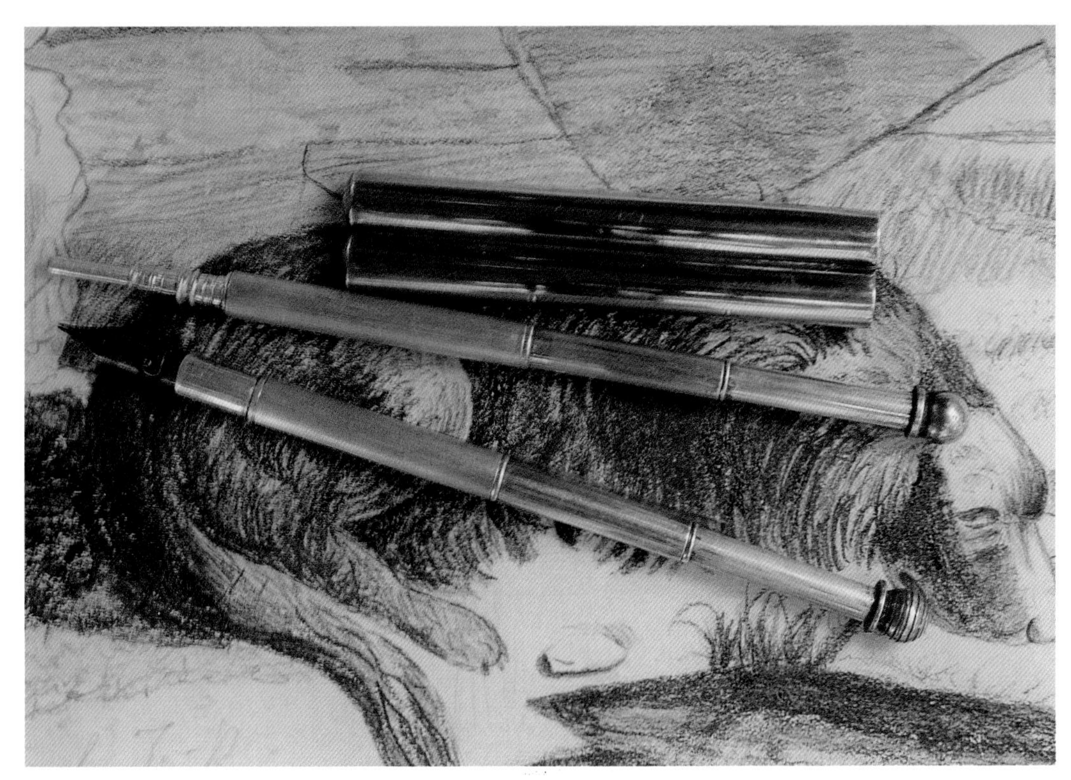

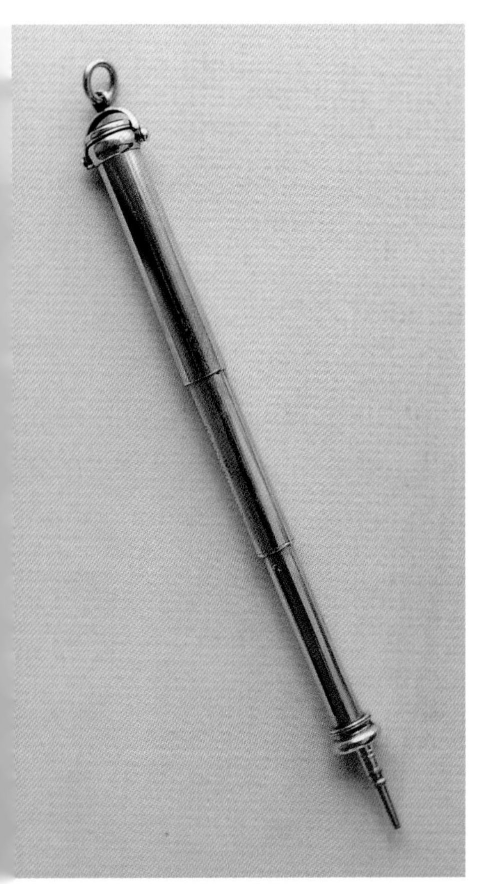

Gold telescoping pen-pencil combo with stirrup bail and bloodstone set in terminal, inscribed "With Much Regard From E. A. P. Middleton, 1881," $250-$300.

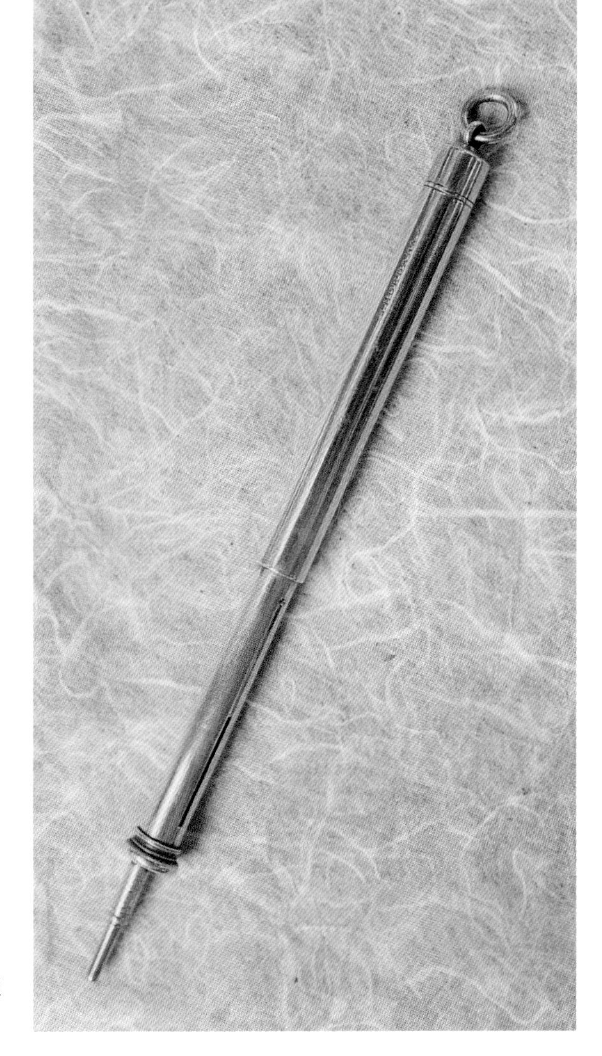

Sterling silver telescoping combo marked S. Mordan & Co., $175-$225.

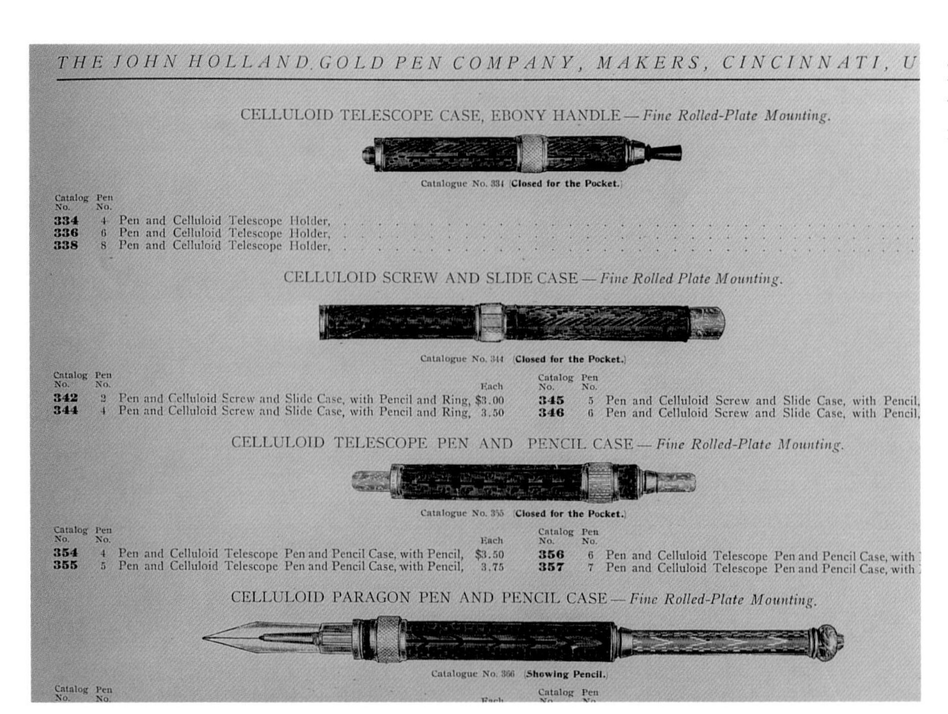

THE JOHN HOLLAND GOLD PEN COMPANY, MAKERS, CINCINNATI, U

CELLULOID TELESCOPE CASE, EBONY HANDLE—*Fine Rolled-Plate Mounting.*

Catalogue No. 334 (Closed for the Pocket.)

Catalog No.	Pen No.	
334	4	Pen and Celluloid Telescope Holder,
336	6	Pen and Celluloid Telescope Holder,
338	8	Pen and Celluloid Telescope Holder,

CELLULOID SCREW AND SLIDE CASE—*Fine Rolled Plate Mounting.*

Catalogue No. 344 (Closed for the Pocket.)

Catalog No.	Pen No.		Each	Catalog No.	Pen No.	
342	2	Pen and Celluloid Screw and Slide Case, with Pencil and Ring,	$3.00	345	5	Pen and Celluloid Screw and Slide Case, with Pencil,
344	4	Pen and Celluloid Screw and Slide Case, with Pencil and Ring,	3.50	346	6	Pen and Celluloid Screw and Slide Case, with Pencil,

CELLULOID TELESCOPE PEN AND PENCIL CASE—*Fine Rolled-Plate Mounting.*

Catalogue No. 355 (Closed for the Pocket.)

Catalog No.	Pen No.		Each	Catalog No.	Pen No.	
354	4	Pen and Celluloid Telescope Pen and Pencil Case, with Pencil,	$3.50	356	6	Pen and Celluloid Telescope Pen and Pencil Case, with
355	5	Pen and Celluloid Telescope Pen and Pencil Case, with Pencil,	3.75	357	7	Pen and Celluloid Telescope Pen and Pencil Case, with

CELLULOID PARAGON PEN AND PENCIL CASE—*Fine Rolled-Plate Mounting.*

Catalogue No. 366 (Showing Pencil.)

Catalog No. Pen No. Each Catalog No. Pen No.

Page from John Holland catalog showing "celluloid telescope pen and pencil case, fine rolled plate mounting."

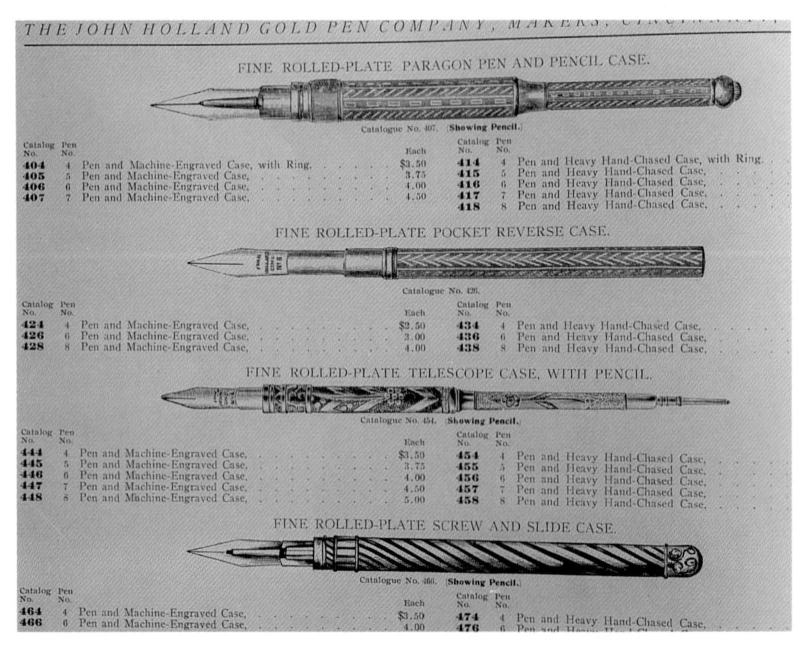

THE JOHN HOLLAND GOLD PEN COMPANY, MAKERS, ...

FINE ROLLED-PLATE PARAGON PEN AND PENCIL CASE.

Catalogue No. 407. (Showing Pencil.)

Catalog No.	Pen No.		Each	Catalog No.	Pen No.	
404	4	Pen and Machine-Engraved Case, with Ring,	$3.50	414	4	Pen and Heavy Hand-Chased Case, with Ring,
405	5	Pen and Machine-Engraved Case,	3.75	415	5	Pen and Heavy Hand-Chased Case,
406	6	Pen and Machine-Engraved Case,	4.00	416	6	Pen and Heavy Hand-Chased Case,
407	7	Pen and Machine-Engraved Case,	4.50	417	7	Pen and Heavy Hand-Chased Case,
				418	8	Pen and Heavy Hand-Chased Case,

FINE ROLLED-PLATE POCKET REVERSE CASE.

Catalogue No. 426.

Catalog No.	Pen No.		Each	Catalog No.	Pen No.	
424	4	Pen and Machine-Engraved Case,	$2.50	434	4	Pen and Heavy Hand-Chased Case,
426	6	Pen and Machine-Engraved Case,	3.00	436	6	Pen and Heavy Hand-Chased Case,
428	8	Pen and Machine-Engraved Case,	4.00	438	8	Pen and Heavy Hand-Chased Case,

FINE ROLLED-PLATE TELESCOPE CASE, WITH PENCIL.

Catalogue No. 454. (Showing Pencil.)

Catalog No.	Pen No.		Each	Catalog No.	Pen No.	
444	4	Pen and Machine-Engraved Case,	$3.50	454	4	Pen and Heavy Hand-Chased Case,
445	5	Pen and Machine-Engraved Case,	3.75	455	5	Pen and Heavy Hand-Chased Case,
446	6	Pen and Machine-Engraved Case,	4.00	456	6	Pen and Heavy Hand-Chased Case,
447	7	Pen and Machine-Engraved Case,	4.50	457	7	Pen and Heavy Hand-Chased Case,
448	8	Pen and Machine-Engraved Case,	5.00	458	8	Pen and Heavy Hand-Chased Case,

FINE ROLLED-PLATE SCREW AND SLIDE CASE.

Catalogue No. 466. (Showing Pencil.)

Catalog No.	Pen No.		Each	Catalog No.	Pen No.	
464	4	Pen and Machine-Engraved Case,	$3.50	474	4	Pen and Heavy Hand-Chased Case,
466	6	Pen and Machine-Engraved Case,	4.00	476	6	[illegible]

Page from John Holland catalog showing four versions of combos.

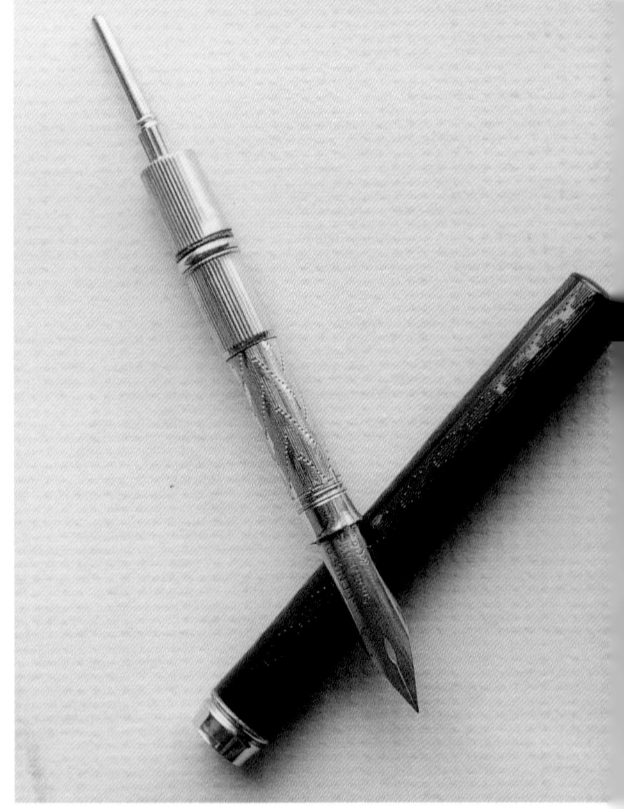

Reverse case combo, chased black hard rubber, Aikin, Lambert Spencerian nib, $75-$100.

Gadrooned cable, rolled gold telescoping combo, slider for pen and twist for pencil, marked Leroy Fairchild, $150-$175. *Photo Courtesy of Chris Odgers.*

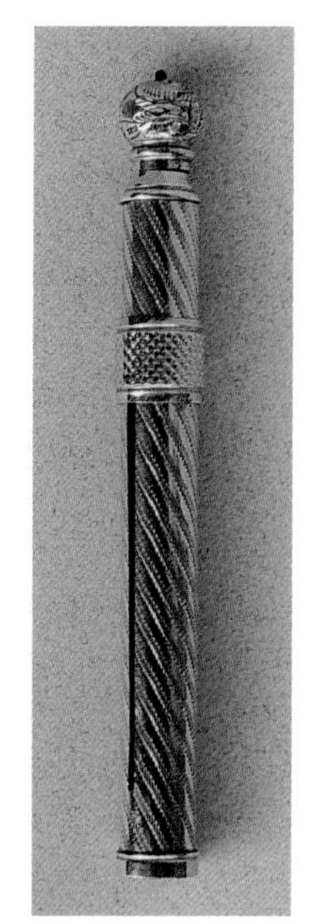

Same piece, closed. *Photo Courtesy of Chris Odgers.*

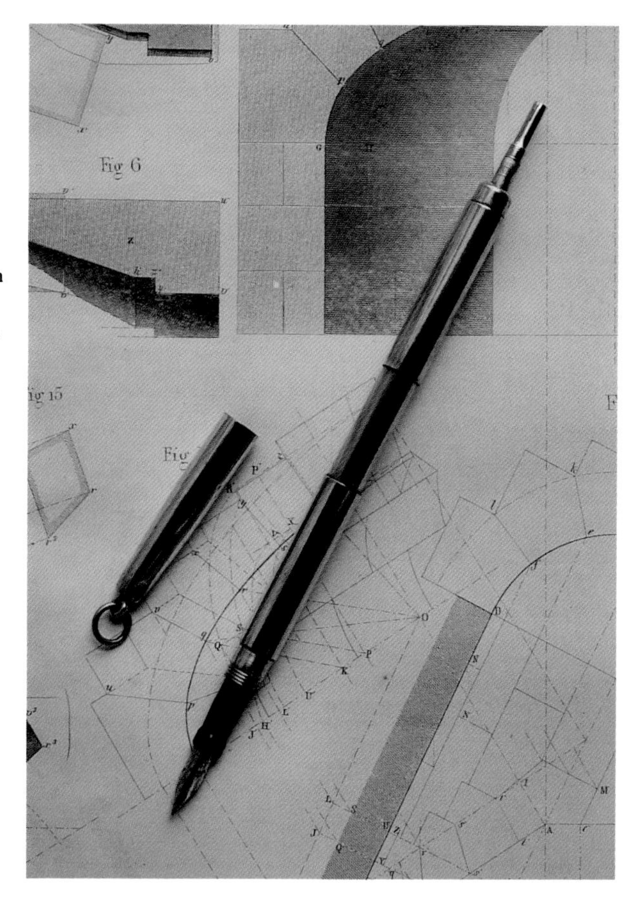

Gold combo, with a telescoping pencil and an eyedropper fill fountain pen, circa 1915. Combos like this were sold by Vickery and Edward Baker (listed as "the Mascot"), $150-$175.

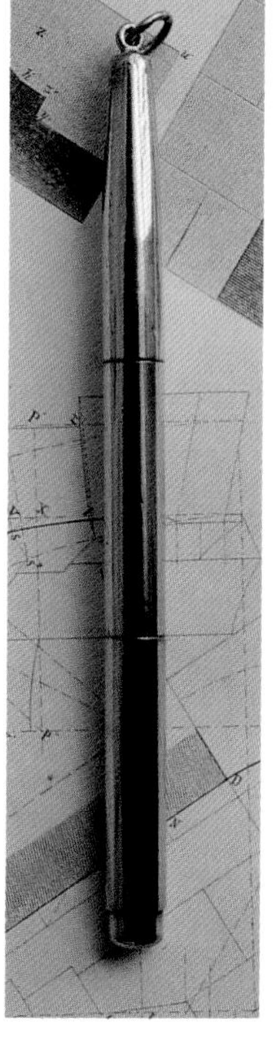

Same combo, as it appears closed.

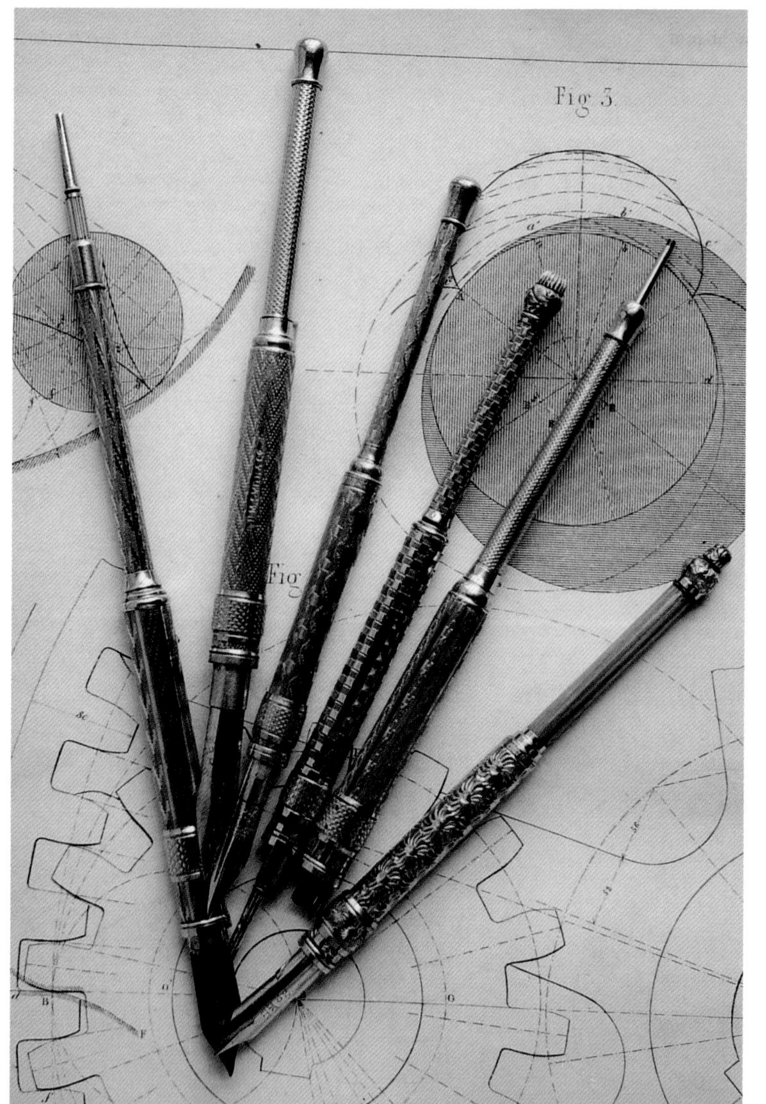

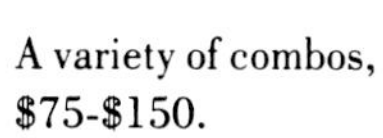

A variety of combos, $75-$150.

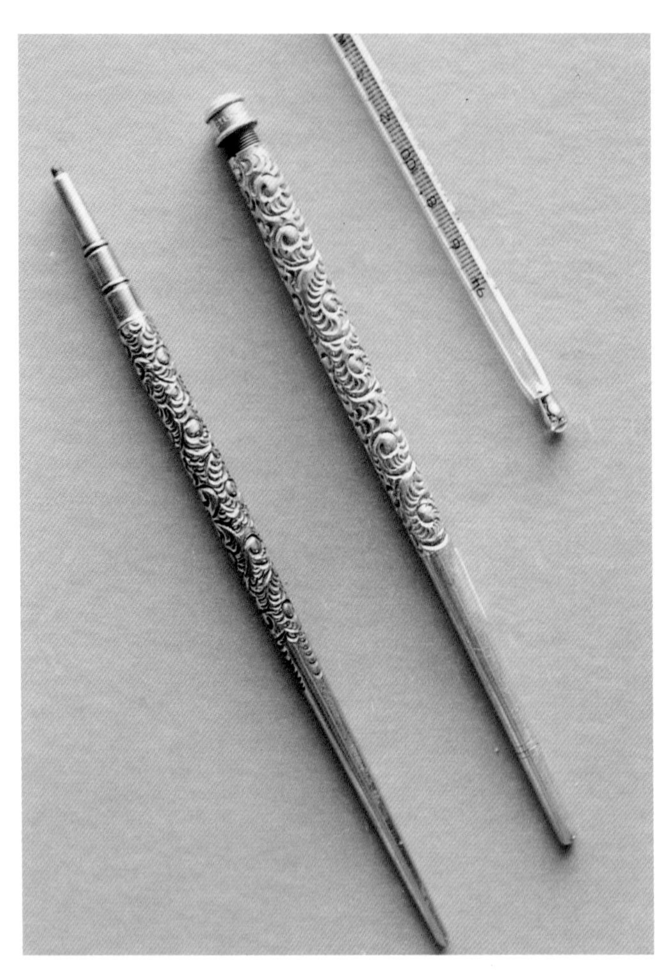

Two silver desk pencils, with chasing. Left: $75-$125; the pencil on the right holds a thermometer in the barrel, $125-$150.

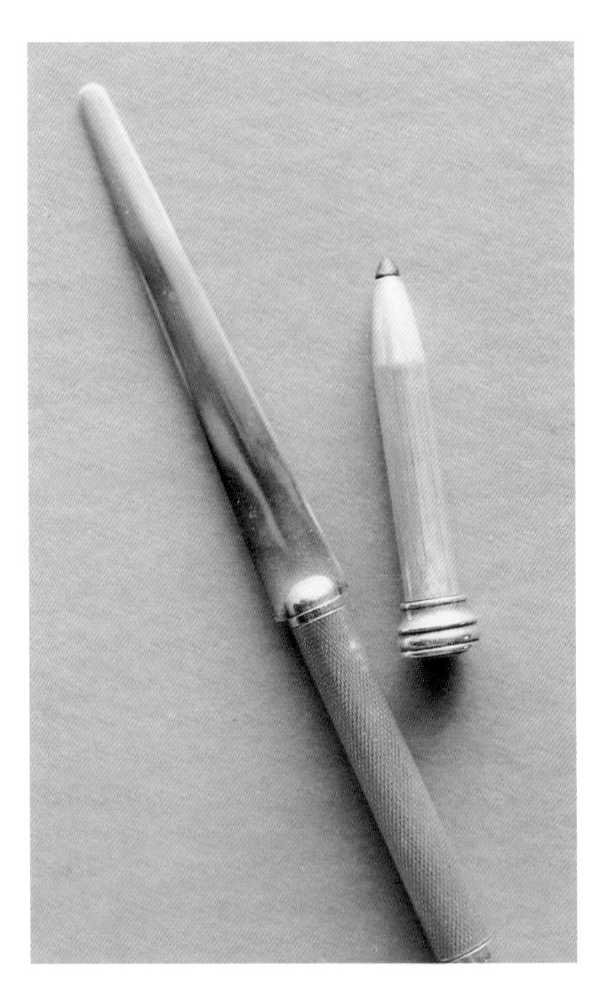

Sterling silver letter opener with a pencil in the handle. The pencil has a blue enamel band around the finial, 6.5", $125-$150.

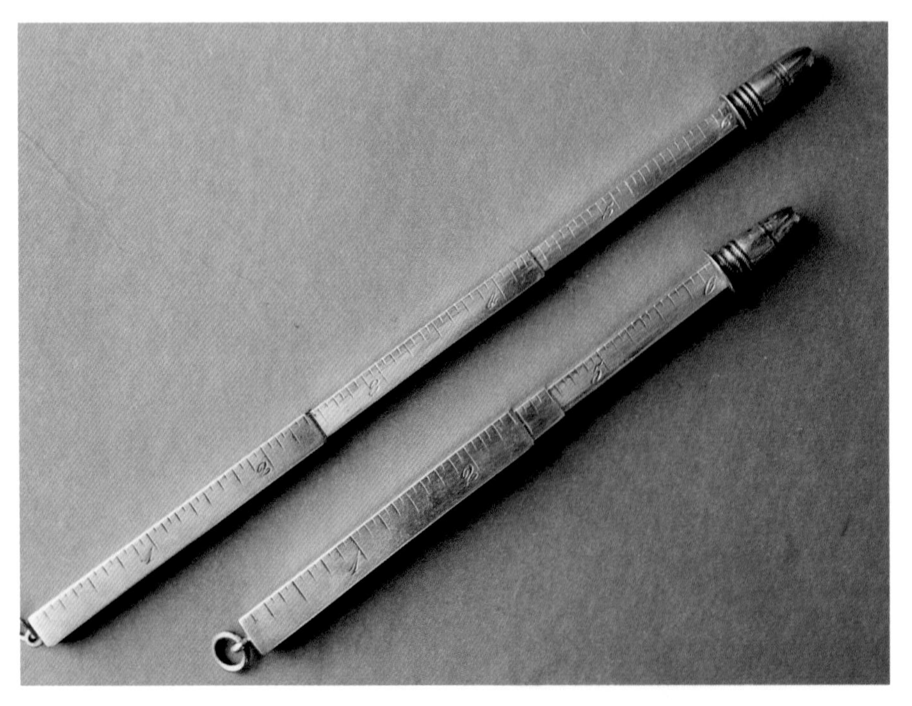

Two sterling silver telescoping ruler pencils, marked Edward Todd, patented January 19, 1897. $150-$175.

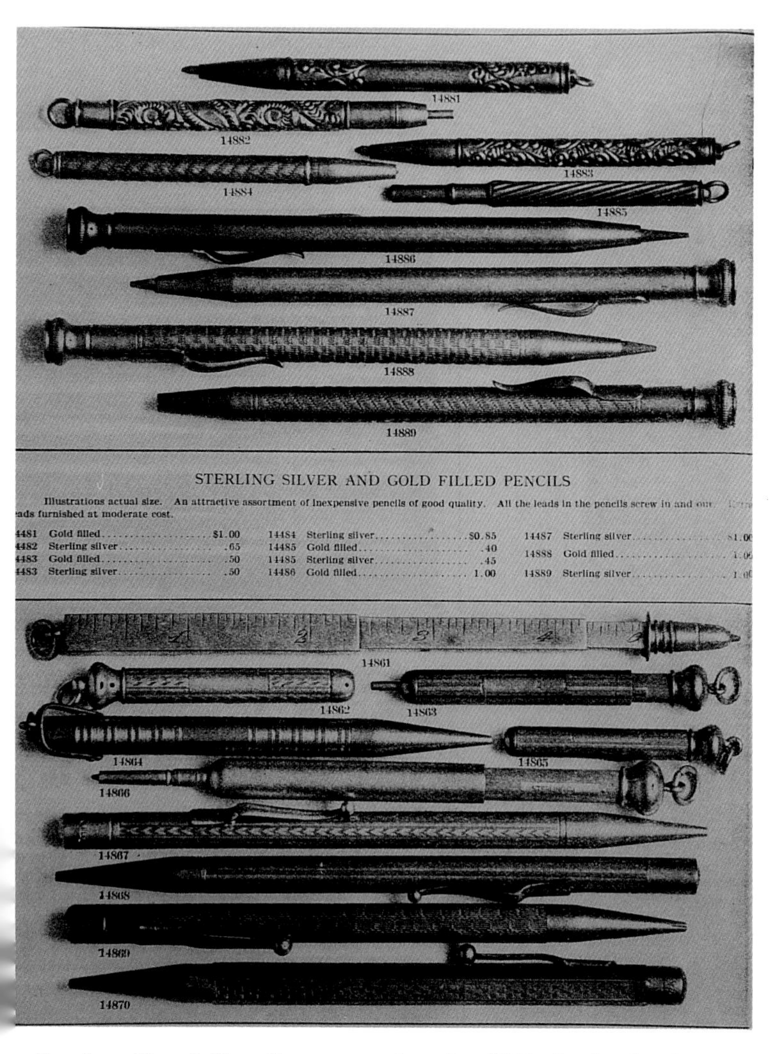

14881
14882
14883
14884
14885
14886
14887
14888
14889

STERLING SILVER AND GOLD FILLED PENCILS

Illustrations actual size. An attractive assortment of inexpensive pencils of good quality. All the leads in the pencils screw in and out.
eads furnished at moderate cost.

4481	Gold filled	$1.00
4482	Sterling silver	.65
4483	Gold filled	.50
4483	Sterling silver	.50
14484	Sterling silver	$0.85
14485	Gold filled	.40
14485	Sterling silver	.45
14486	Gold filled	1.00
14487	Sterling silver	$1.0
14888	Gold filled	1.0
14889	Sterling silver	1.0

14861
14862
14863
14864
14865
14866
14867
14868
14869
14870

Page from Chas. L. Trout Company catalog, circa 1920, showing that the silver telescoping ruler pencils were still being advertised well into the twentieth century.

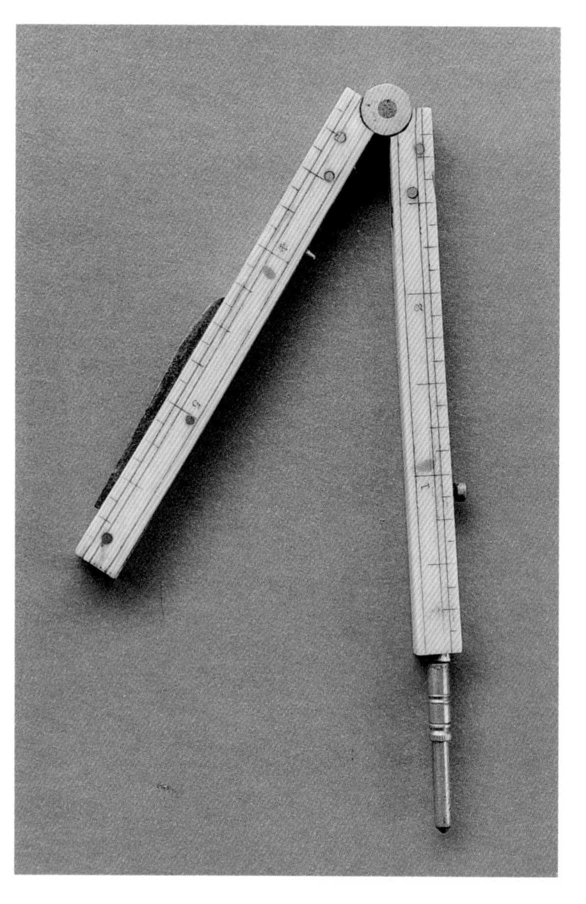

Ivory combination folding rule, pencil and quill blade, mid-1800s, $125-$200.

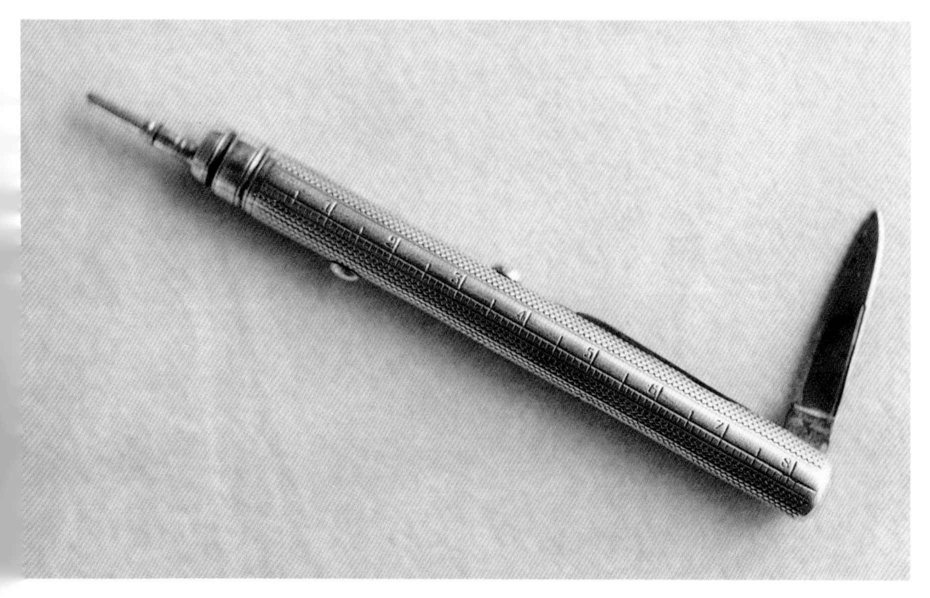

Silver combo with button sliders for pen and pencil, two blades, and ruler. $225-$325.

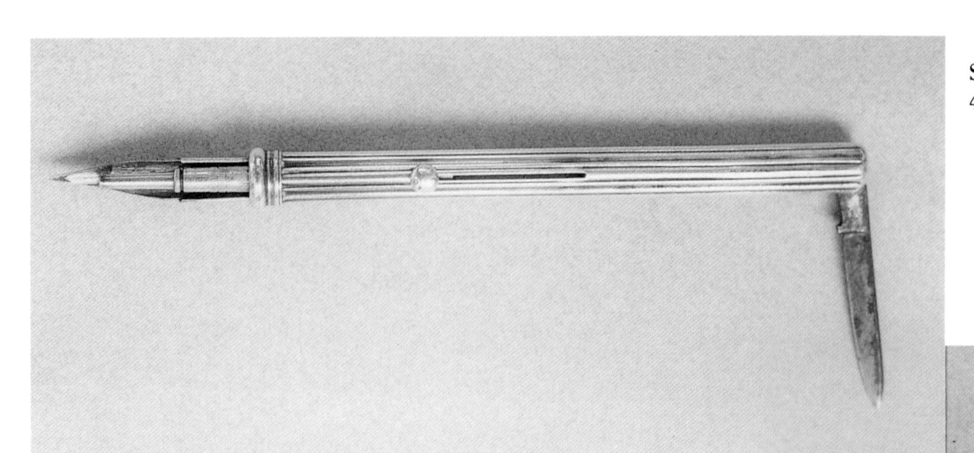

S. Mordan & Co. combo with blade, 3.5" closed 4.5" open, $175-$250.

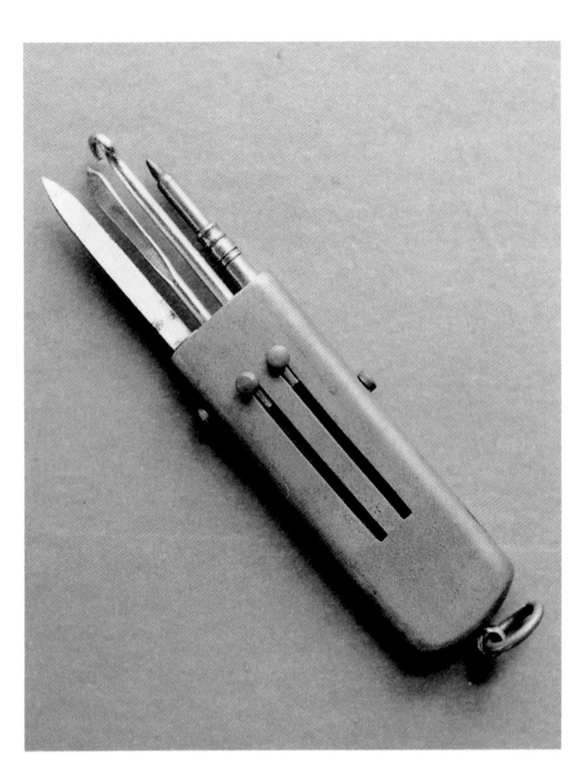

Brass combo with blade, toothpick, buttonhook and pencil, with button sliders, 2", $65-$85.

Sterling silver penknife, pencil combination on watch chain. 1.5" closed, 2.5" open, $125-$150.

Gunmetal and turquoise combo, with blade, toothpick, and pencil. (The buttonhook is missing.) $75-$100.

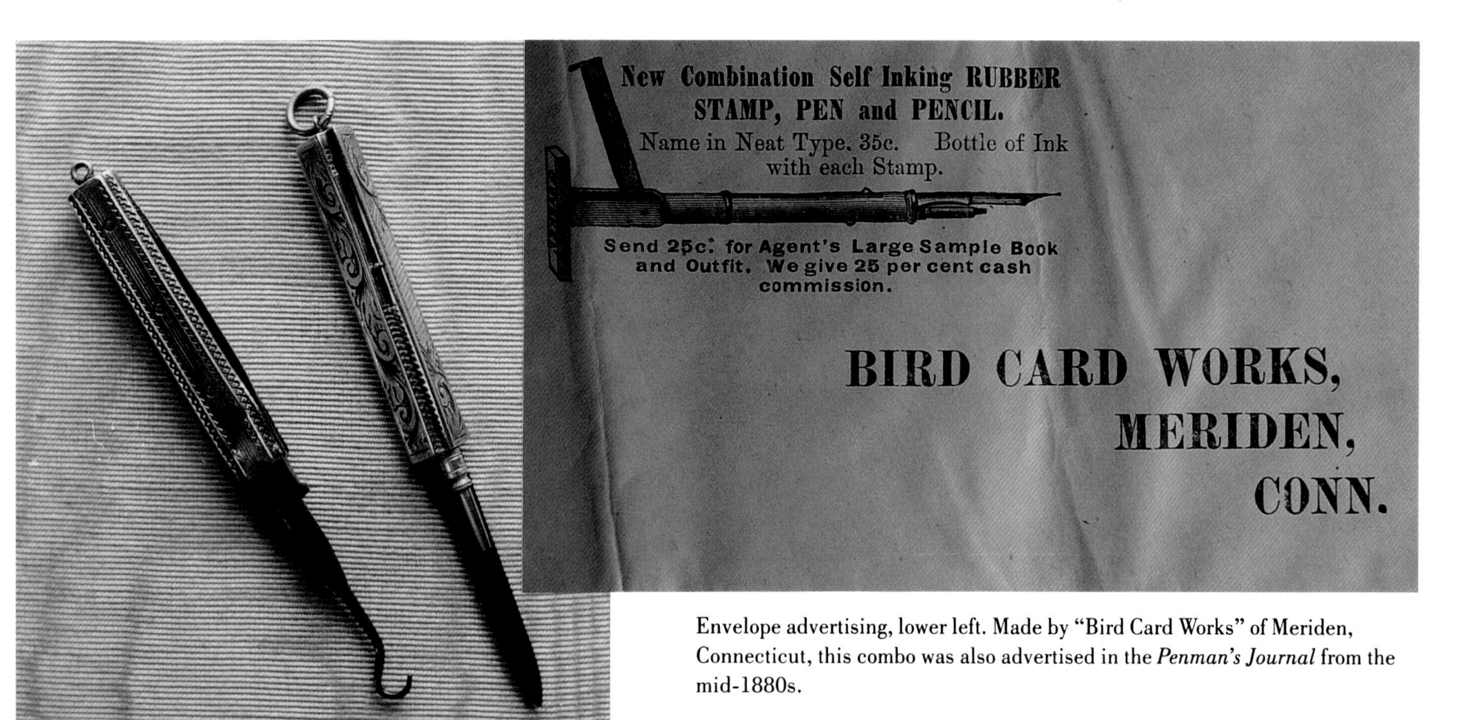

Envelope advertising, lower left. Made by "Bird Card Works" of Meriden, Connecticut, this combo was also advertised in the *Penman's Journal* from the mid-1880s.

Two hand-engraved, sterling silver combos. Left: blade and buttonhook, no pencil, line and dot pattern, $75-$85. Right: triangular pencil, Sheffield blade combination, with a match-striker edge, 1.75" closed, 2.5" open, $125-$150.

Unusual novelty combo, with a self-inking stamp pad and stamp for visiting cards, pen and pencil, made of brass and nickel-plated brass. $35-$50.

Two watch keys. Watch keys were used to wind pocket watches, and often resemble mechanical pencils. Watch keys have a square opening where the graphite would be propelled from a mechanical pencil.

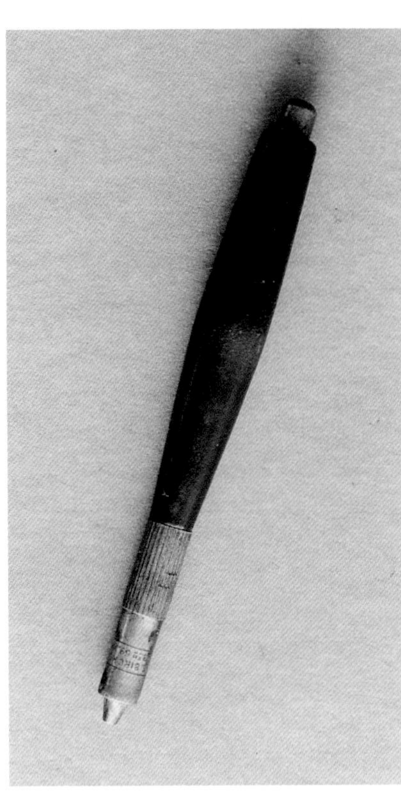

Birch's Patent watch key. There are certain similarities between watch keys and mechanical pencils, but the pencil would have at least an internal collar for the mechanism. (See Aikin, Lambert ad, p. 158.

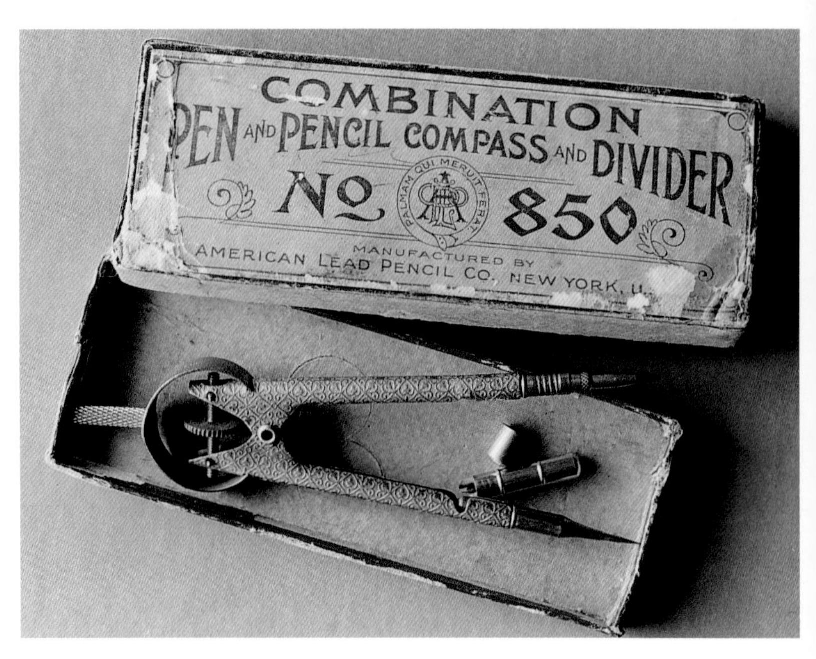

Combination pen, pencil, compass, and divider, made by American Lead Pencil Co., in original box, $35-$45.

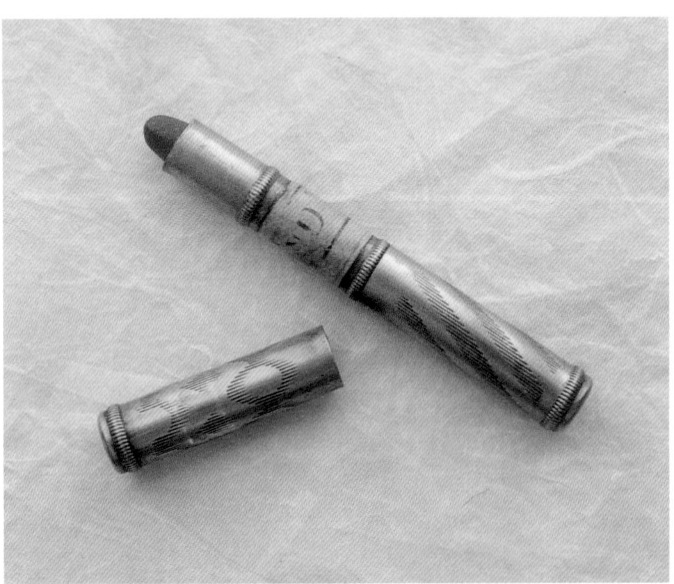

This is an example of eyeliner (kohl) in an engine-turned, brass holder with original price sticker, $5-$8.

Pencil, spare lead compartment in a parasol or cane handle. Without parasol or cane, $125-$175.

Same as above, closed. Faux wood celluloid, 9.25".

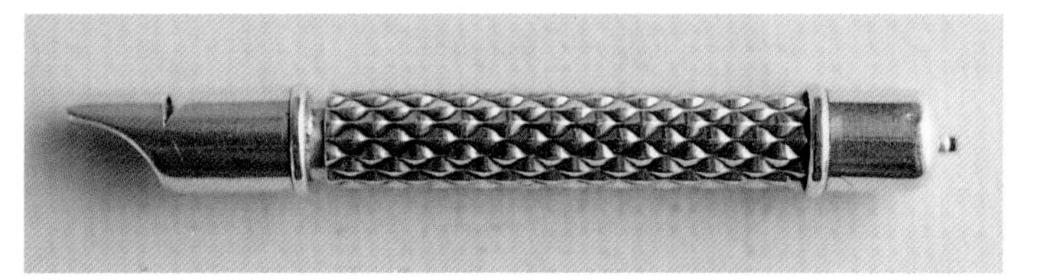

Sterling silver whistle, telescoping pencil. $175-$225.

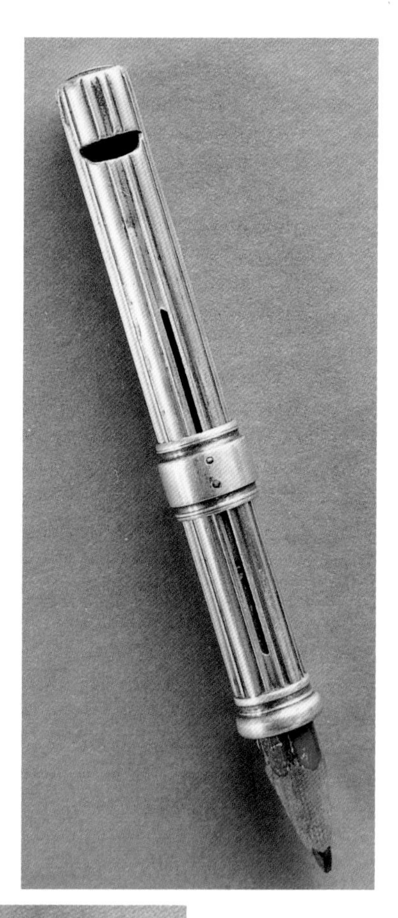

Sterling silver whistle-porte-crayon with ring slider. $150-$175.

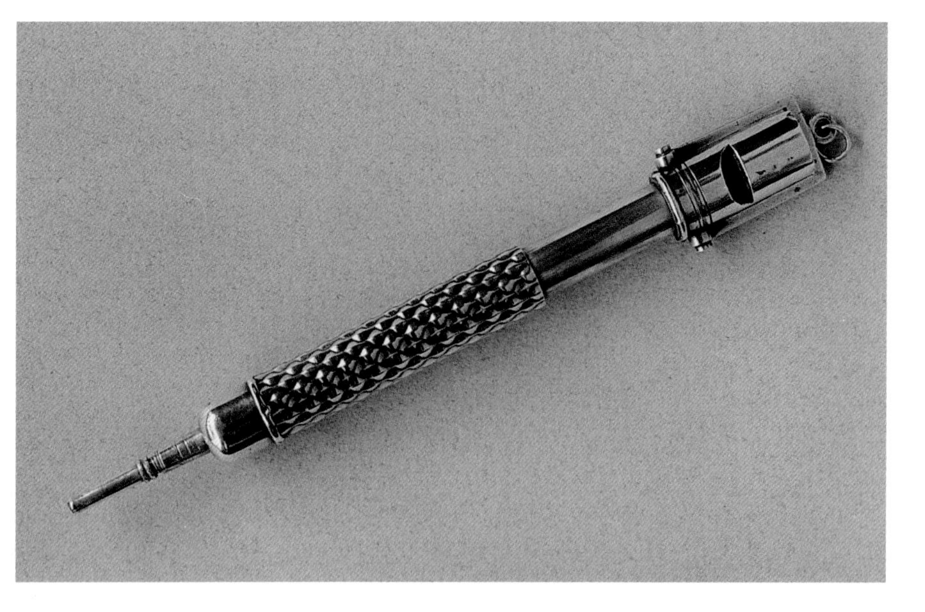

Sterling silver whistle, telescoping pencil, $150-$175. *Photo Courtesy of Chris Odgers.*

Silver-plated match safe, and coin holder, with a pencil and toothpick housed under the lid. $125-$175.

Detail of two pencils with monogram or signet seals, one in citrine, one in amethyst.

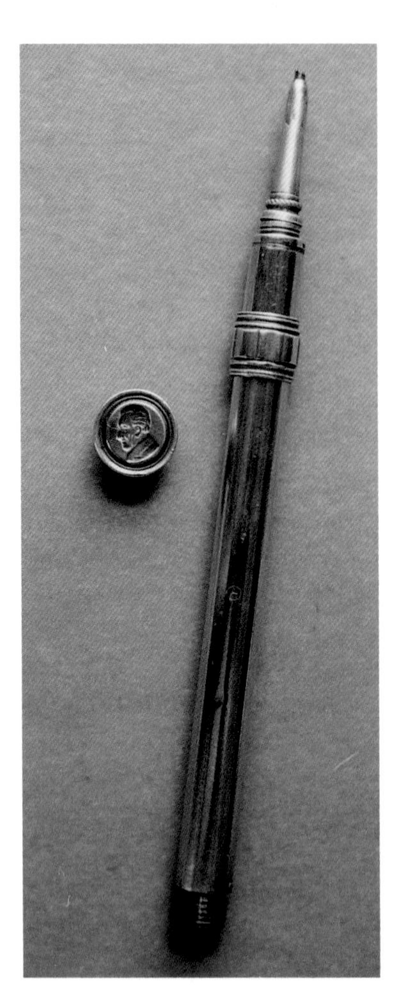

Silver ring-slider pencil, marked "W. M. & Co.," 3" closed, 4" open, with a seal in the shape of a man's profile. $125-$135.

Sterling silver tri-color pencil, with red, black, and blue enameled slider-bands. $150-$175.

Close-up of a hand-engraved tri-color pencil with enameled bands.

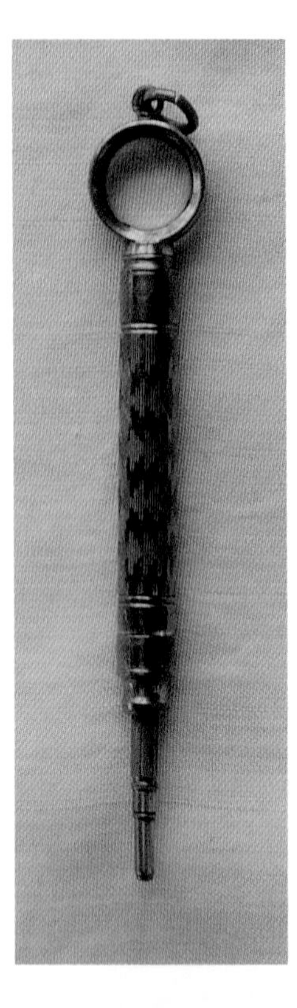

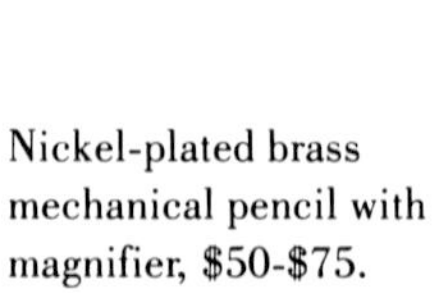

Nickel-plated brass mechanical pencil with magnifier, $50-$75.

Close-up of a silver tri-color pencil, showing how the pins on the button-slider fit into the mechanism, allowing the pencil to be pushed out of and back into the barrel. Hallmarked, "M. Bros."

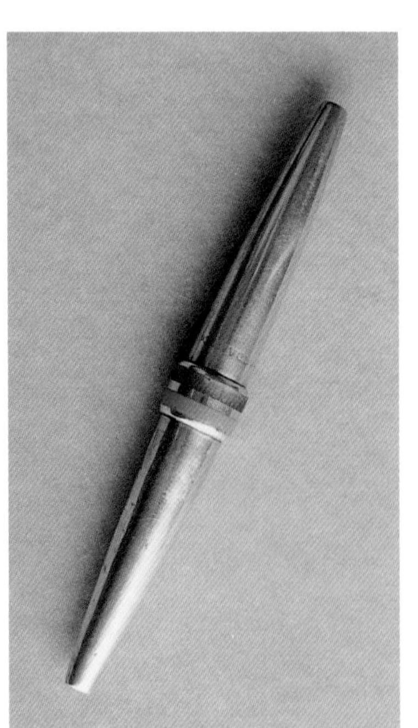

Torpedo-shaped, telescoping bi-color pencil, with two enameled bands indicating the color of the lead. Made by S. Mordan & Co. $150-$175.

Two gunmetal tri-color pencils, with glass button-sliders. $75-$125.

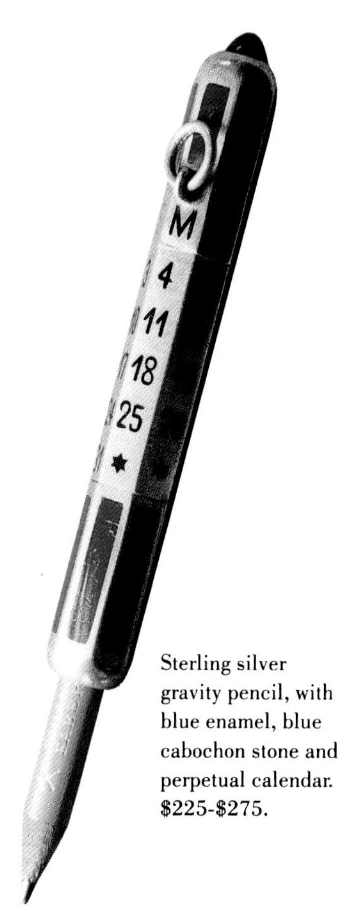

Sterling silver gravity pencil, with blue enamel, blue cabochon stone and perpetual calendar. $225-$275.

Unusual mechanism for a bi-color pencil. The central key changes the color lead.

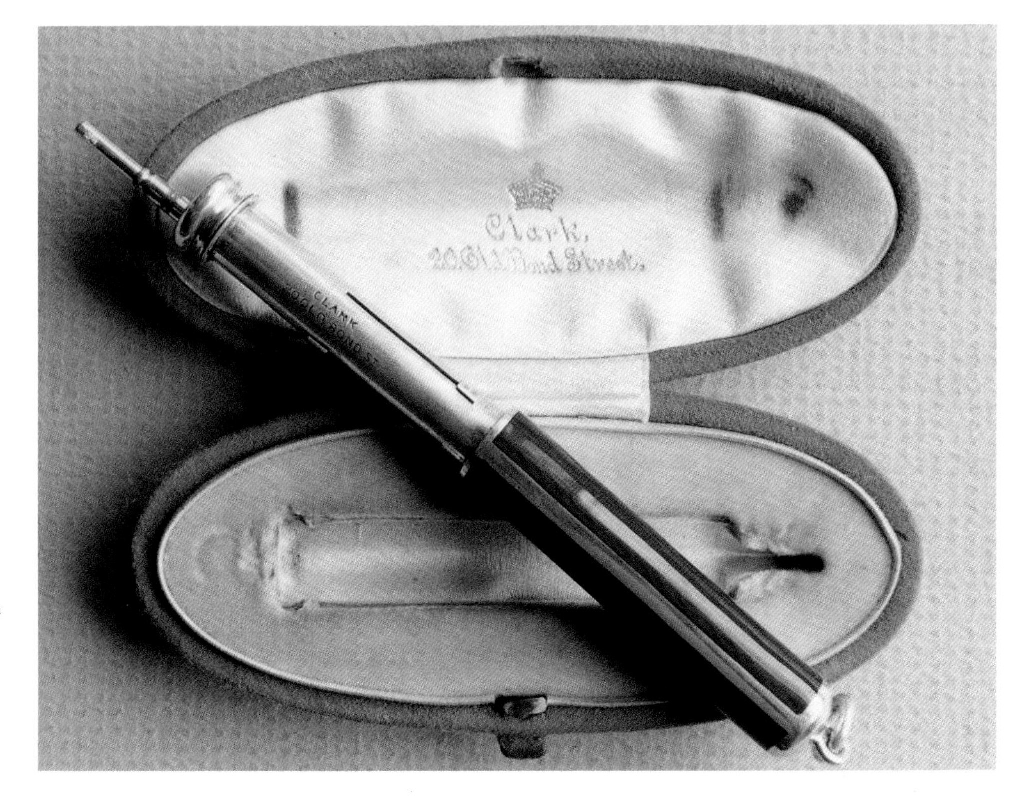

Tri-color telescopic pencil with pin-sliders and three enamel panels, in its original plush-lined case. Marked "Clark, 20 Old Bond Street [London]," $350-$450.

Black hard-rubber advertising pencil (Ingersoll Rand Company). $35-$40.

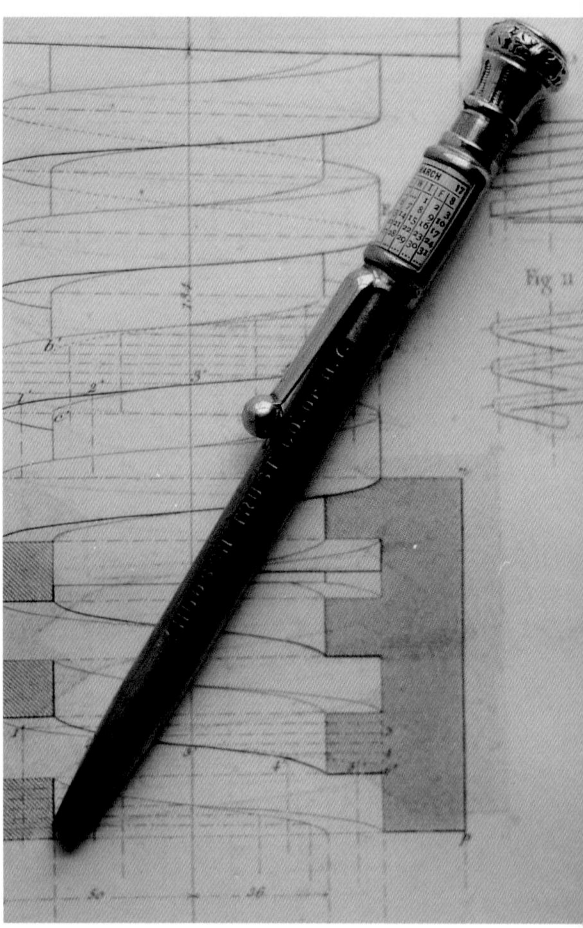

Nickel silver advertising pencil (Hudson Trust Co. of N.Y.) with a changeable calendar. Circa 1917, $25-$35.

Black hard-rubber pencil with original sticker ("For extra supply of leads, unscrew cap holding point downward"), made by Eagle Pencil Co. for The First National Bank, Mechanicville, N.Y., 5.75", $35-$45.

Pencil with perpetual calendar, made of 14kt gold, with inlaid enamel. $325-$350.

Chapter Seven

More Pencils

Sheath, Scabbard, Vest-Pocket, or Flat Cedar Pencil Cases

Sheath pencils (also called *scabbard*, *vest-pocket*, or *flat cedar pencil cases*) have no mechanisms, but the metal cases were often quite lovely. They held a small flat wooden pencil that had to be sharpened with a knife. Distinguished from needle-cases by the location of the ferrule (on the cap rather than on the body, and used to hold the flat pencil in place), sheath pencils were used by both men and women from the late 1800s through the 1920s. Many designers took advantage of the flat surfaces on the sheath pencils to create diminutive works of art. Using the same techniques as those used on mechanical pencil cases, the designs ranged from line engraving to bas-relief (die-stamped or repoussé) to three-dimensional figures. Most were mass-produced, but some of the more artistic were probably created for a small market (some were even signed by the artist who created them). Messages and mottoes were communicated through images: "wine, women, song," the vices, tokens of good luck, etc. Daniel Low, the originator of souvenir spoons, made sheath pencils ornamented with the words "Salem 1622" and a bas-relief image of the Salem Witch riding her broom (the date on the pencil is from the Salem witch trials, not the date the pencil was made).

Postcard from 1903, showing a mechanical and a sheath pencil.

Flat cedar pencils made by A. W. Faber and Eberhard Faber, used inside vest pocket (also called flat or sheath) pencil-cases.

Left: brass, and lacquered brass vest-pocket pencil, $20-$25; right: scabbard shaped vest-pocket pencil, sterling silver, marked "S. M. & Co., Cross, London, 1904." 2.75" closed, $35-$45.

Left to right: sterling silver, etched vest-pocket pencil, $50-$75; hand-engraved vest-pocket pencil, silver, $50-$60; needle case. Notice that the ferrule for the pencil is what secures the cedar pencil, whereas the flange on the needle case is attached to the body rather than the cap.

Gold-colored metal, hand-engraved dragon. $75-$100.

Left to right: gold-colored metal, scene in black lacquer of Mt. Fuji, $125; repoussé silver, $50-$75; silver pencil with lilies of the valley, $75-$125; sterling silver with lilies of the valley, $85-$125;silver floral, $50-$75.

Sterling silver filigree, marked W. C., $50-$75.

Sterling silver vest-pocket pencil with etched design of a griffin, 3" closed, $75-$85.

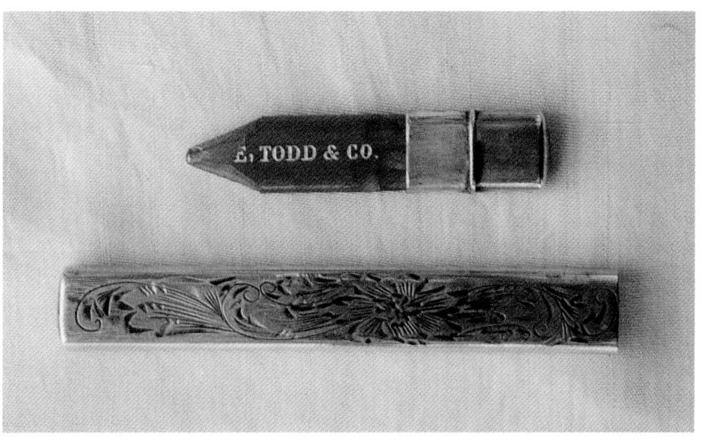

Hand-engraved floral design, perhaps made by Mabie, Todd. $50-$75.

Page from S. Kind & Sons catalog, 1915, showing a variety of metal pencils.

Left to right: small vest-pocket pencil marked "Edward Todd," $40-$50; enamel and silver, $60-$70; hand-engraved vest-pocket pencil, hallmarked Birmingham, 1904, W. V. & Co., $50-$75.

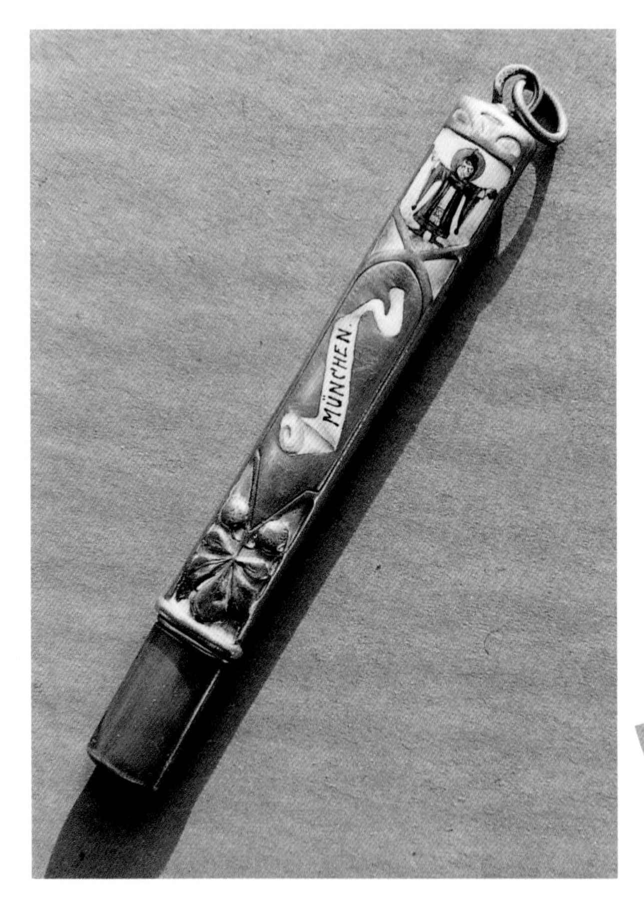

Silver vest pocket pencil, souvenir of München, Germany, with hand-painted enamel. $75-$85.

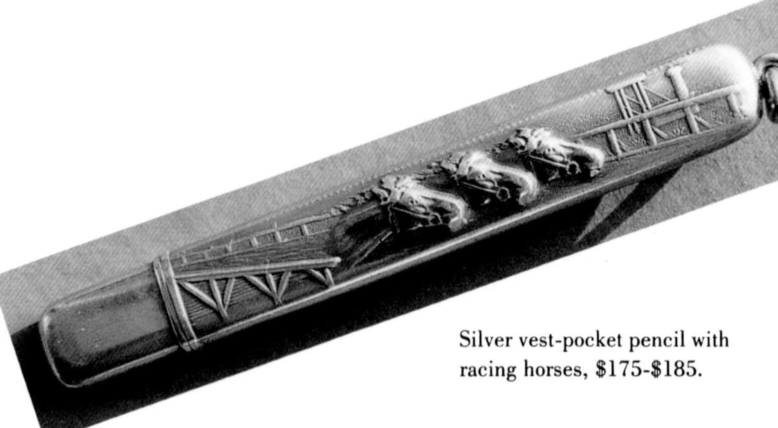

Silver vest-pocket pencil with racing horses, $175-$185.

Sterling silver vest-pocket pencil with a hand-engraved poinsettia design, $50-$75.

Hand-engraved vest-pocket pencil with engraving on the ferrule, photographed on a gutta-percha ink stand.

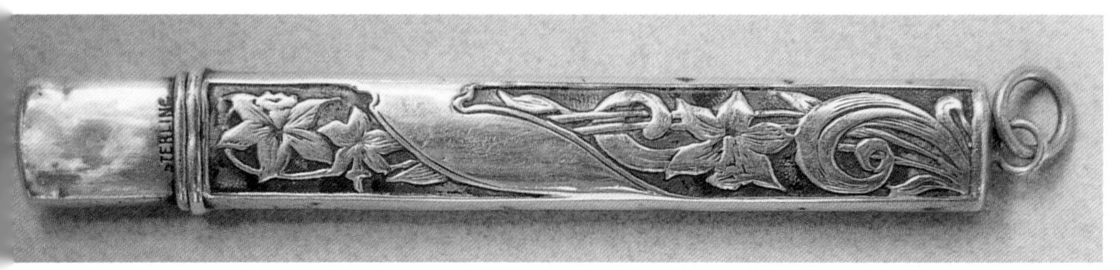

Sterling silver vest-pocket pencil with daffodils, $60-$75.

Repoussé floral vest-pocket pencil, $75-$95.

Silver vest-pocket pencil and matching penknife, $135-$150 set.

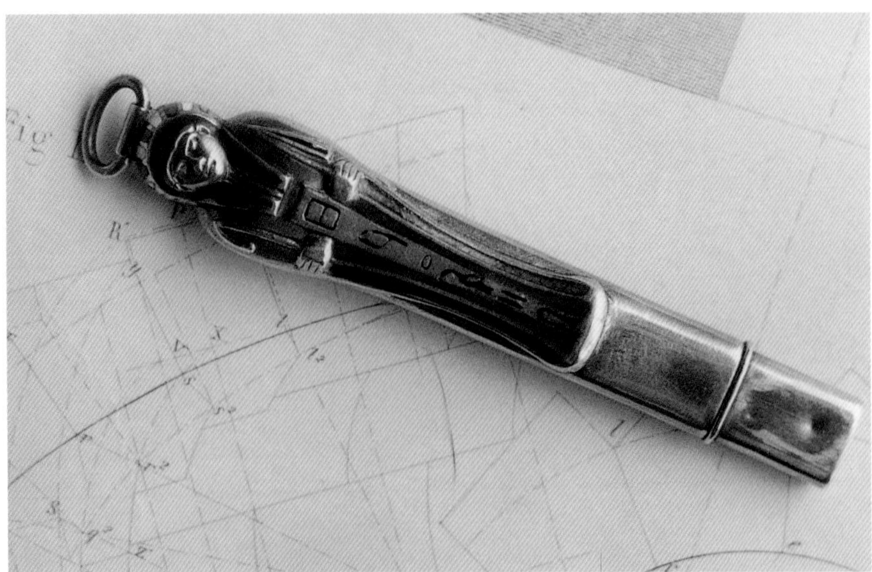

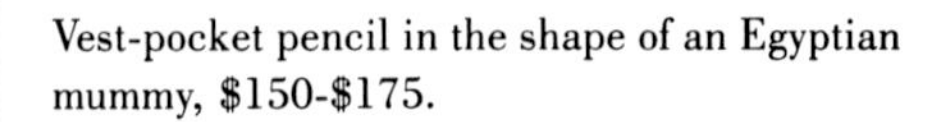

Vest-pocket pencil in the shape of an Egyptian mummy, $150-$175.

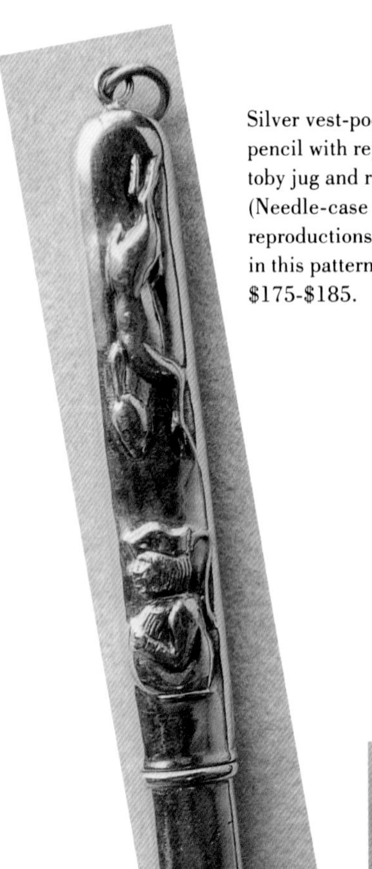

Silver vest-pocket pencil with repoussé toby jug and rabbit. (Needle-case reproductions exist in this pattern.) $175-$185.

Silver vest-pocket pencil with an etched and engraved image of a woman dancing with scarves, $125.

Silver vest-pocket pencil with repoussé dragon. (Needle-case reproductions exist in this pattern.) $135-$150.

Silver vest-pocket pencil with chrysanthemums, $150-$175.

Silver vest-pocket pencil with repoussé gracefully posed woman, $175-$200.

Gold vest-pocket pencil with repoussé woman, $175-$200.

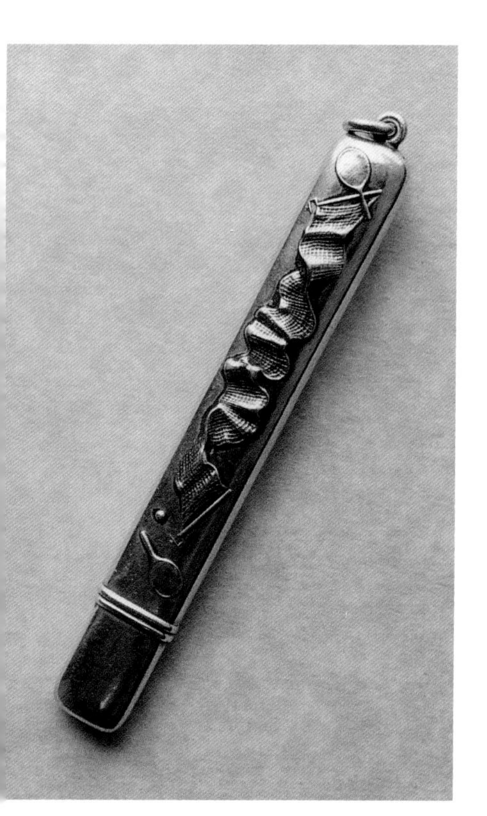

Silver vest-pocket pencil with repoussé net and tennis rackets, $175-$185.

Reverse side of above.

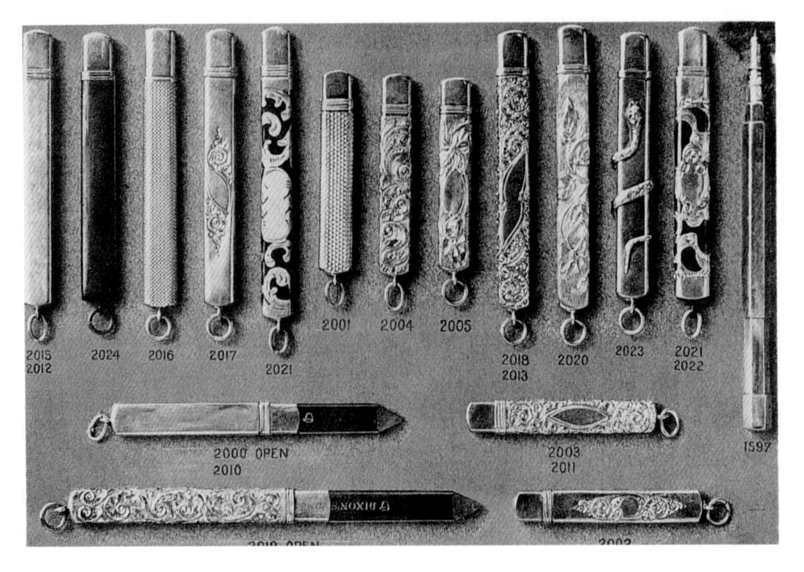

Page from nineteenth century catalog showing vest pocket, or flat pencils.

Silver vest-pocket pencil made by Daniel Low with the Salem Witch (1692 is the date of the Salem Witch trials, not the date the pencil was made), $150-$175.

Sterling silver pencil holders, 3"- 4" long. The chased tops usually held an eraser, which is sometimes missing. Some had stones set in the finial (left). $65-$95.

Pencil Extenders and Point Protectors

Pencil *extenders* or *point protectors* (called *pencil tips* by Unger Bros.) were likewise used to lengthen the wooden pencil, in more than one way. They could be placed over the point of the pencil to keep the pocket clean and the tip from breaking and could be attached to the end of the pencil to make it long enough to hold onto when there was nothing more than a stub left. The extenders sometimes housed a pencil sharpener or an eraser. These items were also offered at both ends of the market. Some were made of silver, and some were made in brass or nickel.

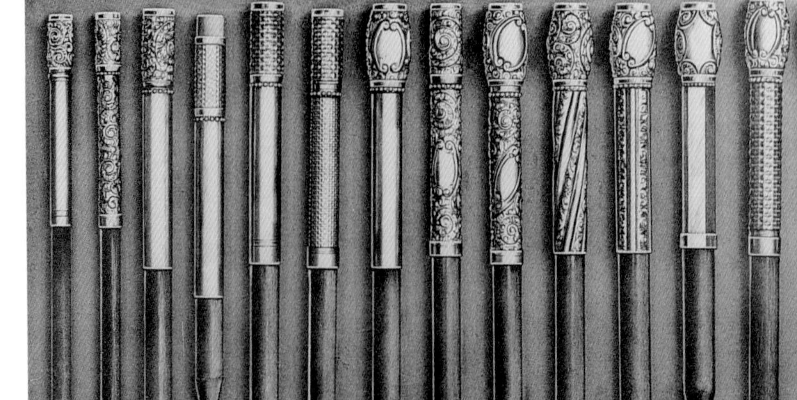

Page from a nineteenth century catalog showing silver and gold pencil holders (also called pencil extenders or pencil protectors).

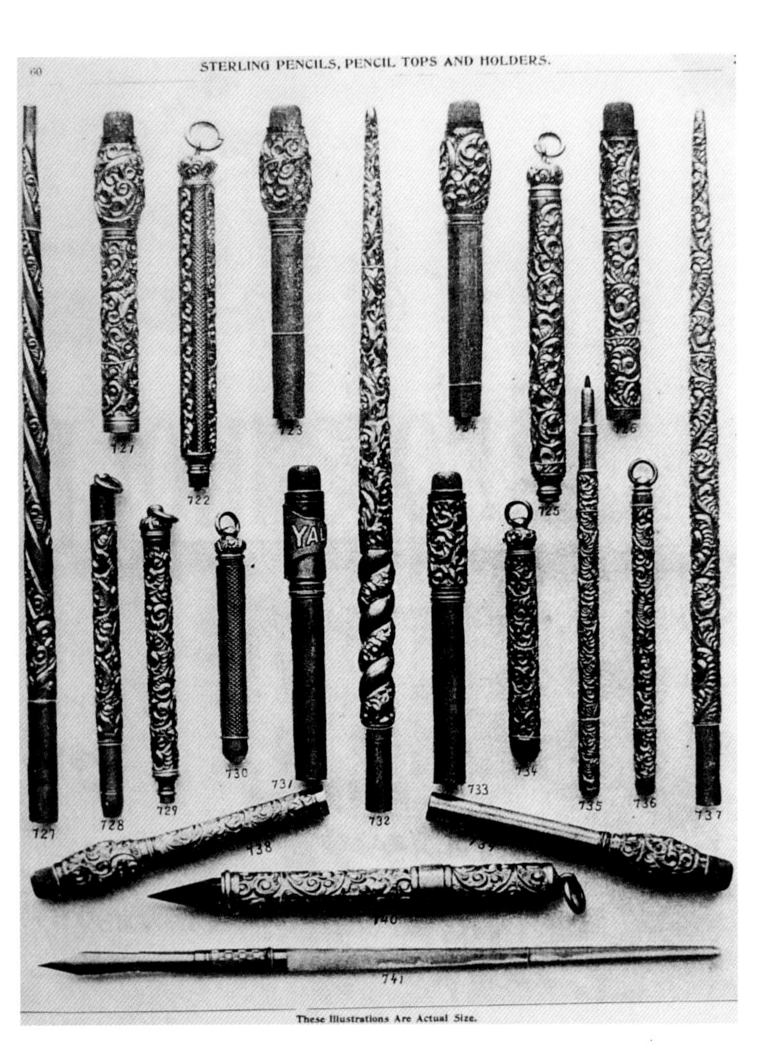

60 STERLING PENCILS, PENCIL TOPS AND HOLDERS.

721 722 723 724 725 726 727 728 729 730 731 732 733 734 735 736 737 738 739 740 741

These Illustrations Are Actual Size.

Page from E. S. Johnson catalog, showing "pencil tops and holders."

Silver pencil holders, $45-$65.

Three pencil extenders, one with sharpener built in. $65-95.

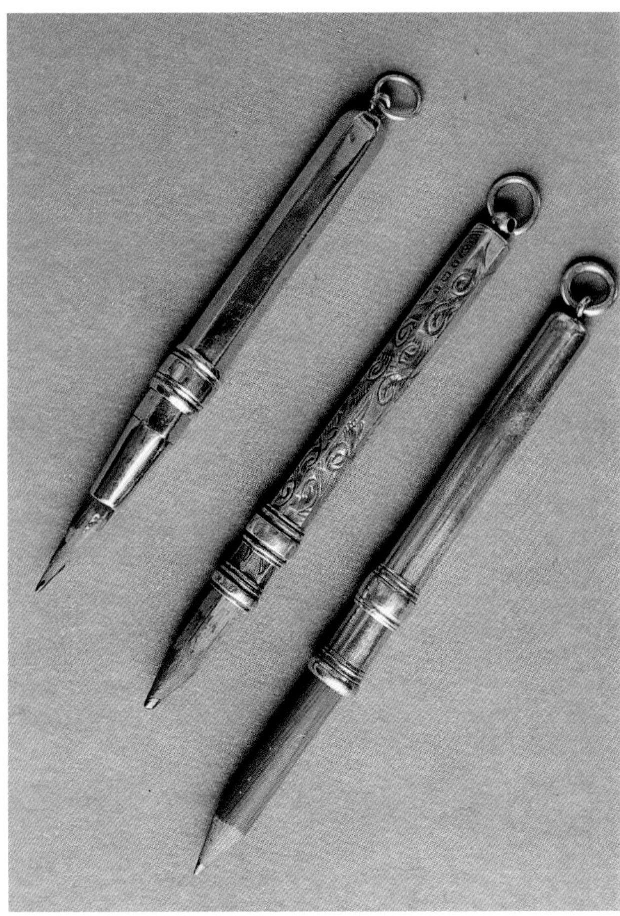

Silver (plain round and hexagonal hand-engraved), and hexagonal gold wood pencil holders, or porte-crayons, $45-$75.

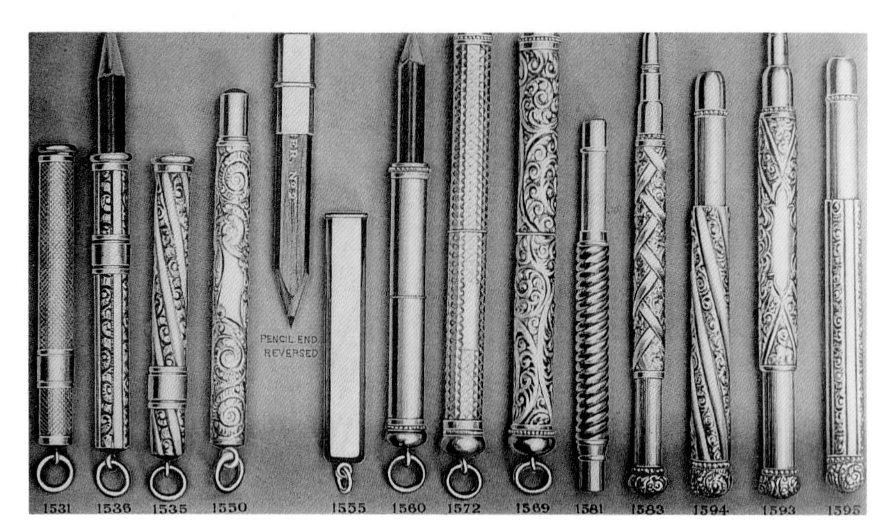

Page from John Holland catalog, showing "round wood pencil holders and business pencils."

Aide Mémoire

Pencils were sometimes made to fit into or accompany a small notebook, called *aide mémoire*. These would hang from a ring or belt chatelaine, or from a sautoir. Sometimes used as a dance card, sometimes made with ivory or celluloid pages marked with the days of the week, these pieces would serve as convenient memo pads for a woman at the end of the nineteenth or beginning of the twentieth centuries.

Sterling silver aide mémoire with repoussé flowers and Native American. Unmarked. $225-$300.

Ivory aide mémoire, with ivory pages marked with the days of the week, with Sunday being conspicuously absent (no lists of chores for Sunday). $175-$225.

Repoussé silver aide mémoire, $225-$300.

Aide mémoire, silver, made by Unger Brothers, of cherubs pulling on a wishbone, $250-$350.

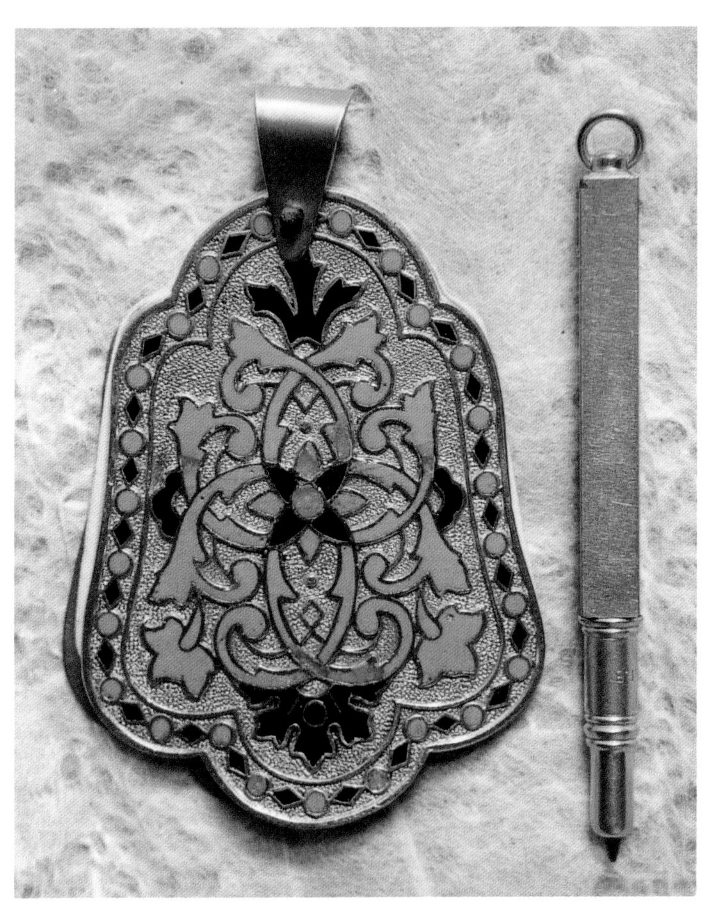

Left: Gold colored metal and enamel inlay, pencil with roman finish on gold, $225-$300.

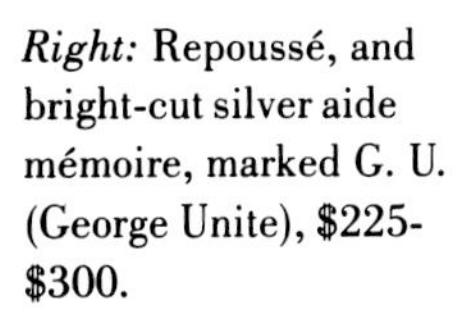

Right: Repoussé, and bright-cut silver aide mémoire, marked G. U. (George Unite), $225-$300.

Two silver aide mémoires, marked Webster Company. $75-$150.

Nickel silver aide mémoire or dance card with wood and bone pencil, on a ring chatelaine, $65-$75.

White metal aide mémoire, with a tiny photograph of Washington's Headquarters in Valley Forge. $65-$90.

Beautiful art nouveau eraser and brush. $325-$375.

Chapter Eight

Pencil-Case Makers

In a one hundred year period (from about 1822 until the 1920s), pencil-case makers created some extraordinary instruments. During this time, the mechanical pencil made the transition from new invention to commonplace object. The inventors and manufacturers (sometimes one in the same) created remarkably simple as well as highly complex writing implements that often doubled as jewelry. They developed a market that they kept active through constant revisions, improvements, and variations in their product. They were among the people who took advantage of the Industrial Revolution to increase their productivity. They were pioneers in the use of advertising. Oddly, they have received little recognition in our time despite their many accomplishments.

While there is much yet to be discovered about these individuals and the companies they ran (and it is hoped that this information leads to more knowledge about them), a brief history of pencil-case makers follows.

Thomas Addison: Thomas Addison of New York patented "Addison's Improved Ever-Pointed Pencil-Case" in 1838. In patent No. 736 (note the very low number) dated May 10, Addison describes his idea: "In a pencil case invented several years ago, the use of which was abandoned on account of not producing the desired effect and on account of its liability to get out of order...In the pencil case now submitted these defects are removed, and in the following manner...Making a spiral channel or screw in the inner or outer tube in combination with a seat in the lower end of either to receive the stop of the pencil tube to prevent the pencil or pen shoving back while writing, as before described." An example of Addison's Improved Ever-Pointed Pencil-Case is shown here.

Silver pencil with engine turning, marked "Addison's Patent" (American), circa 1838, $150-$175.

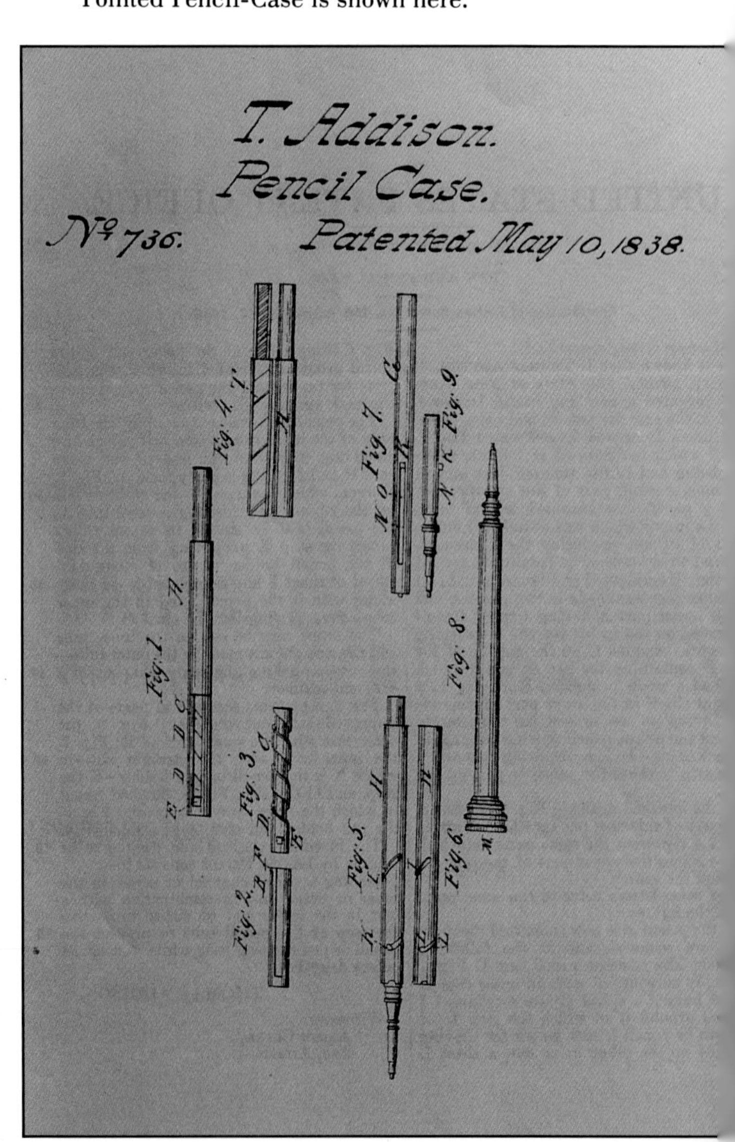

Addison's patent drawing for the pencil-case he designed in 1838.

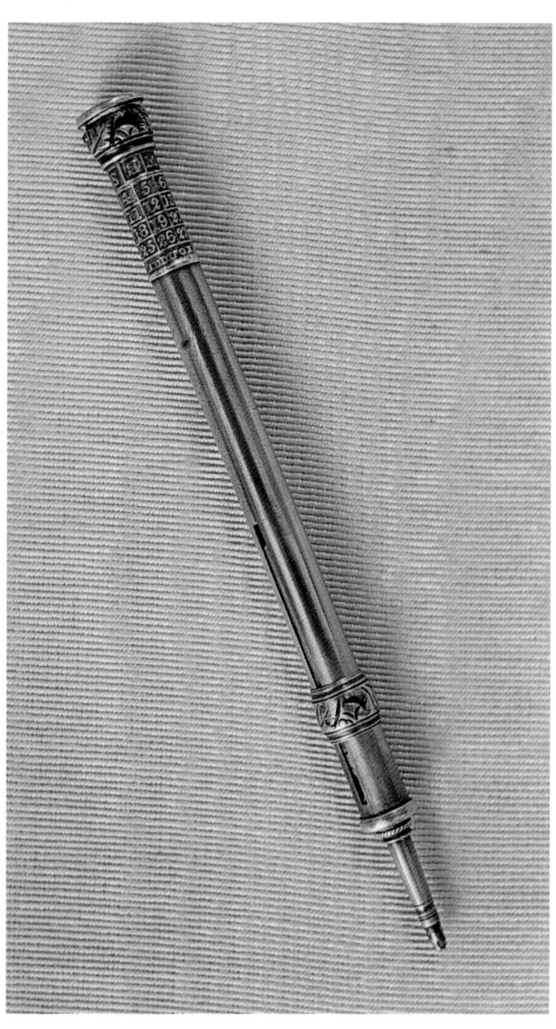

"The Perpetual Pencil," made by American Lead Pencil Co., $20-$30.

Silver pencil with slider-ring made by Addison, with a perpetual calendar. $75-$125.

Printing block for an American Pencil Co. ad.

Ad from *The Cosmopolitan*, November 3, 1903, showing how the perpetual pencil worked.

Aikin, Lambert & Co.: One of the more successful American pen and pencil-case manufacturers was Aikin, Lambert & Company.

A beautifully hand-engraved pen nib (see illustration) establishes that a company called J. C. Aikin & Co. existed prior to the partnership of James Cornelius Aikin and Henry A. Lambert. The nib is marked J. C. Aikin New York (No. 6) and is in a case with the same name, along with the word "Manufrs" embossed on it. Very few pieces bear this mark, as once Aikin and Lambert became partners in 1864 their company grew and prospered into the twentieth century (eventually becoming a part of Waterman).[1]

James C. Aikin was born in Poughkeepsie, New York, on July 14, 1840. He was educated at the University of Rochester and served as a member of the 7th Reg't N.Y.N.G. in the Civil War. He established his first business in New York City in 1859. In addition to being the president of Aikin, Lambert & Co. and Aikin, Lambert Jewelry Co., he also served as president of the Jeweler's Safety Fund Ins. Co. and Met. Burglar Alarm Co., and the director of Maiden Lane Savings Bank, Protective Union. He wrote two family genealogies, and traveled throughout the United States, South America, and Europe. He never married and is not listed as having any offspring. He lived in New York at 418 Central Park West.

Henry Lambert, who is described as a pen manufacturer and jeweler, was born in Woodbury, Connecticut, on August 7, 1837. He married Sarah G. Corliss in Lowell, Massachusetts, on September 3, 1867, and had four children (Wallace C., Helen M., John H., and Fred DeForest). Although he was a clerk in the drug business from 1853-1862, his travels led him throughout the United States and he eventually became involved in the gold pen business. He became partners with Aikin, and served as treasurer and director of Aikin, Lambert & Co. and Aikin, Lambert Jewelry Co.

During the late 1870s and early 1880s it seems there was another partner, J. B. Shea. Little is known about him at this point.

Aikin, Lambert & Co. advertised extensively in the *Jewelers Weekly* and *The Jewelers' Circular and Horological Review*, publications to the jewelry trade. Less frequently and with less impact, Aikin, Lambert Jewelry Co. was also advertised, but sometimes as importers of diamonds and jobbers in watches, jewelry, chains, etc., as opposed to manufacturers.

In an ad from the early 1880s, Aikin, Lambert & Co. advertise "'Novelties' in pencil goods" and describe "an unprecedented demand, which establishes the sale of these goods as staples, and as being suited to any season of the year." These pencils are "inlaid with pearl and gold, in the form of vines, flowers, birds, etc., on celluloid of assorted colors, in imitation of malachite, tortoise shell, agate variegated marble..." A branch in Chicago was located at 113 East Madison Street. Another ad, from 1881, illustrates three figural magic pencils. One is an acorn. The other two are heads of men: one round faced with a tall hat, squat collar, and bow tie; and the other with his hair parted in the middle, wearing a monocle and high collared shirt. The characters are not identified in the ad, which leads one to believe that they were easily recognized in their day. Another ad which depicts a figural pencil is entitled "Bill 'Possum." Describing how popular the teddy bear already was, the 'possum was predicted to be the next favorite animal symbol. Apparently meant to caricature William Taft, this particular political emblem has not withstood the test of time. Obscured by the passage of time, the reference is somewhat unclear, but it seems to show Taft as an unpopular plump figure, lurking about in darkness, not terribly comfortable in the (lime) light.

James Cornelius Aikin (Am.), "President Aikin-Lambert Jewelry Company and Aikin, Lambert & Co., Gold Pens and Pencils," from *Notable New Yorkers of 1896-1899*.

Hand-engraved nib marked "J. C. Aikin & Co.," with pen staff in original box, $100-$125.

Crossed quills, like those used on the Aikin, Lambert & Co. exhibit in 1886. (Approx. 3"x 4", $50-$75.)

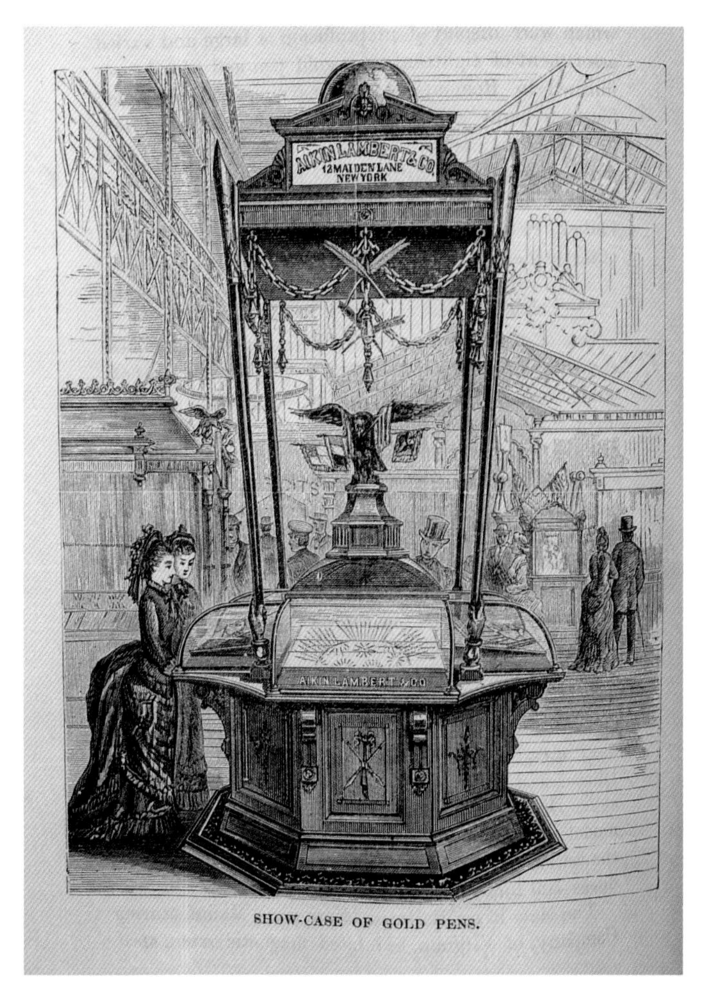

Illustration from *The Centennial Exposition* showing the Aikin, Lambert & Co. exhibit of gold pens and pencils. The globe-topped canopy is held aloft with what appear to be seven-foot high dip pens, with crossed quills ornamenting two sides of the display.

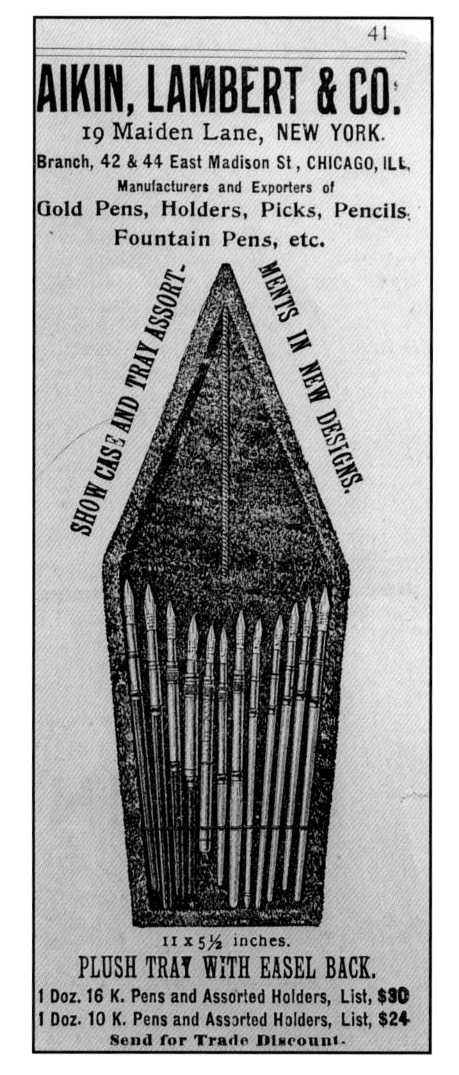

Aikin, Lambert & Co. ad from the *Jeweler's Keystone*, c. 1880.

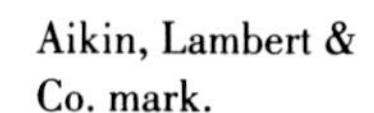

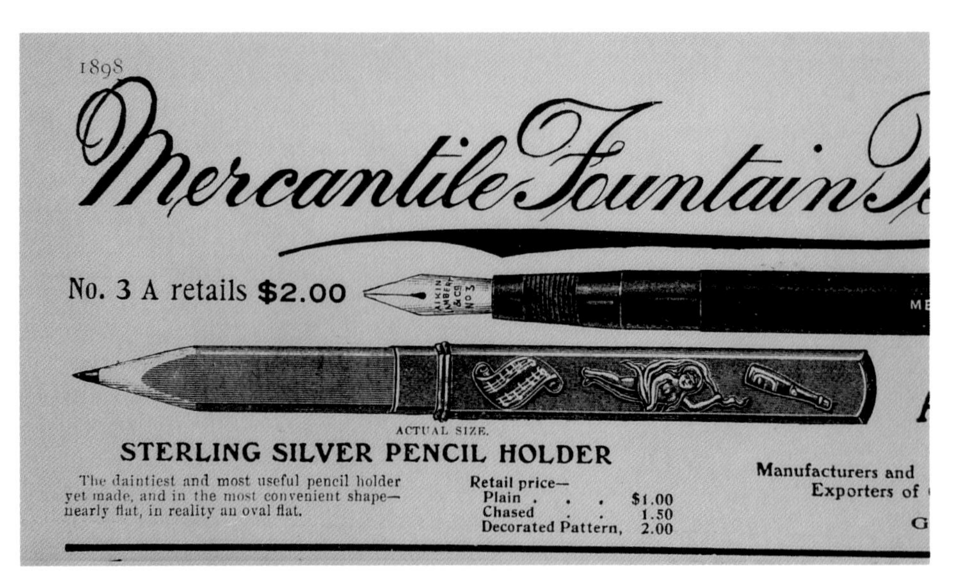

Another Aikin, Lamber & Co. ad from the same magazine.

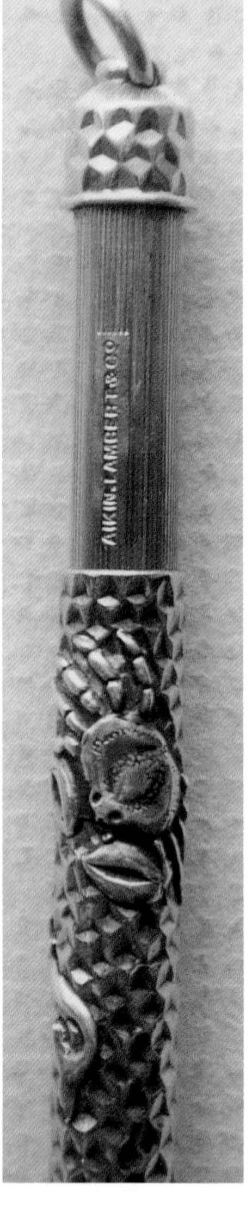

Aikin, Lambert & Co. mark.

Combination pen-pencil, rolled gold, in original box, $125-$150.

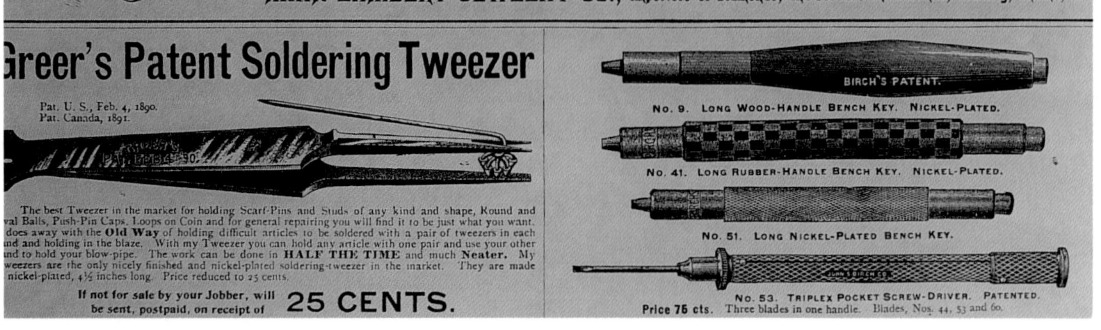

Aikin, Lambert & Co. ad (*The Keystone*, 1894), which includes both the pen-pencil company and Aikin, Lambert Jewelry Co. Also note the ad for Birch's Patent watch keys.

Aikin, Lambert & Co. ad, circa 1870.

Aikin, Lambert Jewelry Co. ad, circa 1870.

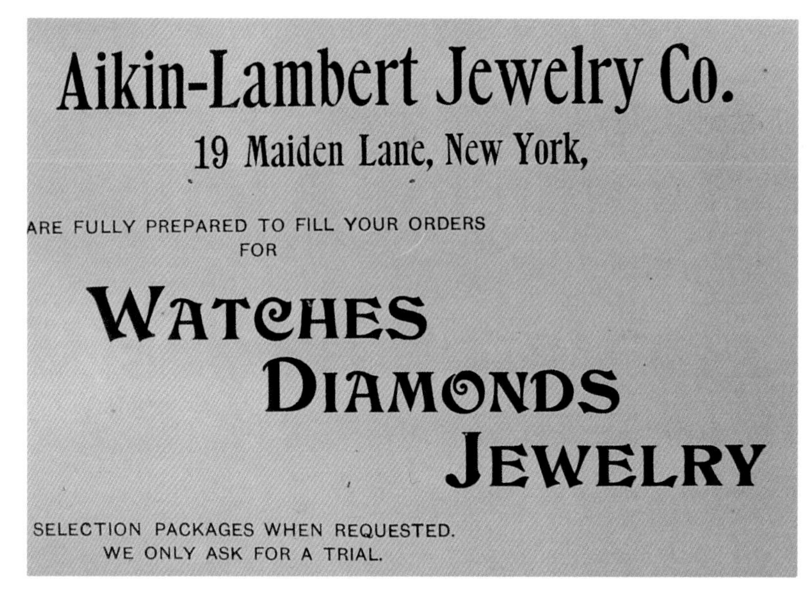

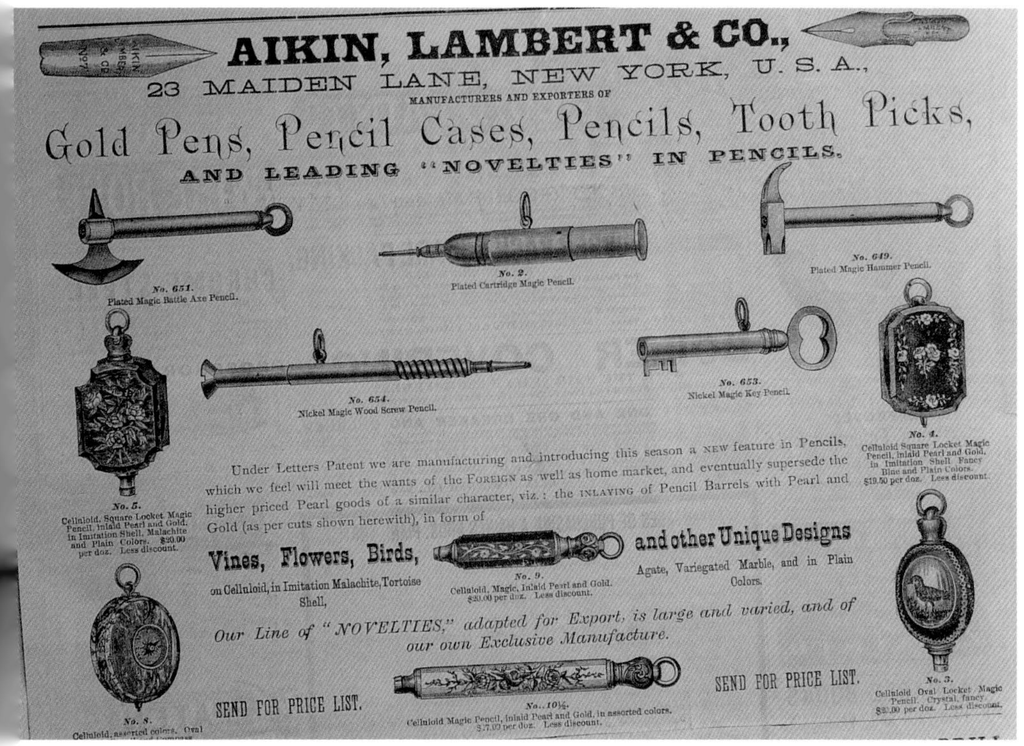

Remarkable ad for Aikin, Lambert & Co. "Gold pens, Pencil Cases, Pencils, Tooth Picks, and Leading 'Novelties' in Pencils." Several figural pencils are shown here, including an axe, hammer, and key.

J. C. Aikin. H. A. Lambert. J. B. Shea.

AIKIN, LAMBERT & CO.,

MANUFACTURERS OF GOLD PENS,

Pen and Pencil Cases, Pencils, Tooth-picks, and "Novelties" in Pencil Goods.

No. 23 Maiden Lane, New York,

Would call the attention of the Trade to our large and complete line of Pen Goods in all styles and varieties, suitable for the Winter and early Spring demand.

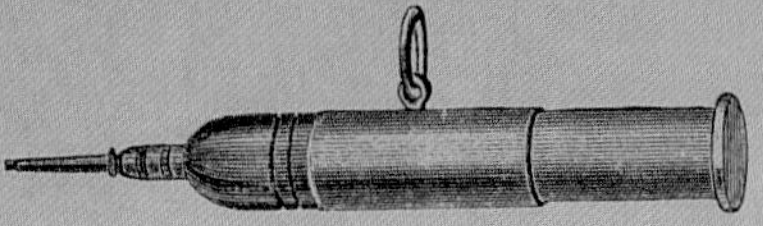

☞ Our introduction last season of Pencils in NEW AND ENTIRELY NOVEL DESIGNS was marked by an unprecedented demand, which establishes the sale of these goods as STAPLES, and as being suited to any season of the year.

The Magic Charms (as per cuts shown below), inlaid with pearl and gold, in form of vines, flowers, birds, etc., on celluloid of assorted colors, in imitation of malachite, tortoise shell, agate variegated marble, etc., are the LATEST and most novel pencils in the market.

Send for circular and new list. **Branch, No. 113 East Madison Street, Chicago.**

Aikin, Lambert & Co. ad, circa 1880.

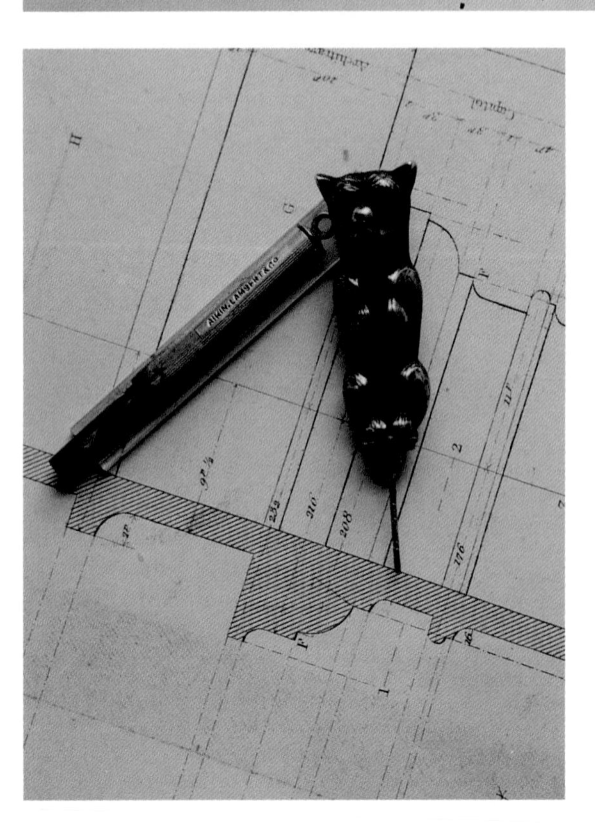

Figural pencil in the shape of a dog, marked "Aikin, Lambert & Co.," disassembled to show mechanism.

Wonderful figural pencil made by Aikin, Lambert & Co. (see Chapter 5).

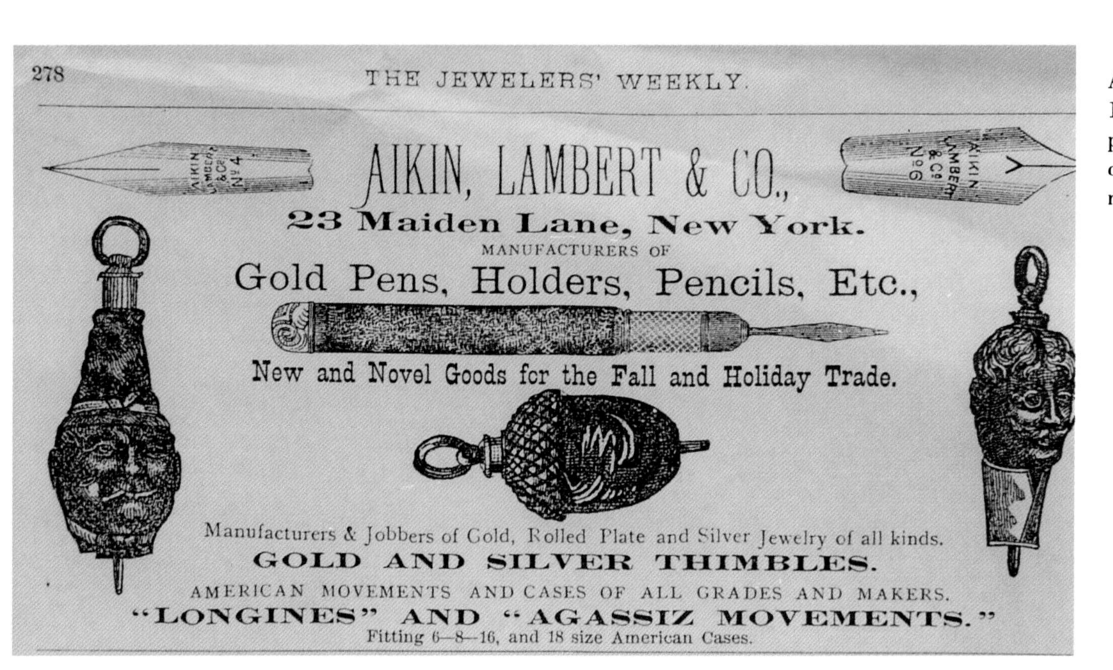

Ad from *The Jewelers' Weekly*, 1881, showing the previous pencil. Unfortunately, the characters are not described or named.

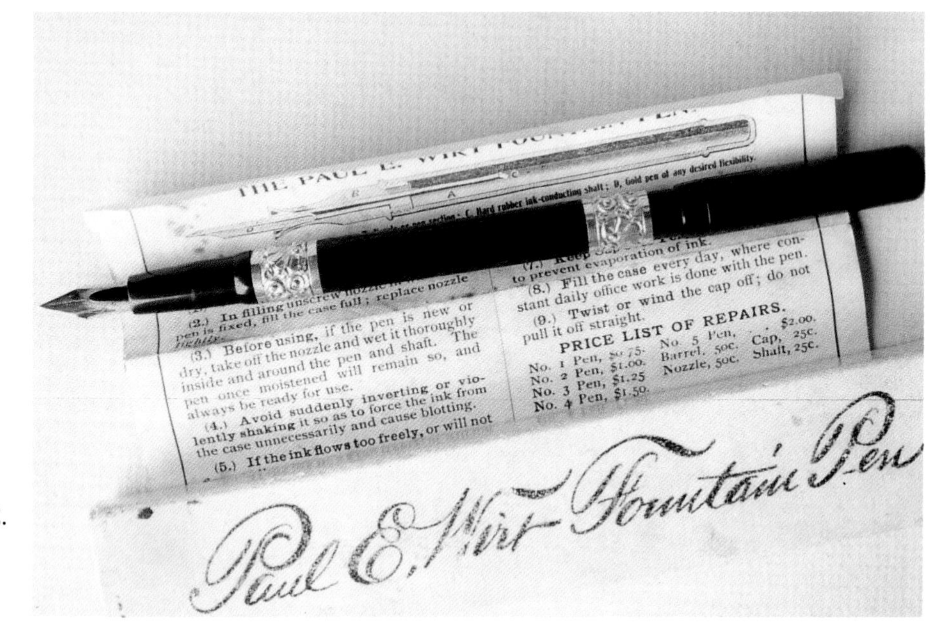

Aikin, Lambert & Co. were authorized dealers of Paul E. Wirt fountain pens.

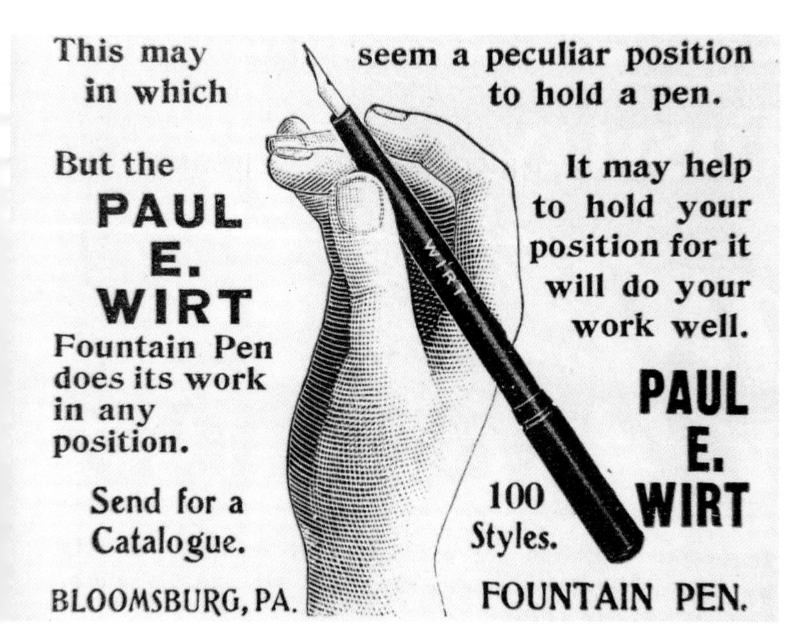

Ad for Paul E. Wirt fountain pens, circa 1890.

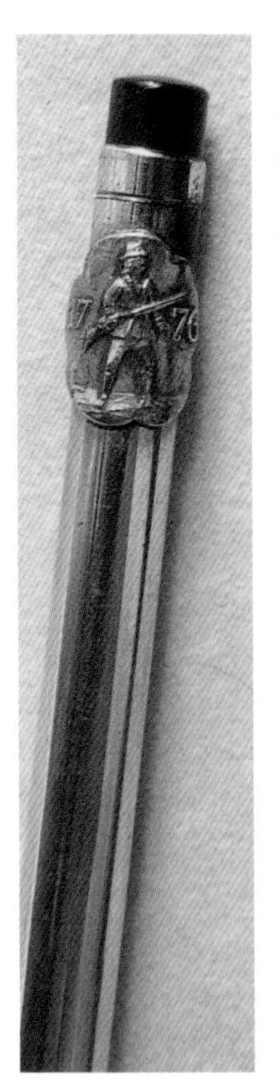

Detail showing pencil made by Aikin, Lambert & Co. and Hardtmuth's Koh-i-noor for the American Centennial in 1876.

Hand-engraved silver pencil made by Aikin, Lambert & Co. and Hardtmuth's Koh-i-noor, with a sterling snake clip. The L. E. Waterman Pen Co. in London sold pencils made by Hardtmuth until World War I. Nearly 5", $150-$175.

"BILL 'POSSUM"

The advent of the successor to the "Teddy Bear" finds the well-known Aikin-Lambert pencils leading the idea.

The presentation of this line in Washington was an immediate success. The large number of styles in Gold and Silver with the 'Possum mounting are now ready for shipment. Dealers will find them to be fast sellers, affording a liberal margin of profit.

Only one of the large line of Aikin-Lambert novelties that are popular for their utility and fineness.

Manufacturers of the well-known Mercantile Fountain Pen.

Send for Catalogues

Aikin-Lambert Co. 15-19 Maiden Lane NEW YORK

Ad for "Bill 'Possum" pencils, circa 1909. President William Howard Taft was nicknamed "Bill 'Possum," but the image never caught on the way the "Teddy Bear" did.

WE WILL RESUME OUR OLD ADDRESS

One year ago, we left our headquarters of twenty years' standing to make room for the handsome structure illustrated opposite.

On May 1st, we will occupy our handsome, new quarters, and extend to the trade and our friends in general, a cordial invitation to visit us, and see the growth and development in the distribution of the following well-known lines, the output of our large and well-equipped factories.

A. L. Co. Gold Pens
Mercantile Fountain Pens
Mercantile Self-Fillers
Beacon Stylographic Pens
Gold and Silver Pen Holders
Gold and Silver Pencils
Universal Outfits
Gold and Silver Penknives
Cigar Piercers
Etc., etc.

Send for complete catalogues to-day.

Aikin-Lambert Co. 15-19 Maiden Lane, NEW YORK Established (1864)

Ad from the early twentieth century, showing products made by Aikin, Lambert & Co.

American Graphite Company, Mines, and Works: located in Ticonderoga, New York. They mined and sold plumbago, "standard unequalled grades...grades for special uses prepared to order..." Cyrus Butler was president, and the company had an office located at 24 Cliff Street, New York (letterhead, 1878).

American Lead Pencil Company: Edward Weissenborn, manufacturing partner, and Hecht Brothers, selling partners. Makers of inexpensive mechanical pencils, including "the Perpetual Pencil," and pencil extenders. They were located at 491-493 Broadway, New York, with their factory and pencil-lead works in Hoboken, New Jersey (letterhead, 1903).

Asprey: was a high-end retailer of Mordan pencils in London. The firm is still in existence.

S. Mordan & Co. drop-action pencil, marked Asprey.

Baird-North Company: gold and silversmiths who sold mechanical pencils (and other pieces) to hang from chatelaines. Located in Salem, Massachusetts (catalog, circa 1880).

Edward Baker: Pencil-case maker (London), established in the mid-nineteenth century. Manufactured gold and silver pencil-cases, as well as "penholders, pocket knives, toothpicks, cigar piercers, cigar cutters, cigar and cigarette holders and tubes, spirit lamp extinguishers, fountain pens, gold pen nibs, fountain pen chatelaine cases and many patented and registered novelties" (catalog, 1915). Baker sold "Baker's patent Pointer pencil-cases," a wide array of telescopic pencils, including enameled, removable sheath, and pen-pencil combinations.

Banks and Company: established in 1832 at Greta Pencil Works, Keswick, Cumberland. They manufactured black lead pencils and black lead. A nineteenth-century advertisement describes the process:

Plumbago, or graphite, commonly known as "black lead," is in reality a pure form of carbon, and is found deposited in a natural condition in varying sized lumps, from a few ounces to several pounds in weight, in what the miners term "sops," or "pockets," at irregular distances below the surface of the ground. It is also found at Seathwaite, in Borrowdale, Cumberland, and also in Bohemia, Ceylon and elsewhere. That found at Seathwaite, in Borrowdale, is considered the finest quality for pencil-making and is principally used by us in our extensive works where all the varying processes may be witnessed... The plumbago is prepared by reducing it to powder and grinding it in a wet condition, with binding materials, between two large mill-stones, until it is reduced to a perfectly smooth paste; it is then partially dried and pressed by Hydraulic Machinery, through agate dies, of the required shape and size... The resulting long threads of plumbago are cut into lengths of 7 inches or more, and are rendered of the necessary hardness by the process of tempering, and then by heating to redness in large iron pots in furnaces...

Wilson and Maney Benton: successors to B. T. Benton and Brother. Manufacturers of gold pens, gold and silver pen holders, pencils, lockets, and jewelry. Offices located at 4 Jewelers Exchange, Gilsey Building, 169 Broadway, New York. Their manufactory was located in the Deans Building, Providence, Rhode Island (letterhead, 1857).

Berolzheimer, Illfeder, and Company: manufacturers and sole importers of The Eagle Pencils, patented April 3, 1860. Located at 118 William Street, New York (letterhead, 1863).

Joseph Bramah: Inventor of the swivel nib holder, water closet, Bramah lock, etc. Influenced Sampson Mordan.

Joshua Butler: Joshua Butler & Co. was also known as Butler & Wise (c.1835-6), Thomas Wise (1836-9), Wise & Nash (1839-49), and Eliezer Nash (1849-?). Butler himself retired in 1839, and the business continued at 30 Coppice Row, Clerkenwell, with Wise and Nash as partners. They were listed as "manufacturers of Butler & Co. ever-pointed pencils and tortoiseshell workers... silversmiths and pencil-case makers." At the Crystal Palace Exhibition in 1851 Nash, now working without a partner, showed "Pencil-cases: Engraved, coloured gold, and set with turquoise; engine-turned bright gold, engraved, elongated; triangular; engine-turned hexagon, and engraved round silver...Set of tortoiseshell instruments, gold inlaid and mounted, comprising of paper-folder, pen-knife, pen-holder, pencil, and desk seal The elastic palladium point, or lead-holder, an improvement applied to pencils, was invented by Mr. Joshua Butler and has been in use for 20 years." (John Culme, *The Directory of Gold and Silversmiths, Jeweller's and Allied Trades.*)

Detail of silver pencil marked "Butler & Co." (English).

More elaborate pencil made by Butler & Co., London, with cast foliate slider and terminal.

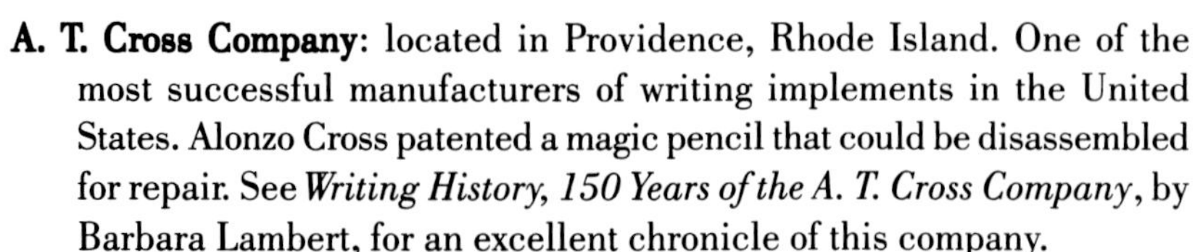

A. T. Cross Company: located in Providence, Rhode Island. One of the most successful manufacturers of writing implements in the United States. Alonzo Cross patented a magic pencil that could be disassembled for repair. See *Writing History, 150 Years of the A. T. Cross Company*, by Barbara Lambert, for an excellent chronicle of this company.

A. T. Cross was established in 1846, and is still in existence today. Founded by Richard Cross and Edward Bradbury (who were trained in the pencil-case making trade in Birmingham, England), the company was eventually run by Richard's son, Alonzo Townsend Cross. Alonzo was inventive and ambitious, patenting pencil and pencil mechanisms, as well as patenting a stylographic fountain pen. He also recognized the potential of the steam engine, and utilized them in his manufactory (he also was involved in a firm called Shipman Engine Export Company).

A. T. Cross Company maintains an excellent archive, where an interested collector may make arrangements to view many of their early products. Among these are novelty pencils in the shapes of screws, cut nails, keys, brooms, railroad lanterns, cannons, matchsticks, umbrellas, and cigars. Also, there are examples of their "Alwrite" pencils, as well as their famous black banded pencils, magic pencils, stylographic pens and fountain pens. The modern facility in Providence is bustling with activity, and the fact that it is the only of the nineteenth century pencil-case companies to still be in existence (without having been merged into another company) is a tribute to its founders.

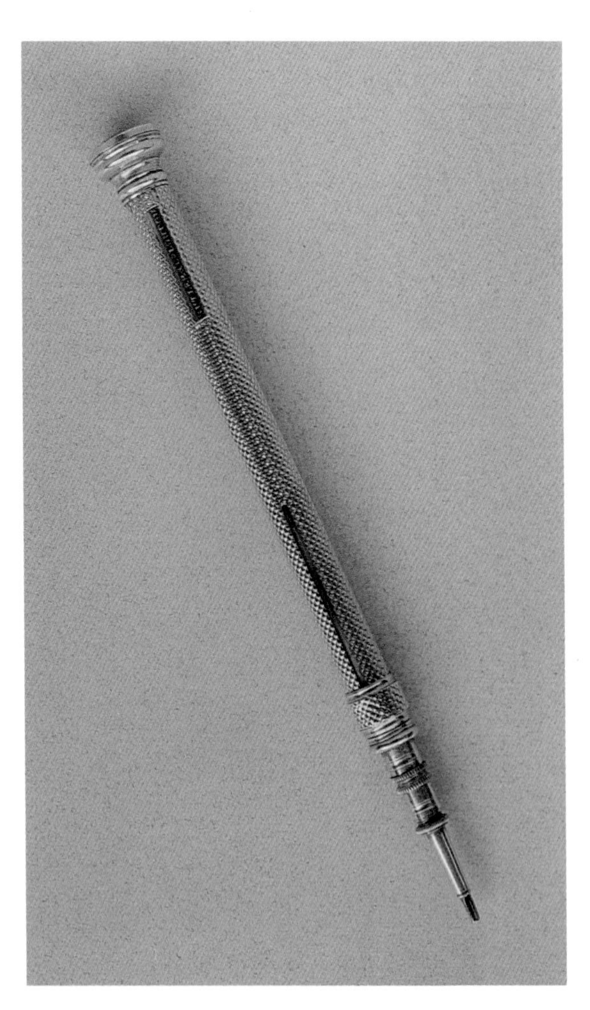

Barleycorn pattern, mounted with lapis lazuli, 18ct gold mechanical pencil made by Butler. $225-$300. *Photo Courtesy of Chris Odgers.*

Thin, silver pencil made by Butler & Co. with a cast button-slider, $150-$175.

Two novelty pencils shaped like square nails, one advertising "Hemlock Lumber and Lath," $75-$125.

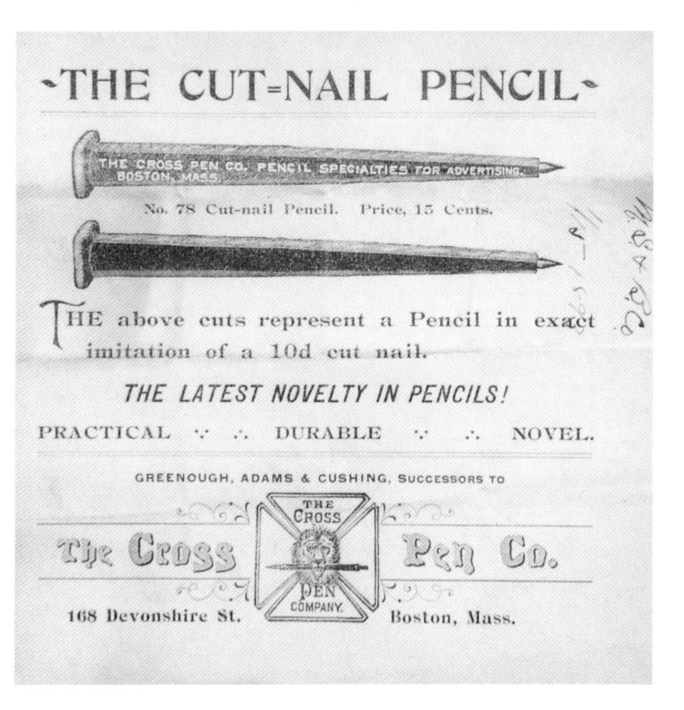

Early ad for A. T. Cross Co. (called the Cross Pen Co.), showing cut-nail pencils, "The Latest Novelty in Pencils!"

Two silver pencils made by A. T. Cross. $50-$75.

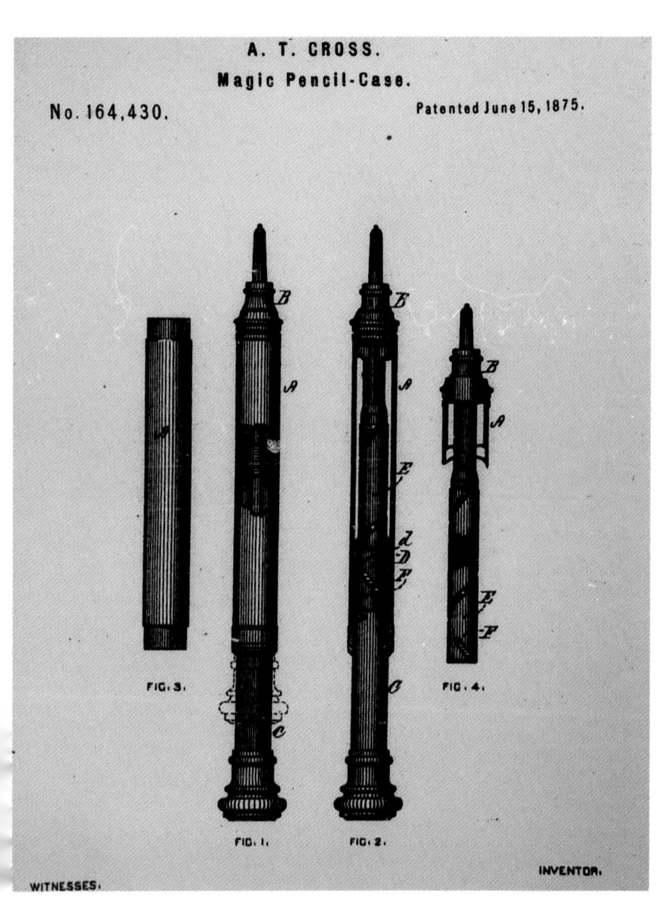

Alonzo Cross patented a mechanical pencil that could be taken apart for repair, indicating that some of the mechanisms were already in need of attention only a few years after they were made.

Close-up of the A. T. Cross logo.

Cutter, Tower, and Company: importers, manufacturers, and dealers in superior gold and steel pens, pen holders, pencils, seal presses, blank books, and stationery. They were listed as general agents for Fosters' Patent Improved Pencil Sharpener, and were located at 7 Beekman Street, New York City, and 17 & 19 Cornhill, Boston, in 1858. J. M. Cutter, L. L. Tower, and S. A. Tower were partners in 1858. I. L. Kidder joined them in 1860 (letterheads, 1858, 1860).

W. F. Doll Manufacturing Company: established in 1876. Advertised "from factory to pocket, saves you two profits." Sold gun-metal items made of "the steel recovered from the wreck of the Battleship Maine..." as well as figural pencils (cannon, screw, and railroad lamp) made by A. T. Cross (catalog, circa 1890).

Eagle Pencil Company: The products offered by Eagle Pencil Company were practical, varied, and often innovative. One of the earliest pencil manufacturers in the United States, the company profited from the knowledge and expertise of the founder's European predecessors.

Daniel Berolzheimer came to America in 1856 in order to create an office in New York City for the purpose of selling the products made by Berolzheimer & Illfeder (manufacturers of lead pencils in Fuerth, Bavaria). When Daniel died, his son Henry took over the business and it was he who built the original factory in Yonkers, New York in 1869. Emil Berolzheimer, who was also born in Fuerth, came to America to succeed his father as chief executive of the company in 1883 and, in 1885, incorporated it as Eagle Pencil Company. He remained president of the firm until he died in 1922. His son, Alfred, succeeded him.

Alfred studied at Exeter Academy and graduated from Harvard in 1913. For a brief period he worked for a New York investment firm. However, by 1915 he was an employee of Eagle Pencil Co., where he became a director in 1920, treasurer in 1922 and vice president in 1925.

Considered by some to be the "largest pencil manufacturing concern in the world," the company manufactured "pencils, steel pens, pen holders, mechanical pencils, fountain pens, compasses, sharpeners, erasers, school sets, metal specialties and various novelties." (James T. White & Company, *The National Cyclopedia of American Biography*, New York, 1921.)

Products made by Eagle Pencil Company are generally relatively inexpensive, but interesting nonetheless. They were often colorful and imaginative, and sometimes stretch the limit by being very large (a pencil may be ten or more inches long and nearly an inch wide), or extremely tiny (three and a half inches long and 1/8 inch wide). Some of their oversized products were called "Just For Fun."

PRIZE MEDALS:
NDON 1862 PARIS, 1867.
New York, Sept 12 1889.
Mr. T. A. Anglin City.
BOUGHT OF THE
EAGLE PENCILS.
HIGHEST AWARD AT CENTENNIAL EXHIBITION FOR CHEAPNESS & GOOD QUALITY.
(EXTRACT FROM JUDGE'S REPORT.)
Eagle Pencil Company.
OFFICE & SALESROOM, 73 FRANKLIN ST.
WORKS 710 TO 732 EAST 14TH ST.
S.B.Folio Terms
We assume no responsibility for packages sent for enclosure. All bills to be settled in 30 days either by Cash or Note.
Shipped by No. Weight

50 Gross 140 Sharp.	63		31 50	
5 " 337		2 47½	12 38	
5 " 92		1 50	7 50	
3 " 202		3 00	9 00	
1 " 553			7 25	67 6

Billhead from Eagle Pencil Company (Am.), 1889.

An "automatic pencil" made by Eagle Pencil Co. When you pushed on the button at the end of the barrel, the graphite would fall into place.

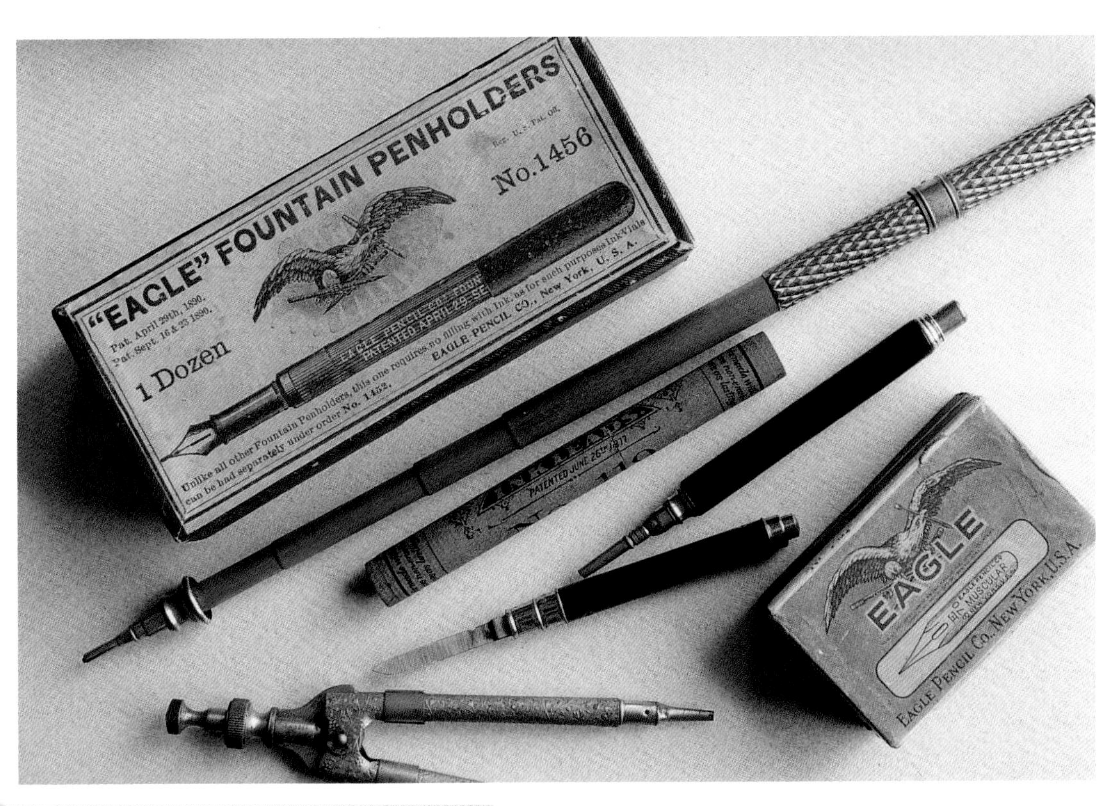

A variety of products made by the versatile Eagle Pencil Co., including one dozen "Fountain Penholders" in their original box ($300), a telescoping pointer pencil for teachers or lecturers ($50-$75), an automatic pencil ($25-$50), a gravity pen-knife ($35-$50), a box of Eagle nibs ($10-$15), an Eagle compass with pencil ($20-$25).

.................................DOZ., .84

UTOMATIC LEAD PENCILS.

Eagle automatic stop gauge, ebony nd nickle-plated case, with blue copy ds; (gross, $7.85)................DOZ., .68

-672. Imported automatic combination and penholder of chased imitation Aus- gold, mounted with assorted birthday a cleverly concealed lever projects an tic pencil or pen as desired; detachable op, on being removed displays an extra of leads; put up 1 dozen in neat black ....DOZ., 1.35

Handsome German silver presentation set, containing pocket penholder with n, pocket pencil, chain pencil and box defills; eac hset in neat box........ DOZ. SETS., 2.10

UTOMATIC PENCIL LEADS.

Challenge copying leads; packed 3 leads nickle case, 2 dozen cases on display pecially adapted for automatic and all ink pencils, 1¾ inches long; (gross, ...DOZ., .35

Eagle indelible leads, for automatic 3 leads in round metal c 1 dozen 1 box; (gross, $8.25).............DOZ., .72

SLATE PENCILS.

One Hundred in Box.

Plain slate pencils, 5½ inches long, ...PER M, .75

wrapped pointed slate pencils—
-F. 5½ inches long...........PER M, .84
-F. 6½ inches long...........PER M, 1.00
wrapped pointed slate pencils—
-G. 5½ inches long...........PER M, .84
-G. 6½ inches long...........PER M, 1.00

Ad from Charles Broadway Rouss catalog, 1907, showing Eagle's "automatic pencil."

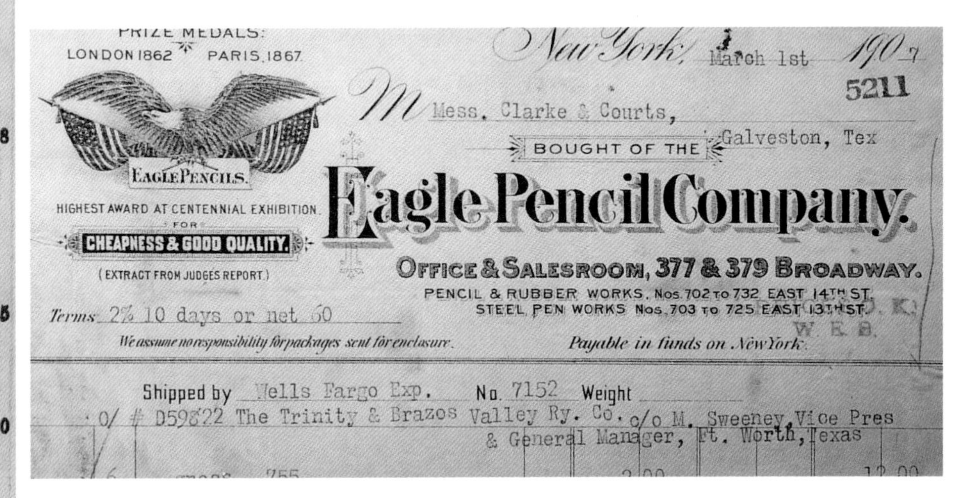
PRIZE MEDALS: LONDON 1862 PARIS 1867

New York, March 1st 1907

5211

M Mess. Clarke & Courts, Galveston, Tex

BOUGHT OF THE

EAGLE PENCILS.

HIGHEST AWARD AT CENTENNIAL EXHIBITION FOR CHEAPNESS & GOOD QUALITY. (EXTRACT FROM JUDGES REPORT.)

Eagle Pencil Company.

OFFICE & SALESROOM, 377 & 379 BROADWAY.

PENCIL & RUBBER WORKS, Nos. 702 to 732 EAST 14TH ST.
STEEL PEN WORKS Nos. 703 to 725 EAST 13TH ST.

Terms 2% 10 days or net 60

We assume no responsibility for packages sent for enclosure. Payable in funds on New York.

Shipped by Wells Fargo Exp. No. 7152 Weight

O/ # D59822 The Trinity & Brazos Valley Ry. Co. c/o M. Sweeney, Vice Pres & General Manager, Ft. Worth, Texas

Billhead from Eagle Pencil Company, 1907.

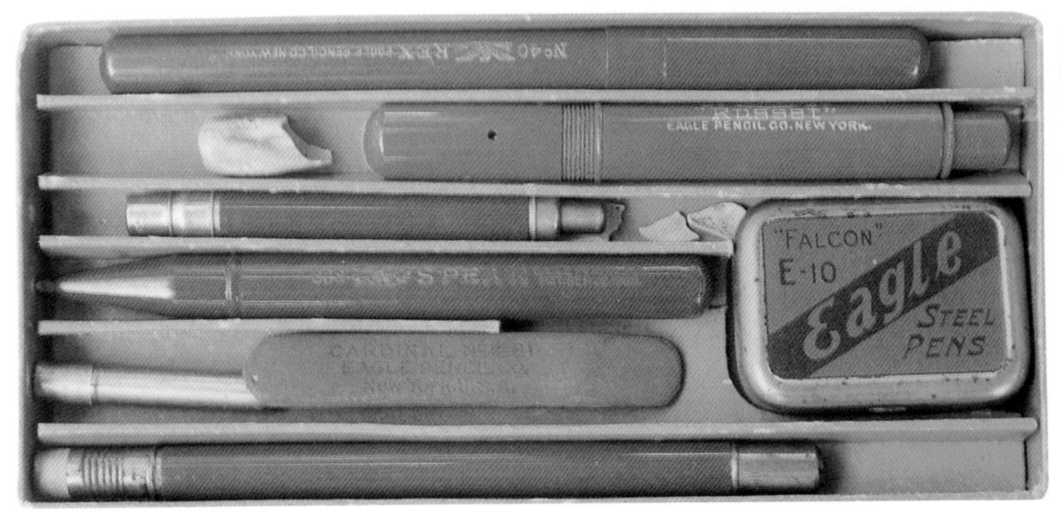

Eagle assortment in original box: fountain pen with glass cartridge, stylographic pen, gravity pen-knife, "Spear" pencil, eraser, nibs, and pen-pencil holder, all in dark orange ("Russet"). $100-$125.

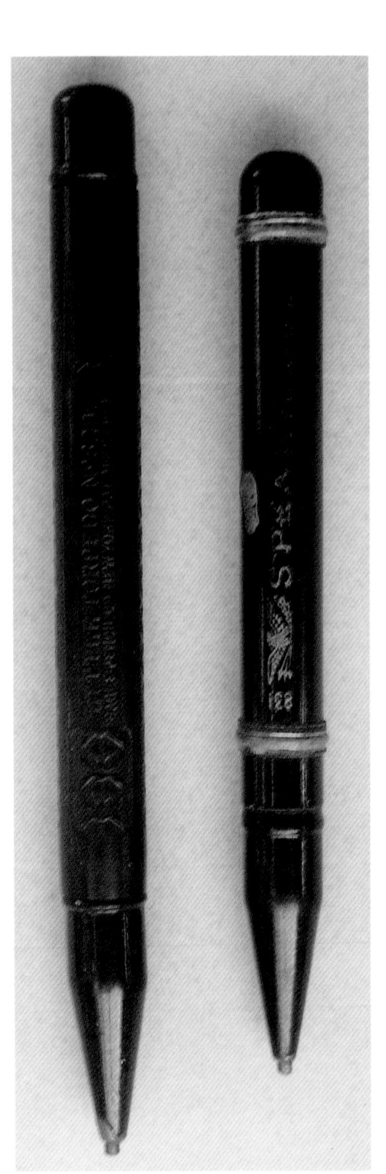

Eagle "Spear" and "Torpedo" pencils, made of lacquered brass and (top) black hard rubber, $25-$30.

Eagle Pencil Co., "Elfin" pencil set. One dozen multi-colored pencils in the original box (each pencil is approximately 4" x 1/8"), $40-$60.

Eagle Pencil Co. "America For All" pencil from 1917. Red, white and blue painted wood, with a metal hat on the end. 8" x 3/4", $45-$75. Photographed on an early painting of a canoe on birch bark.

Ad showing the variety of products offered by Eagle Pencil Company (one of the few firms that chose to maintain the use of the word "pencil" in their name, even after they started making fountain pens; although they later also used the name "EPEN Co."

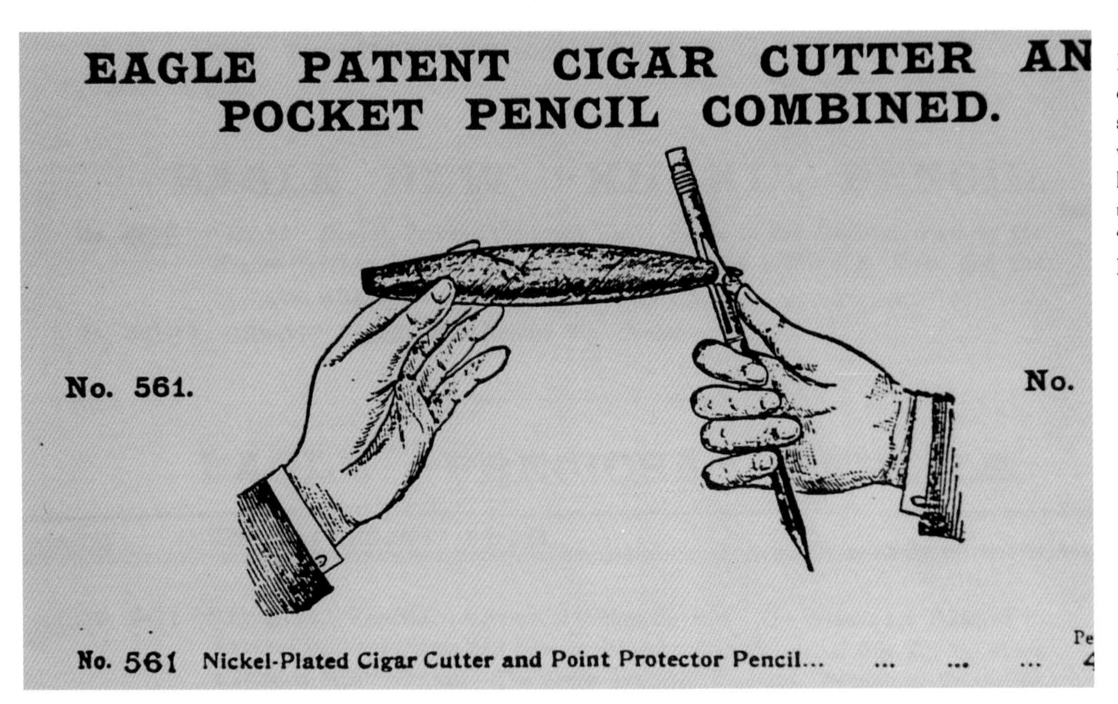

Eagle did not generally make expensive products (although some of their pens and pencils were made of sterling silver), but they certainly made unusual ones. Here is an ad for "Eagle Patent Cigar Cutter and Pocket Pencil Combined."

EAGLE PENCIL COMPANY.

STERLING SILVER AND ROLLED GOLD MAGIC PENCILS.—Contd.

492S

Per Dozen.

No. 492s Magic Pencil, Half Chased Sterling Silver... ... ... Large size 84/-

493S 328S

No. 493s Magic Pencil, Full Chased Solid Silver ... ... Large size 84/-
No. 328P Magic Pencil, Full Chased Fine Rolled Gold Plate, same pattern as 493s. Large size 126/-

494S SILVER 327P PLATED

No. 494s Magic Pencil, Plain Sterling Silver ... ... Large size 60/-
No. 327P Magic Pencil, Plain Fine Rolled Gold Plate, same pattern as 494s .. Large size 120/-

EAGLE PENCILS, WITH STERLING SILVER PROTECTORS. ON ATTRACTIVE CARDS.

450S

Per Dozen.

No. 450s Eagle Pencil, Round, fitted with an Embossed Sterling Silver Protector and Rubber 30/-

451S

No. 451s Eagle Pencil, Hexagon, fitted with Embossed Sterling Silver Protector and Rubber... 36/-

452S

No. 452s Eagle Pencil, Round, fitted with massive Sterling Silver Protector and Jewelled with Stone at end 54/-

453S

No. 453s Eagle Pencil, Round, fitted with Sterling Silver Protector and Rubber 54/-

457S

No. 457s Eagle Pencil, Round, fitted with Sterling Silver Protector and Paper Cutter combined 30/-

Sterling silver and rolled gold pencils and pencil extenders made by Eagle in the late 1800s.

Four pencils made by Eagle, all with glass "stones" set in the finials. $5-$25.

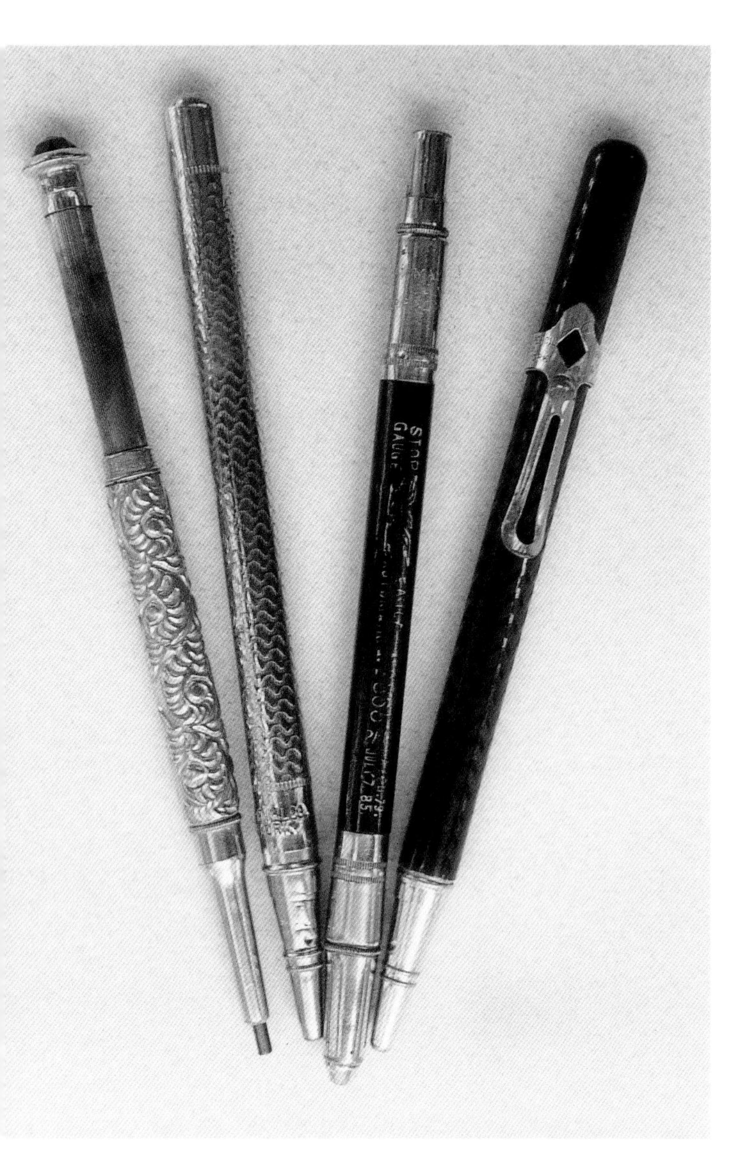

More examples of nineteenth century pencils made by Eagle Pencil Company. $10-$25.

Combination pen-pencils made by Eagle in black hard rubber and nickel, $75-$125.

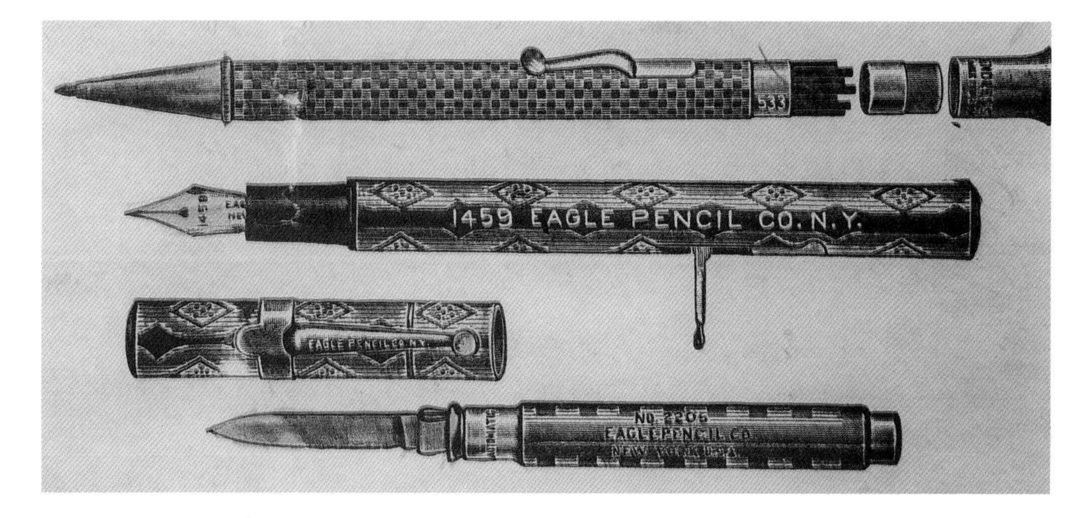

Ad for Eagle pencil, fountain pen and gravity penknife, circa 1890.

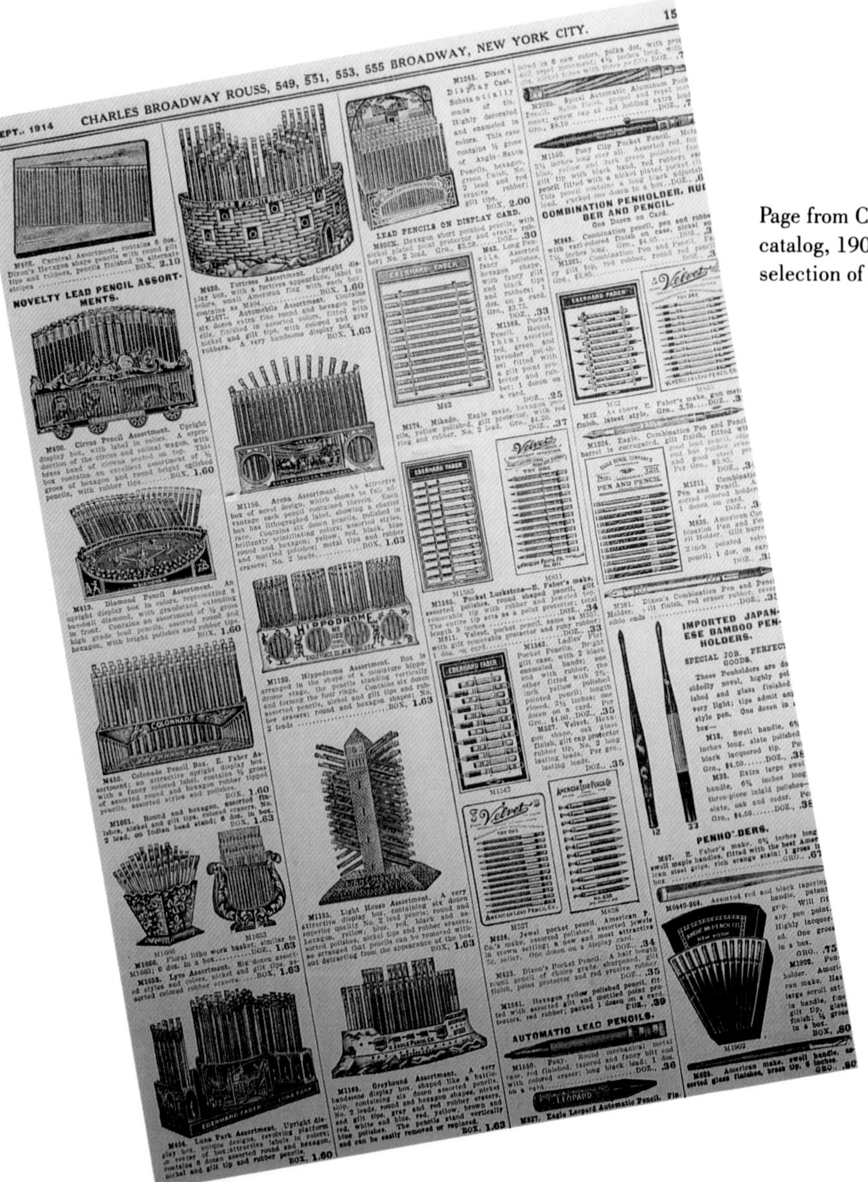

SEPT., 1914 CHARLES BROADWAY ROUSS, 549, 551, 553, 555 BROADWAY, NEW YORK CITY. 15

NOVELTY LEAD PENCIL ASSORTMENTS.

LEAD PENCILS ON DISPLAY CARD.

COMBINATION PENHOLDER, RUBBER AND PENCIL.

IMPORTED JAPANESE BAMBOO PENHOLDERS.

PENHOLDERS.

AUTOMATIC LEAD PENCILS.

Page from Charles Broadway Rouss catalog, 1907, with an interesting selection of Eagle pencils.

Blotter, advertising Eagle Comfort Pencil. $15-$25.

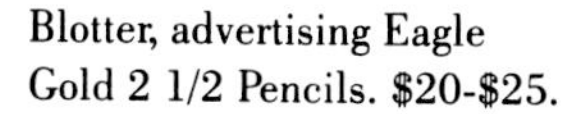

Blotter, advertising Eagle Gold 2 1/2 Pencils. $20-$25.

Blotter, Eagle Water Color pencils. $15-$25.

Blotter, Eagle Simplex pencils. $15-$25.

Eagle Simplex pencil.
$20-$25.

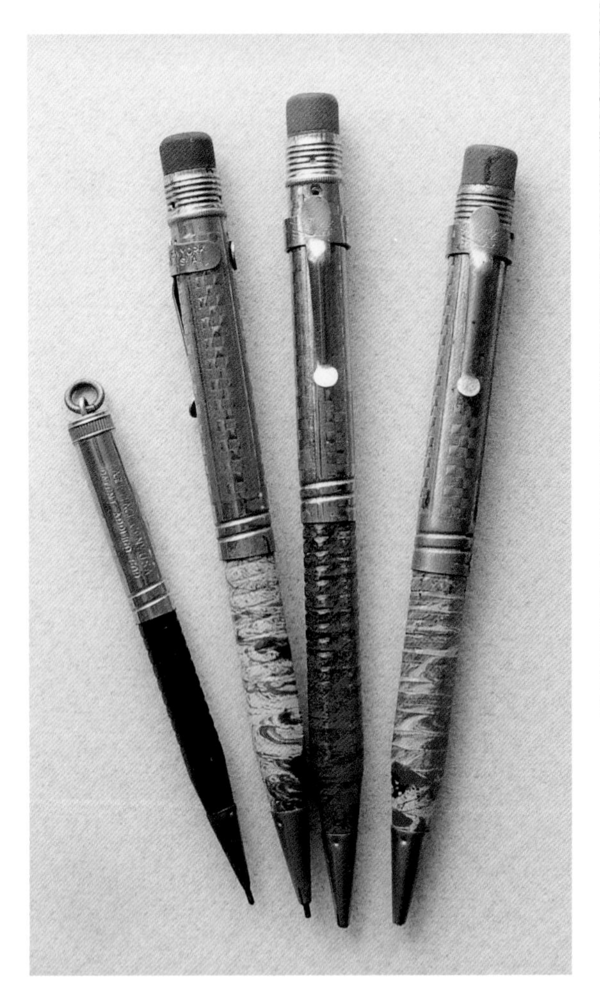

Eagle Simplex pencils were colorful and easy to use.

Eagle Pointer pencil, with storage for extra leads in the barrel end of the tip. $25-$30.

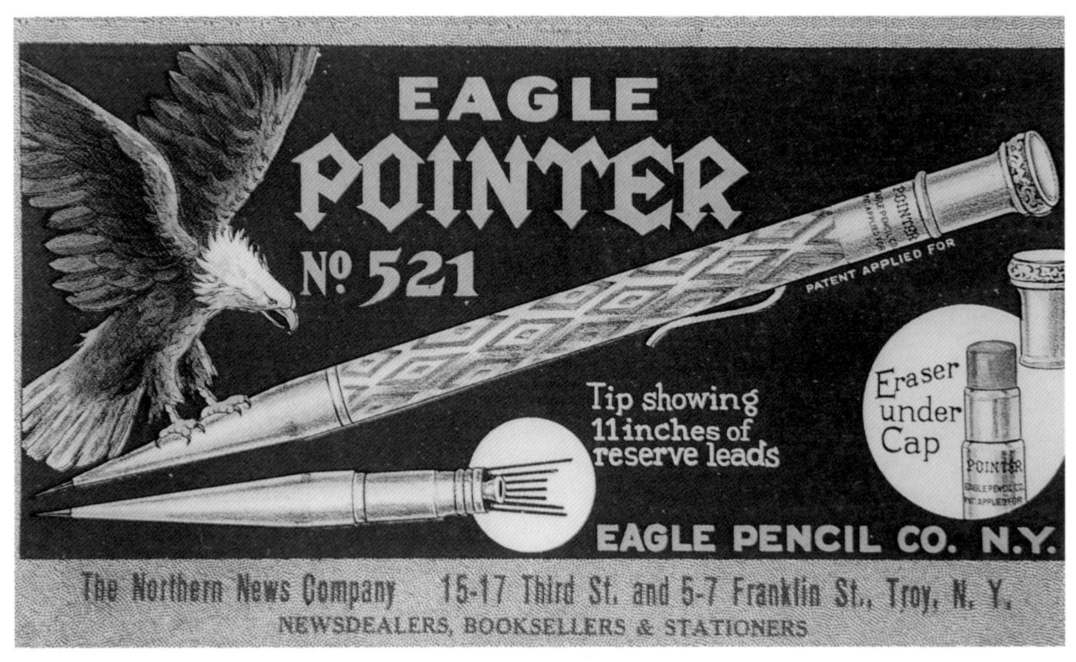

Blotter, advertising Eagle Pointer pencil.$15-$25.

Large advertising pencil, on top of an Eagle Pencil Company publication, "The Largest Pencil Company in the World."

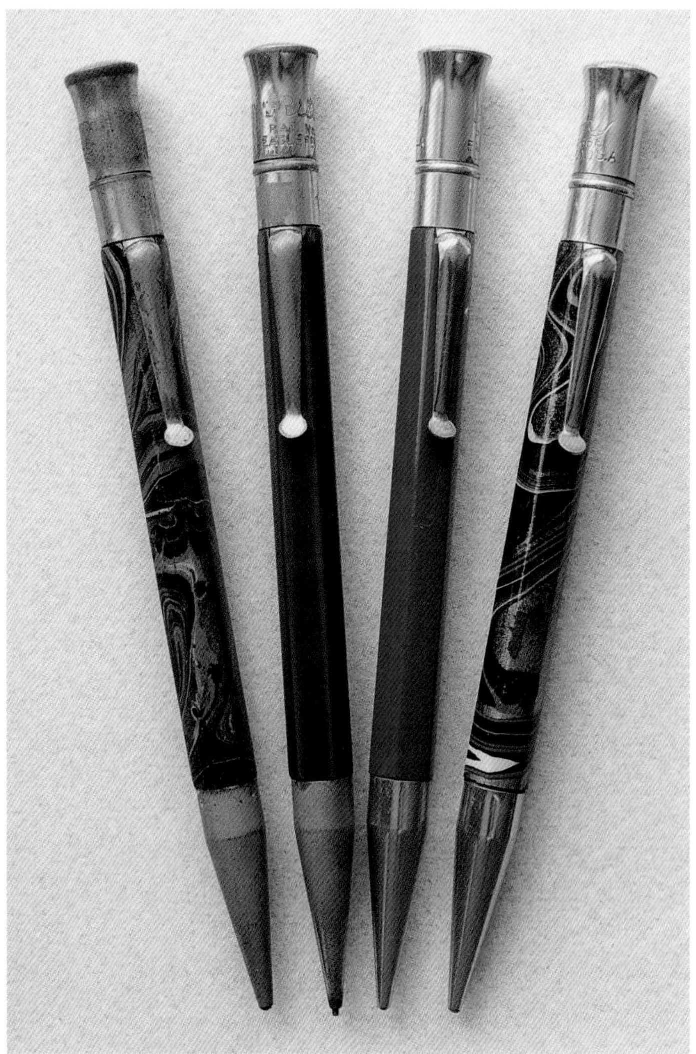

More colorful Eagle pencils, with marbleized and solid lacquer paint over brass, $35-$50.

A three-pack of Eagle Vanity pencils, in their original wrapper, $35-$50.

Aluminum folding cup, advertising Eagle Pencil Co. Cups like this were often contained in school sets. $25-$50.

Wm. S. Eastburn: "specialty manufacturer of the Eureka or magic pencil-case. Also, all styles of gold and silver pen and pencil cases. Southwest corner Corlandt and Greenwich Streets, New York" (letterhead, no date).

A. W. Faber: one of the first products to be labeled with its manufacturer's name was the pencil made by A. W. Faber.

Wm. S. Eastburn (Am.), manufacturer of "Eureka" magic pencil-cases, circa 1870.

Wood, bone, and metal pencil made by A. W. Faber (German) using graphite from the Alibert mine in Siberia. $10-15.

Eberhard Faber: originally with A. W. Faber, as was his brother, Johann. He formed his own company in New York City, and was listed as "sole Agent for A. W. Faber's Lead Pencils" in 1859. "It may, indeed, well be asked, to what this firm owe their uncontradicted control of the world's pencil trade; but the answer is easy as it is plain. Their unparalleled reputation is due solely to the uniform superiority of their manufacture, their unwavering adherence to the principle that *show* shall be *in proportion* to *quality*, and not *take the place of it*; to the unvarying straightforwardness and fairness which characterize all their dealings, and make their name the best guarantee of superiority that any article can bear; and, finally, to that system of branch houses and connected agencies, which enables them to carry out the same principles in every country with which they deal to any extent..." (p. 79, *John C. Gobright, The New York Sketch Book and Merchants' Guide*). This claim is countered in 1877 (p. 125, Edward Bruce, *The Century: Its Fruits and Its Festivals*) by the following statement: "Pens, pencils, and school machinery... make a show that excludes all apparent need for foreign assistance. It is not many years since our chief supplies of this description came from abroad. Now, metallic pens of every grade, pencils rapidly gaining on Faber, and better than two-thirds of those sold under that name, instruments as good, though not yet so cheap, as the German..."

By the end of the nineteenth century, both Eberhard Faber and Dixon Crucible company were proclaiming that they were the oldest producers of pencils in the United states.

Brass case for a wooden pencil, made by Eberhard Faber. The pencil is capped by a ferrule, which holds the eraser and screws into the case. $5-$8.

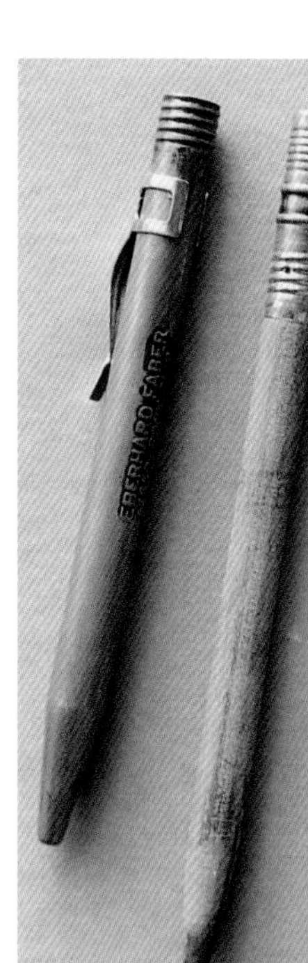

Same as above, open.

Nickel "magazine" pencil made by Eberhard Faber, early twentieth century. If the tip of this pencil is unscrewed, one finds prongs holding the graphite. This is very much like the pencils made of bone and ivory one hundred years earlier. 5", $5-$10.

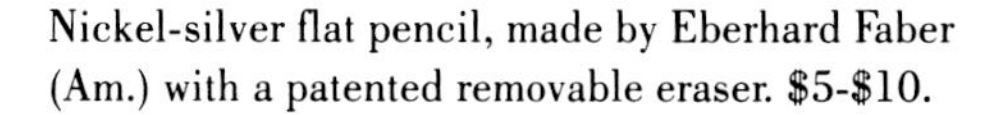

Nickel-silver flat pencil, made by Eberhard Faber (Am.) with a patented removable eraser. $5-$10.

Leroy W. Fairchild and Company (also known as Fairchild-Johnson): Manufacturers of gold pens and pencil-cases. They won prizes for their products in 1847, 1848, 1849, 1850, 1853, 1867, and at the Centennial Exhibition in Philadelphia in 1876. Their office was at 110 William Street, New York, and the company was owned by Leroy Fairchild and James White (letterhead, 1877). Also made lovely figural pencils (p. 95).

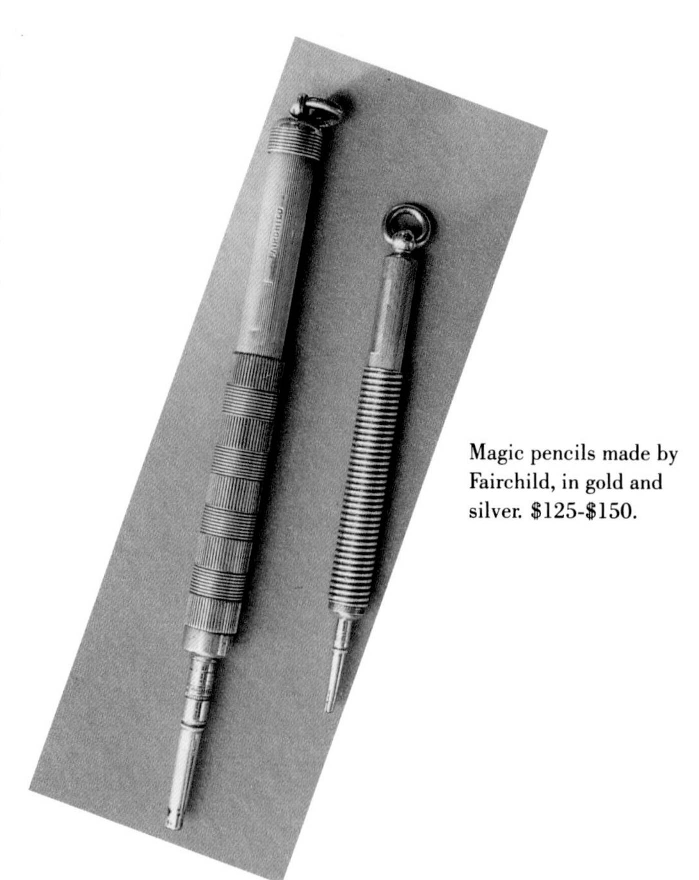

Magic pencils made by Fairchild, in gold and silver. $125-$150.

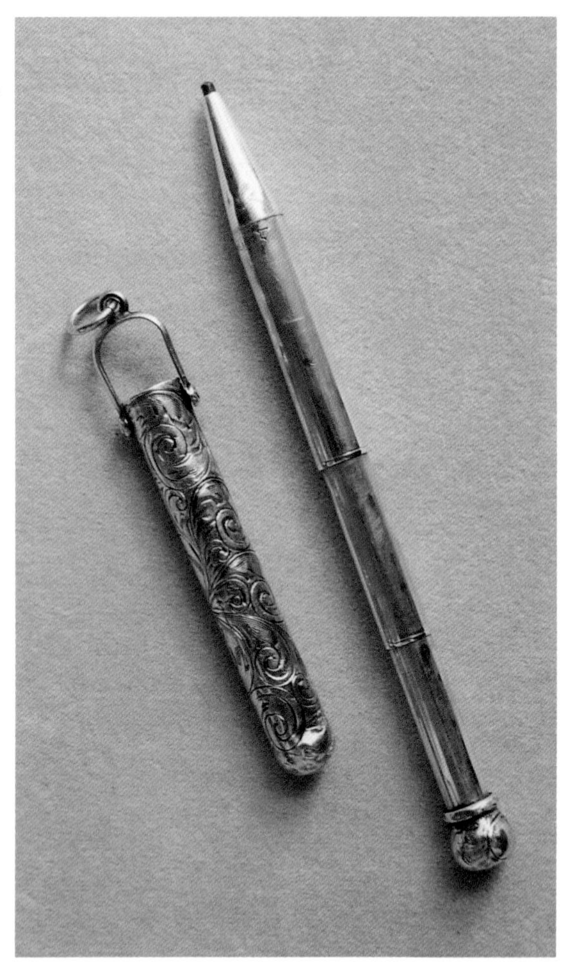

Collapsible or telescopic pencil and hand-engraved sheath, made of sterling silver by Leroy Fairchild (Am.). The mark, a shield with an "F" in it, is located near the tip of the pencil. 2.5" in sheath, 4.5" open, $100-$125.

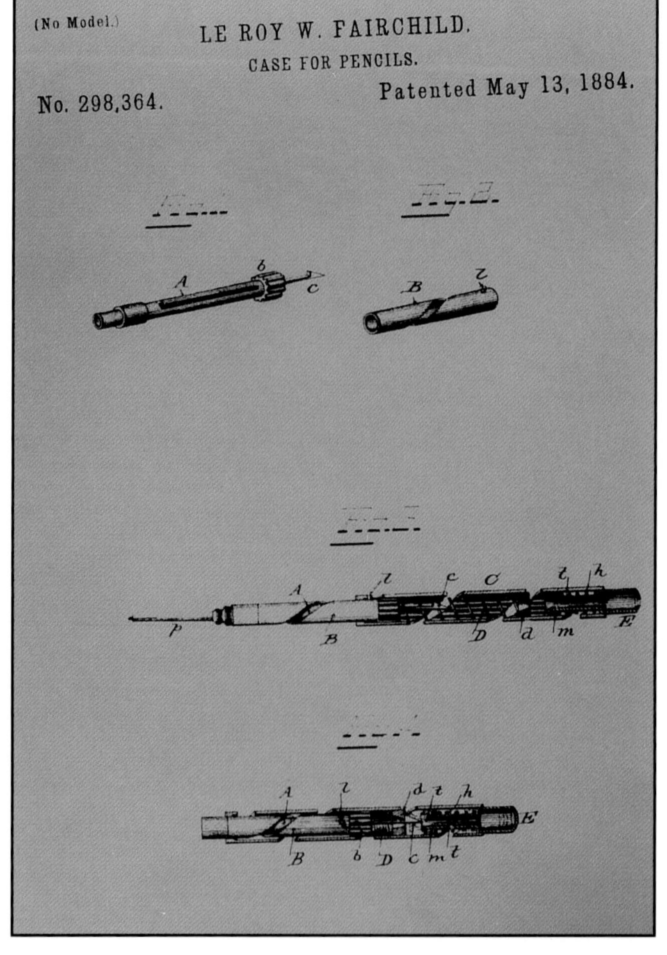

(No Model.) LE ROY W. FAIRCHILD.

CASE FOR PENCILS.

No. 298,364. Patented May 13, 1884.

Leroy Fairchild patent, "case for pencils," 1884. "Pencil-cases" or "cases for pencils" were the actual mechanical pencil (sometimes just the shell, but often the mechanisms as well; the terms were apparently used interchangeably).

D. F. Foley: manufacturers of "gold pens, holders, pencils, toothpicks, match boxes, &c." Located at 23 Maiden Lane (letterhead, 1888).

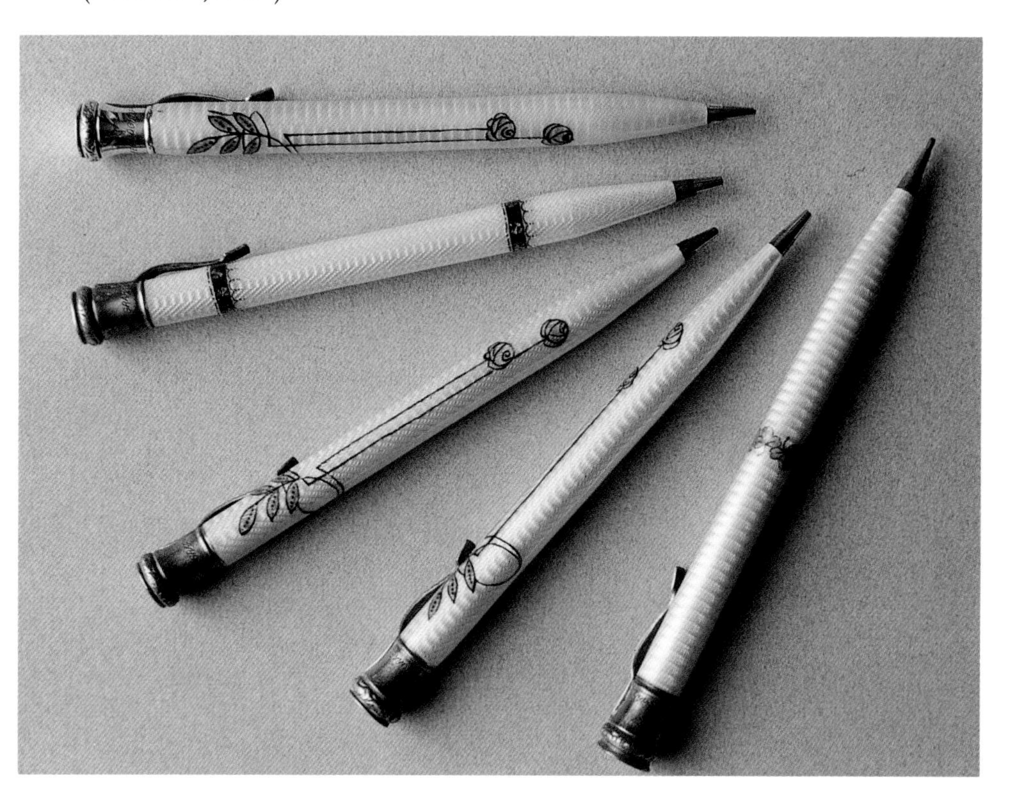

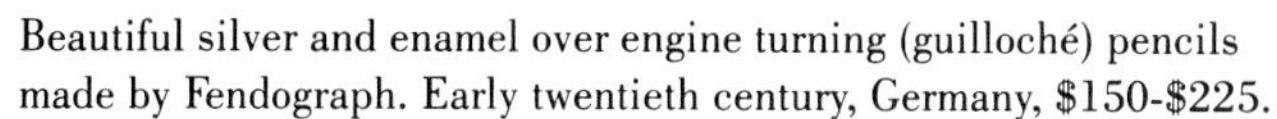

Beautiful silver and enamel over engine turning (guilloché) pencils made by Fendograph. Early twentieth century, Germany, $150-$225.

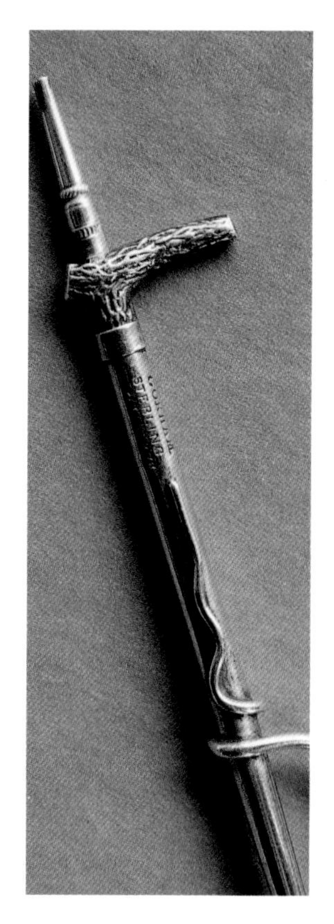

Sterling silver pencil-buttonhook, in the shape of a buggy whip. Marked "Gorham, patented Oct. 8, [18]78." See Chapter 5.

John Foley: according to a trade card, sold "Best Quality Only" gold and silver pencil cases and were located at 167 Broadway, New York. Also advertised as "the Oldest Gold Pen House in the United States, established 1848." Foley manufactured "solid silver screw and extension pen and pencil cases, gold plated telescope pen and pencil cases, gold plated slide or reverse holders, gold mounted reverse pen and pencil cases...screw and magic..." Foley is also known for his history of writing implements, entitled *An Interesting History: Gold Pens, Who Invented Them, When and Where*, published in 1875. In addition, Foley, by his discovery and exposure of fraud perpetrated by the notorious Tweed and the Tammany Ring, caused the Board of Supervisors (New York City) to be abolished in 1870.

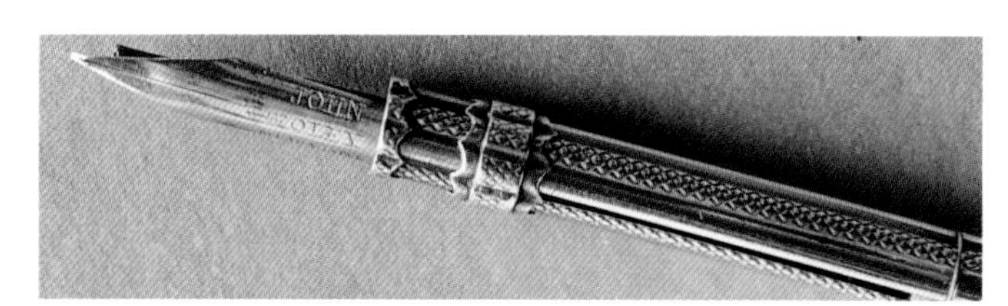

Fluted silver pen-pencil combo, with nib made by John Foley of New York.

Foster and Bailey (F & B): of Providence, Rhode Island. Makers of novelties, including pencils.

Charles Goodyear: Invented the process of "vulcanizing" rubber. Prior to his invention, no one could figure out a way to refine natural rubber so that it would maintain its shape (so that it wouldn't melt or decompose in hot weather, for example). The Goodyear patent, number 3,633, is important to pencil-case manufacture because black hard rubber became a popular material to use on pencil barrels.

Goodyear experimented with rubber for nearly fourteen years before he was able to create a formula for a reliable product. During this time, he and his family lived in extreme poverty, having considerable debt from the expenses incurred by Goodyear's experiments, and little to no income during this period. He and his family sold all their possessions and lived primarily on the kindness of their family and friends, a situation that soon became untenable. When Goodyear finally did discover the vulcanizing process, it took him two years to convince anyone that he had actually succeeded. Goodyear earned little from his patent, as he was heavily in debt and sold most of the licenses and royalties for his invention for far less than he could have if his situation had been different.

Throughout the remainder of his life he continued to experiment with uses for rubber, and asserted that he had discovered more than five hundred. However, he died in 1860 at the age of sixty, leaving his second wife and six children with a debt of two hundred thousand dollars.

Gowland Brothers Ltd.: (English) Established in 1834. Thomas was listed as a watchmaker, James and George as proprietors. Although their name is stamped on pencils, it is believed that they were retailers, not manufacturers, of pencils. They did, however, manufacture jewelry and make watches. They were also described as dealers of precious stones. It is possible that they sold pencils manufactured by Mabie Todd & Co.

Gold porte-crayon marked "Gowland Bros." with "Mordan arrow."

Same as above, showing detail and M. T. & Co. mark, 2", see Chapter 5.

Stephen M. Griswold: manufacturers of gold pens, gold and silver pen and pencil cases (letterhead, 1858).

Grossberger and Kurz, Bleistifte: of Nuremberg, registered trademarks with the U. S. Patent Office. They used an anchor symbol for their "anchor pencil," and a pencil and key for their pencil-cases. They won prize medals for their products in 1862, 1872, 1873, and 1874.

John Hague and Samuel Maycock: John Hague was one of the earliest Americans to patent an ever-pointed pencil. Although nothing is yet known about Hague's training or education, his patent describing improvements on the ever-pointed pencil indicates that he was both familiar with other mechanisms and well enough trained to invent and manufacture his own pencil-cases. Patent No.1,291 (notice the low number) was awarded to him on August 16, 1839. In his specifications, Hague states "The nature of my improvement consists in the mode of protruding the point by a middle outside tube... On pushing the middle tube Fig. 3 downward the pencil point is protruded and ready for use. The pencil point is stationary when held in a position for use, on account of being pinned to the tube Fig. 3, and operates different to other cases..." An example of Hague's ever-pointed pencil is shown here with virtually the same ornamentation as in his patent drawing. In one photograph the pencil is shown closed, with the external tube covering the point. The other photograph shows the pencil as it looks when it's ready for use.

Apparently very successful businessmen, he and his partner Samuel Maycock had coins minted in 1837, hailing the two as "S. Maycock & Co. Everpointed Pencilcase Manufacturers Society Hall Place NY." Called *store cards* or *hard time's tokens*, these coins were created originally as a form of advertising for stores, businesses, etc. However, they were also used as currency during the Panic of 1837. The same size as the large U.S. pennies and made in nearly two-thirds the quantity of the actual pennies minted in 1837, hard times tokens were used interchangeably with other American currency, both by customers and the general public. There is speculation that they were perhaps also considered status symbols; businesses doing well enough could afford to make such coins while, ironically, the government couldn't. (Russell Rulau, *Standard Catalog of United States Tokens.*)

"Trade card" coins, or "hard times tokens" from 1837 (both sides are shown in this photograph). "S. Maycock & Co., 35 City Hall Place, Everpointed Pencil Case Manufacturers, Saml. Maycock, John Hague." $25-$50.

John Hague patented pencil, engine-turned silver. Marked, with the August 16 patent date. Shown closed, $125-$175.

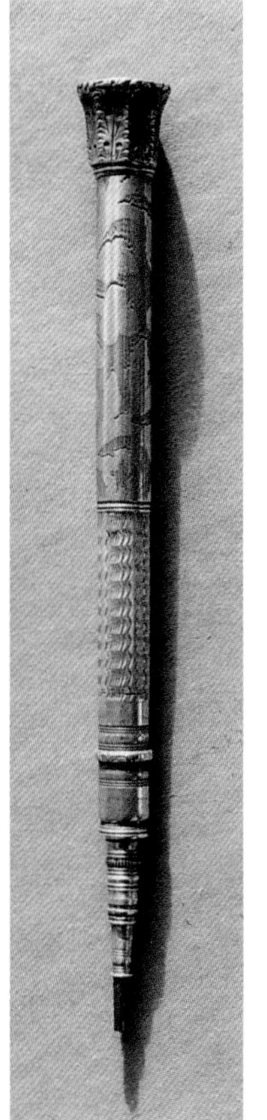

Silver mechanical pencil, marked "John Hague Patent." A thin-gauge metal sleeve, which is connected to the pencil mechanism by a pin, slides down to expose the pencil. The invention is actually a variation on the slider-ring, but instead of using a slider-ring or button to advance the pencil, a metal sleeve is what moves back and forth.

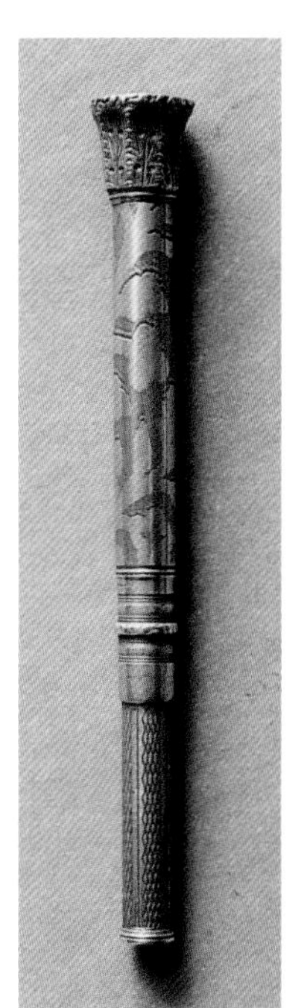

Same as above, closed. Silver with engine turning and hand engraved, cast finial. $150-$175.

Hague pencil, shown open.

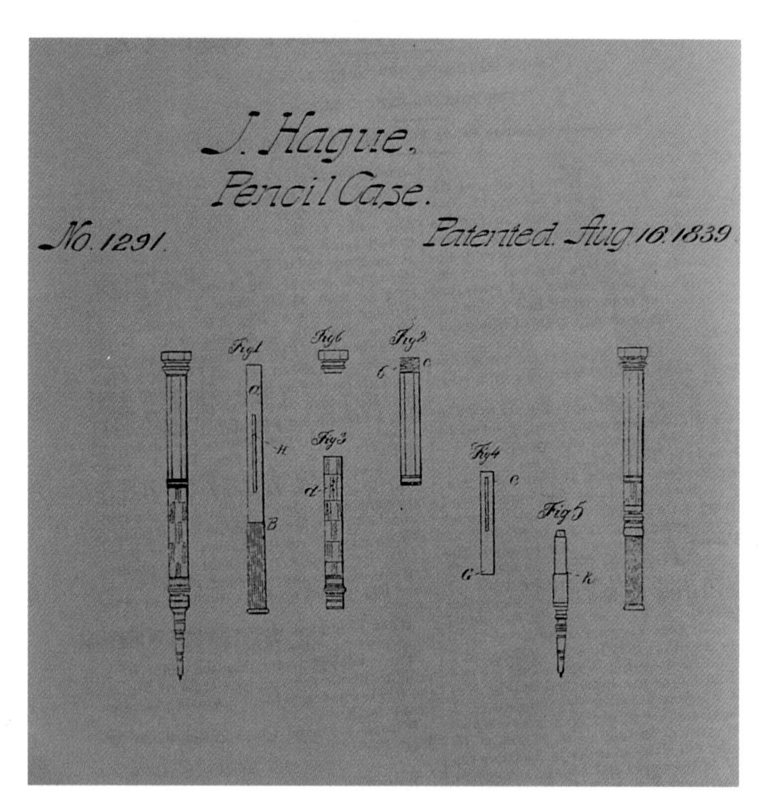

John Hague patent, number 1,291, August 16, 1839.

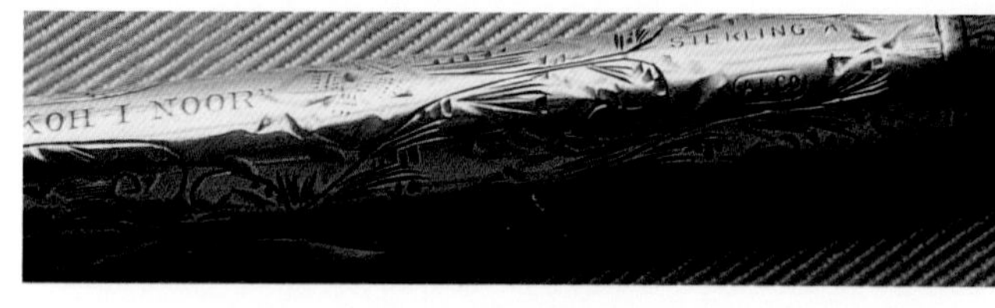

Hand-engraved silver pencil, marked "Koh-i-noor," "Sterling A.," and "A. L. & Co." (Aikin, Lambert).

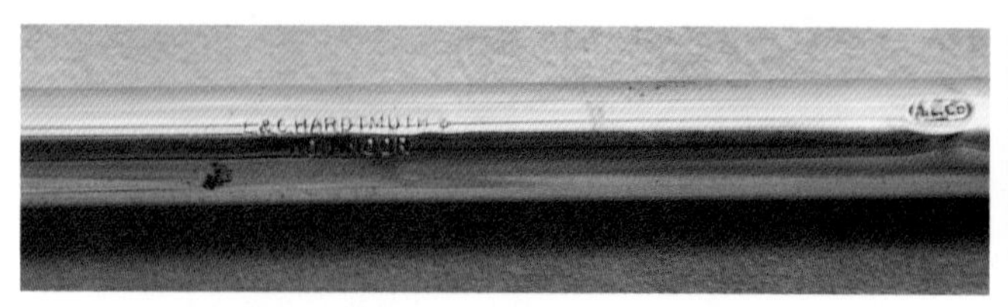

This pencil was made for the American Centennial, in 1886. Aikin, Lambert & Co. made the rolled-gold case for the L. & C. Hardtmuth "Koh-i-noor" mechanism. It is believed that Aikin & Lambert also did metal work for L. E. Waterman (who sold "Koh-i-noor" pencils from their London store). See "Aikin, Lambert & Co." at the beginning of this chapter.

L. and C. Hardtmuth: established in Austria, 1790, and made propelling and repelling, ever-pointed, and point-protector pencils; manufacturers of "Koh-i-noor" Pencils. They provided pencils used in gold, rolled gold, and silver pencil-cases made by Aikin, Lambert, and were agents for Waterman pens sold in London in the early twentieth century. According to Stephen Hull (page 11, *The Marestin Magazine*, issue number 2, 1996), in 1914, the " 'Alien Enemies' campaign by Da La Rue forces Waterman to terminate Hardtmuth's agency agreement."

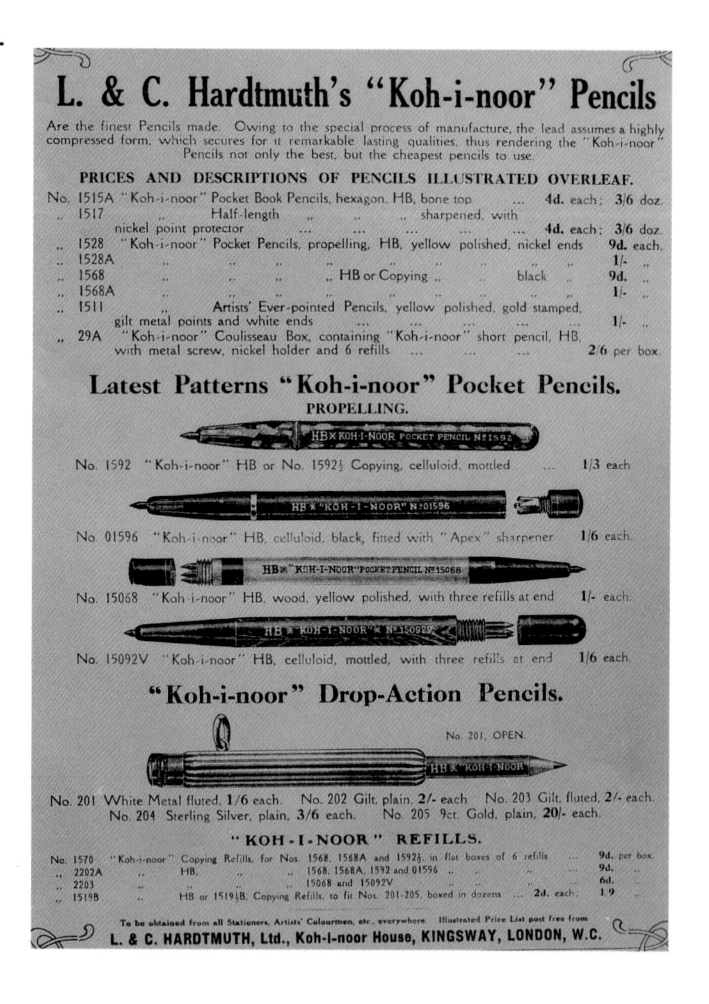

Ad for L. & C. Hardtmuth's "Koh-i-noor" pencils (Germany), early twentieth century.

Haskins' Brothers: inventors and manufacturers of "Imbedded, Pointed, Round Finished, and Rolled Spring Fine Gold Pens. Also, Gold and Silver Pen and Pencil Cases" (letterhead). They were located in Shutesbury in 1866.

C. A. Haskins and Company: importers, manufacturers, and dealers in superior gold and steel pens, pencils, blank books, and stationery. The partners C. A. Haskins, J. M. Cutter, and L. L. tower were located at 23 Cornhill, Boston (letterhead, 1852).

John Hawkins: British inventor of the ever-pointed pencil who co-patented the design with Sampson Mordan. He later developed a gold nib that had an iridium point and also sold the rights to that invention. Hawkins eventually immigrated to the United States (a narrative of Hawkins's invention is included in John Foley's *An Interesting History.*)

Geo. W. Heath: manufacturers of "Tribune" fountain pens, mechanical pencils, and a #0504 pearl filigree pen that retailed for $6.50. They had an office located at Canal Street and West Broadway, New York, and their manufactory was located in Newark, New Jersey. Cross & Beguelin were their "special agents."

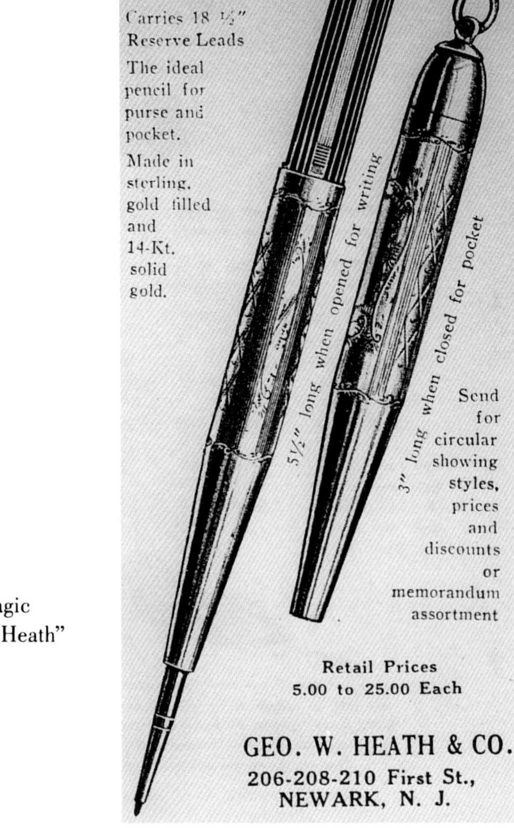

Ad from a New York City directory, circa 1910, showing Heath magic pencils. "The ideal pencil for purse and pocket."

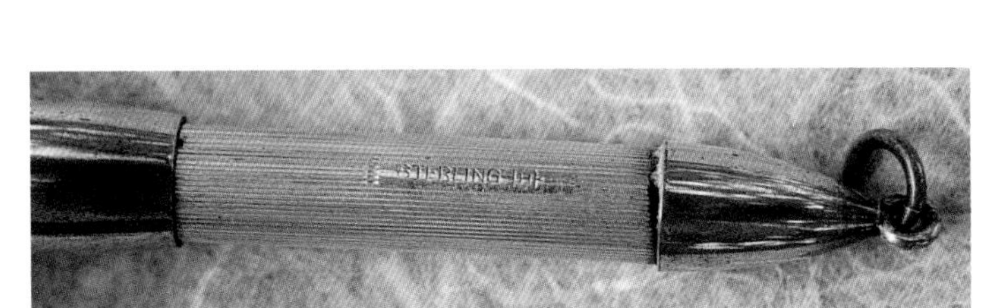

Sterling silver magic pencil, with the "Heath" (Am.) mark.

Hand-engraved silver pencil, torpedo shape. Geo. W. Heath, $125-$175.

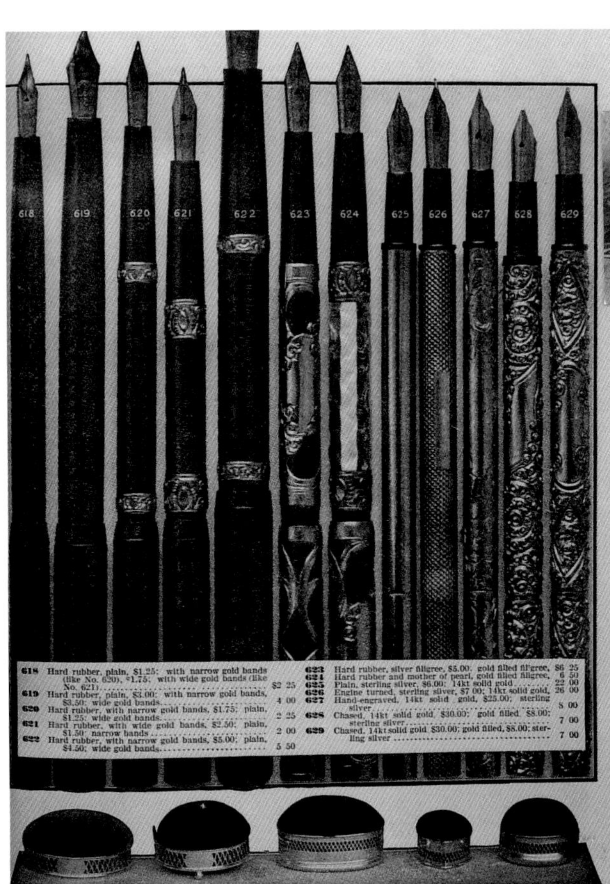

Fountain pens made by Heath, sold by S. Kind & Sons, circa 1910.

W. S. Hicks: American manufacturer of high quality mechanical pencils. Hicks held many patents. Made mechanisms for pencils sold by Tiffany's.

In March, 1871, Hicks patented a type of magic pencil: "My invention relates to that class of pencil-cases which is intended for pocket use; and it consists in the novel construction of the case, whereby pulling backward on the head, the point is made to protrude, and more especially whereby the outer shell, which is held fast in the hand during the operation, is mounted loosely upon the body of the case as hereinafter more fully explained...Having thus described my invention, what I claim is...An extension pencil-case having its outer shell, A, arranged to turn loosely on its body, B, and independently of the other parts of the pencil or case, substantially as described...The arrangement of the tubes d and e and the tubes a and b, the latter having reverse spiral slots, all sliding telescopically within one another, as set forth." The Hicks owl pencil, illustrated here, is a perfect example of that patent, and has the date Mar. 21, [18] '71 inscribed on the ferrule.

Hicks died in 1890 at the age of 72, at which time the name of his company was changed to William S. Hicks' Sons.

In addition to the large H mark, with a small W and S nestled above and below the crossbar on the H, Hicks also sometimes used an acorn mark.

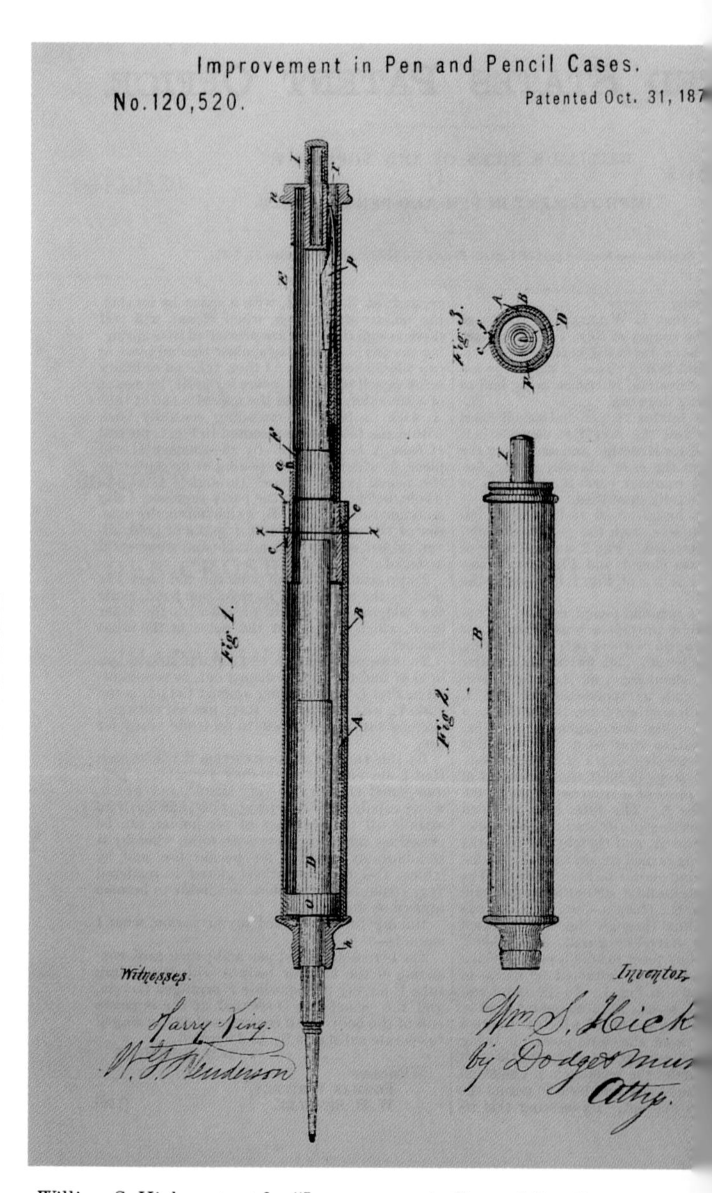

William S. Hicks patent for "Improvement in Pen and Pencil Cases." October 31, 1871.

William S. Hicks patent for "Pen & Pencil Case," September 12, 1865.

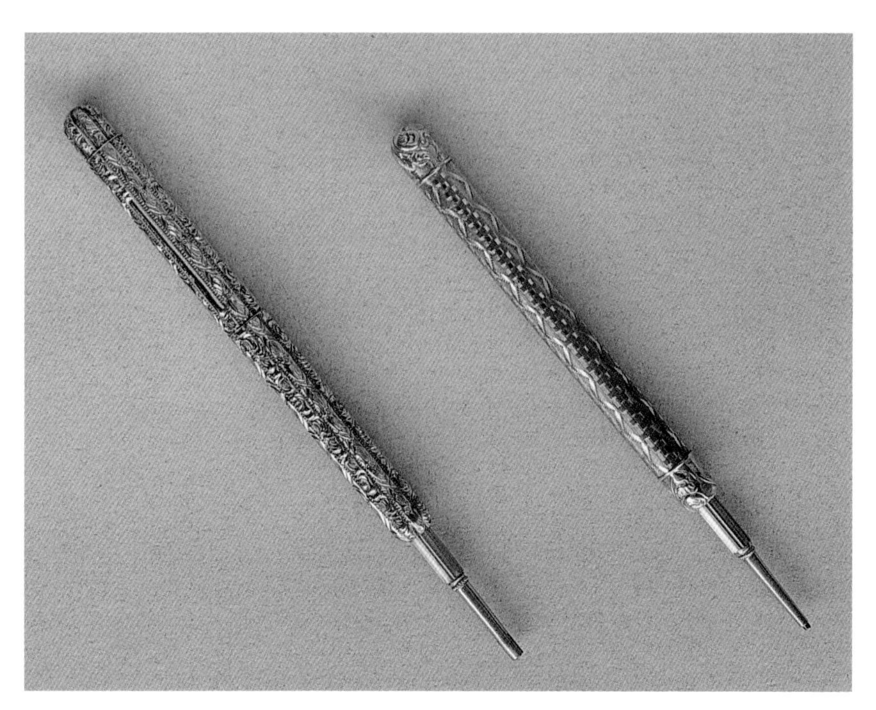

W. S. Hicks pencils made of "solid gold" with twist or screw mechanisms. Top: hand-engraved, $175-$225. Bottom: engine-turned with hand-engraved trim, $175-$200. *Photo Courtesy of Chris Odgers.*

Rolled-gold owl, figural pencil made by W. S. Hicks. The patent *date* (March 21,1871) rather than the patent *number* was marked on the piece, making it more difficult for someone to look up the actual patent and thus make alterations that could be accepted as a new idea, but which were really nearly plagiarized versions of the original. 1" closed, 2.25" open, $375-$500.

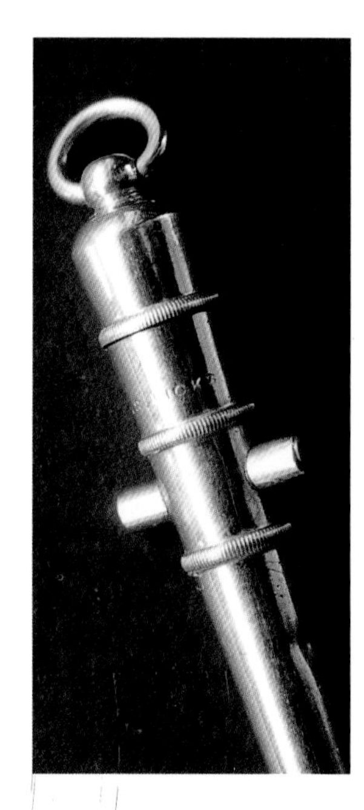

Pencil in the shape of a cannon, made by W. S. Hicks.

Hutcheon Brothers: of New York. Made the "Finerpointe" leads for their Finerpointe pencils, some which were capable of holding graphite that was 3/16" wide. Also made the "Hutch Clutch." They made smaller pencils (including an interesting one, with a small Viking head appliqué), as well as fountain pens.

Frank Hyams Ltd.: was located at 128 New Bond Street, London, and was in business there from 1902 to about 1912. Hyams immigrated to England from New Zealand in 1902 and began to use jade from New Zealand in his jewelry. One representative piece has a telescoping gold pencil, which extends from 1-3/4" when closed to 2-7/8" when opened, is housed in a case that would remain on a chain when the pencil was taken out for use. The upper-most chamber of the pencil is reeded, and a bezel-set piece of translucent green jade is set right above it. The pencil is marked "F. Hyams Ld. 128 New Bond St" and the mark appears to have been hand punched with a linear tool, one side of a letter or number at a time, making the characters somewhat boxy and irregular. The craftsmanship in the piece, however, is outstanding and the mechanism works smoothly to allow the pencil to double its length as it is pulled from its case.

John Jago: Early British instrument maker.

E. S. Johnson: Prolific manufacturer of writing implements in the 1800s. Established in 1848, E. S. Johnson published an "Illustrated Catalogue of Unequaled Gold Pens, Pencil Holders, Pencils, Pen and Pencil Cases, Tooth Picks, Tooth and Ear Picks, &c., in Gold, Silver, Pearl, Ivory, Rubber & Celluloid." This catalog refers to magic pencils as "magic extension pencils."

William Johnson: early pencil-case maker, holding a British patent as early as 1825. Incorporated Bramah type quill holder with an everpointed pencil.

Geo. F. King and Company: blank book manufacturers. Proprietors of King's celebrated "Nonpariel" pens, King's "Nonpariel" office and carpenters' pencils, 38 Hawley Street, Boston (letterhead, 1899).

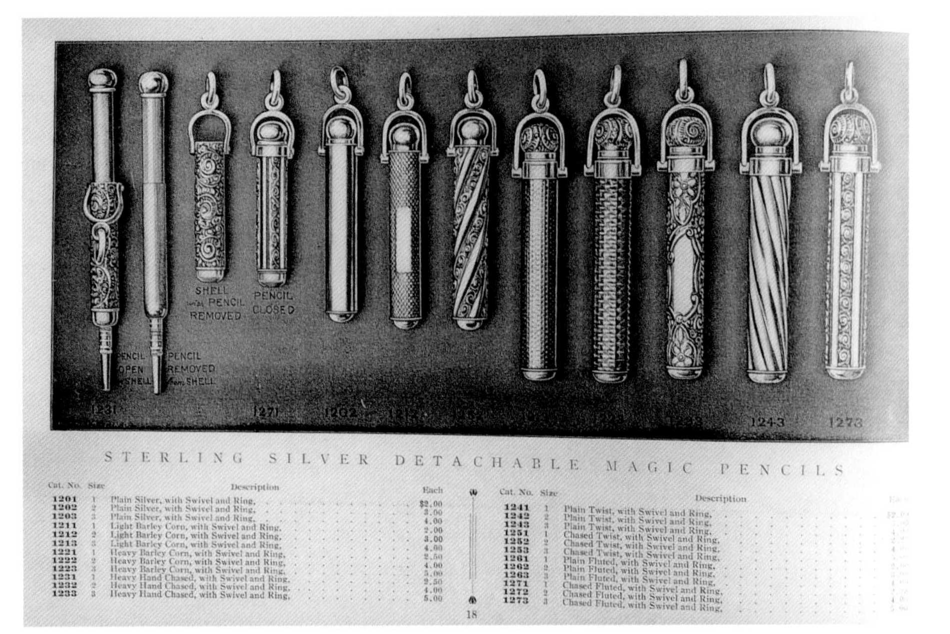

S T E R L I N G   S I L V E R   D E T A C H A B L E   M A G I C   P E N C I L S

Cat. No.	Size	Description	Each
1201	1	Plain Silver, with Swivel and Ring,	$2.00
1202	2	Plain Silver, with Swivel and Ring,	3.00
1203	3	Plain Silver, with Swivel and Ring,	4.00
1211	1	Light Barley Corn, with Swivel and Ring,	2.00
1212	2	Light Barley Corn, with Swivel and Ring,	3.00
1213	3	Light Barley Corn, with Swivel and Ring,	4.00
1221	1	Heavy Barley Corn, with Swivel and Ring,	2.50
1222	2	Heavy Barley Corn, with Swivel and Ring,	4.00
1223	3	Heavy Barley Corn, with Swivel and Ring,	5.00
1231	1	Heavy Hand Chased, with Swivel and Ring,	2.50
1232	2	Heavy Hand Chased, with Swivel and Ring,	4.00
1233	3	Heavy Hand Chased, with Swivel and Ring,	5.00

Cat. No.	Size	Description	Each
1241	1	Plain Twist, with Swivel and Ring,	[illegible]
1242	2	Plain Twist, with Swivel and Ring,	[illegible]
1243	3	Plain Twist, with Swivel and Ring,	[illegible]
1251	1	Chased Twist, with Swivel and Ring,	[illegible]
1252	2	Chased Twist, with Swivel and Ring,	[illegible]
1253	3	Chased Twist, with Swivel and Ring,	[illegible]
1261	1	Plain Fluted, with Swivel and Ring,	[illegible]
1262	2	Plain Fluted, with Swivel and Ring,	[illegible]
1263	3	Plain Fluted, with Swivel and Ring,	[illegible]
1271	1	Chased Fluted, with Swivel and Ring,	[illegible]
1272	2	Chased Fluted, with Swivel and Ring,	[illegible]
1273	3	Chased Fluted, with Swivel and Ring,	[illegible]

18

Page from a John Holland Gold Pen Company, Makers, Cincinnati, U.S.A., showing "sterling silver detachable magic pencils."

Photograph of the original headquarters of Holland Gold Pens.

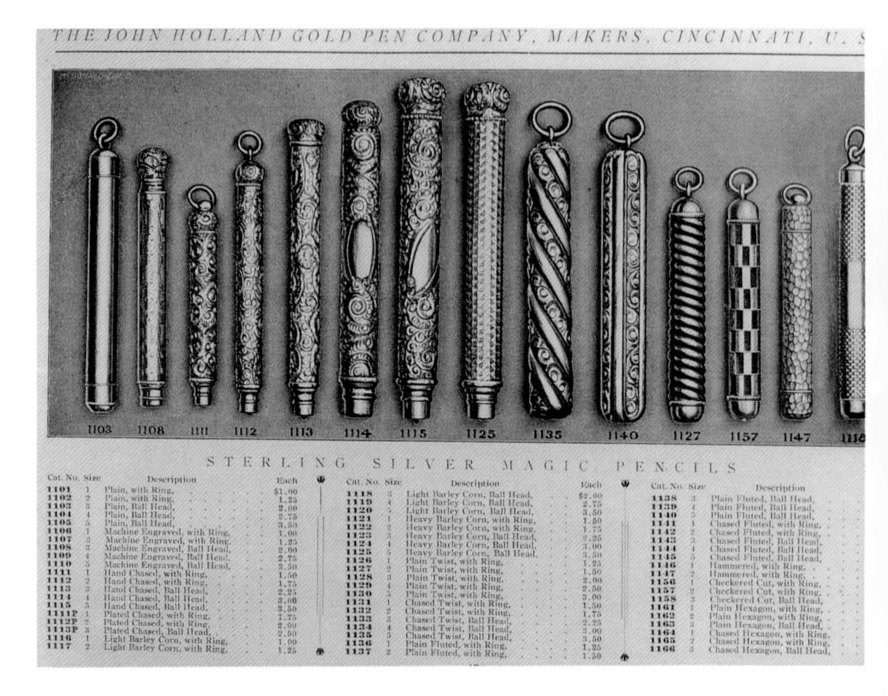

THE JOHN HOLLAND GOLD PEN COMPANY, MAKERS, CINCINNATI, U.

S T E R L I N G   S I L V E R   M A G I C   P E N C I L S

Cat. No.	Size	Description	Each
1101	1	Plain, with Ring,	$1.00
1102	2	Plain, with Ring,	1.25
1103	3	Plain, Ball Head,	2.00
1104	4	Plain, Ball Head,	2.75
1105	5	Plain, Ball Head,	3.50
1106	1	Machine Engraved, with Ring,	1.00
1107	2	Machine Engraved, with Ring,	1.25
1108	3	Machine Engraved, Ball Head,	2.00
1109	4	Machine Engraved, Ball Head,	2.75
1110	5	Machine Engraved, Ball Head,	3.50
1111	1	Hand Chased, with Ring,	1.50
1112	2	Hand Chased, with Ring,	1.75
1113	3	Hand Chased, Ball Head,	2.25
1114	4	Hand Chased, Ball Head,	3.00
1115	5	Hand Chased, Ball Head,	3.50
1111P	1	Plated Chased, with Ring,	1.75
1112P	2	Plated Chased, with Ring,	2.00
1113P	3	Plated Chased, Ball Head,	2.50
1116	1	Light Barley Corn, with Ring,	1.00
1117	2	Light Barley Corn, with Ring,	1.25

Cat. No.	Size	Description	Each
1118	3	Light Barley Corn, Ball Head,	$2.00
1119	4	Light Barley Corn, Ball Head,	2.75
1120	5	Light Barley Corn, Ball Head,	3.50
1121	1	Heavy Barley Corn, with Ring,	1.50
1122	2	Heavy Barley Corn, with Ring,	1.75
1123	3	Heavy Barley Corn, Ball Head,	2.25
1124	4	Heavy Barley Corn, Ball Head,	3.00
1125	5	Heavy Barley Corn, Ball Head,	3.50
1126	1	Plain Twist, with Ring,	1.25
1127	2	Plain Twist, with Ring,	1.50
1128	3	Plain Twist, with Ring,	2.00
1129	4	Plain Twist, with Ring,	2.50
1130	5	Plain Twist, with Ring,	3.00
1131	1	Chased Twist, with Ring,	1.50
1132	2	Chased Twist, with Ring,	1.75
1133	3	Chased Twist, Ball Head,	2.25
1134	4	Chased Twist, Ball Head,	3.00
1135	5	Chased Twist, Ball Head,	3.50
1136	1	Plain Fluted, with Ring,	1.25
1137	2	Plain Fluted, with Ring,	1.50

Cat. No.	Size	Description
1138	3	Plain Fluted, Ball Head,
1139	4	Plain Fluted, Ball Head,
1140	5	Plain Fluted, Ball Head,
1141	1	Chased Fluted, with Ring,
1142	2	Chased Fluted, with Ring,
1143	3	Chased Fluted, Ball Head,
1144	4	Chased Fluted, Ball Head,
1145	5	Chased Fluted, Ball Head,
1146	1	Hammered, with Ring,
1147	2	Hammered, with Ring,
1156	1	Checkered Cut, with Ring,
1157	2	Checkered Cut, with Ring,
1158	3	Checkered Cut, Ball Head,
1161	1	Plain Hexagon, with Ring,
1162	2	Plain Hexagon, with Ring,
1163	3	Plain Hexagon, Ball Head,
1164	1	Chased Hexagon, with Ring,
1165	2	Chased Hexagon, with Ring,
1166	3	Chased Hexagon, Ball Head,

John Holland catalog, with examples of machine engraving, hand-chasing, light and heavy barleycorn, plain twist, chased twist, etc.

Where the Holland Business Is Now Housed.

Photograph of the building later used by John Holland for his business.

The Improved Pencil Co.
Providence R.I. 158 Pine St.
Oct 4/27

Wish you would ship us the following Pencil actions so same will reach us on November 10th 1927 or before. All the actions to be made in a substantial way as per the samples you sent and very little side wobble to the lead and very little end play to the lead as per your samples.

10 gro. 10Kt 1/40 Rolled Gold Plate (yellow color) Pencil Actions .406 outside diameter. action to have a bushing .297 outside diameter.

5 gro 10Kt 1/40 Rolled Gold Plate (yellow color) Pencil Actions .325 outside diameter, action to have a bushing .250 outside diameter

Rec. Nov. 11/27 - 5 gross 10K 1/40 Rolled Plate small Pencils 20.50 Gr. 102.50

9 3/4 Gross 10K 1/40 Rolled Plate Large Pencil 20.50 Gr. 199.88

Page from Holland notebook, 1927, showing an order for pencil "actions" from The Improved Pencil Co. of Providence, Rhode Island.

Photograph of John Holland.

John Holland pencil, with the Hague-style mechanism and an eraser under the cap, in celluloid and rolled gold. $45-$65.

Jacob J. Lownds: He held one of the earliest patents for mechanical pencils issued in the United States: number 32, dated September 22, 1836. In a version of his ingenious invention (one without the pen), the nozzle of the pencil was propelled out of the external case by pulling on the end of the barrel, giving it a quarter turn, and then pushing the end back in. The pencil is retracted the same way. It also includes a compartment to hold extra lead.

Also shown here is an example of Lownds patent used by Edward Todd, in 14-kt. gold, using the pen component as well as the pencil.

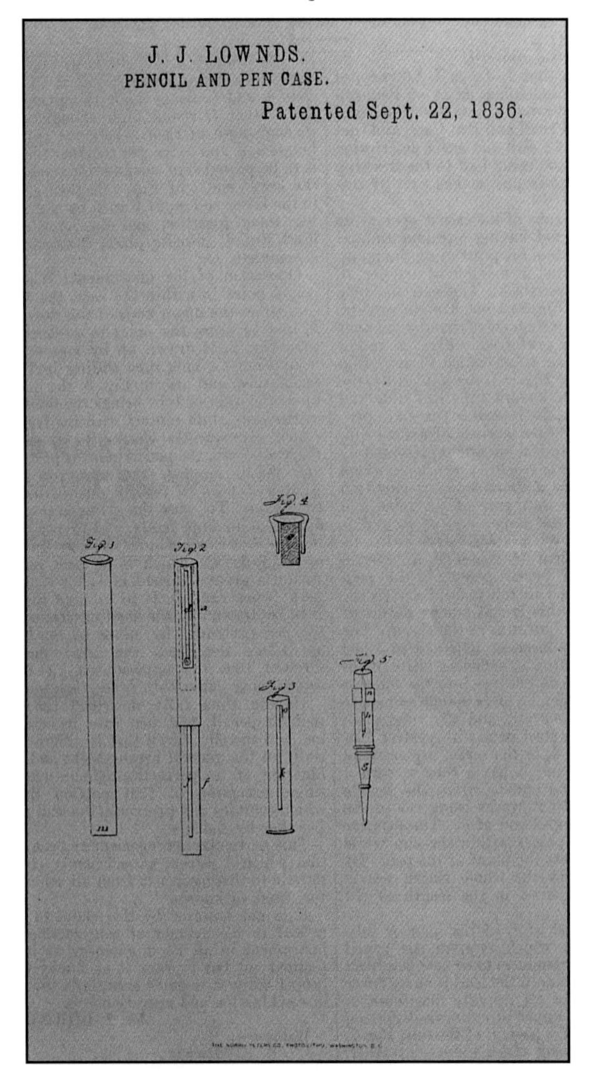

J. J. Lownds' patent for a "Pencil and Pen Case," dated September 22, 1836. This is one of the earliest of the American patents for everpointed pencils.

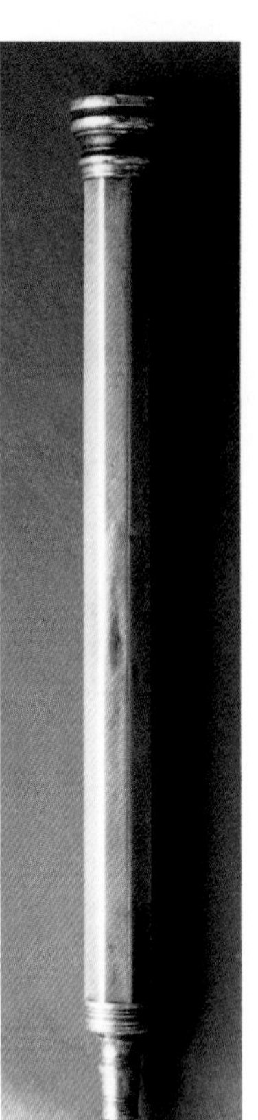

Lownd's patent pencil, detail of maker's mark.

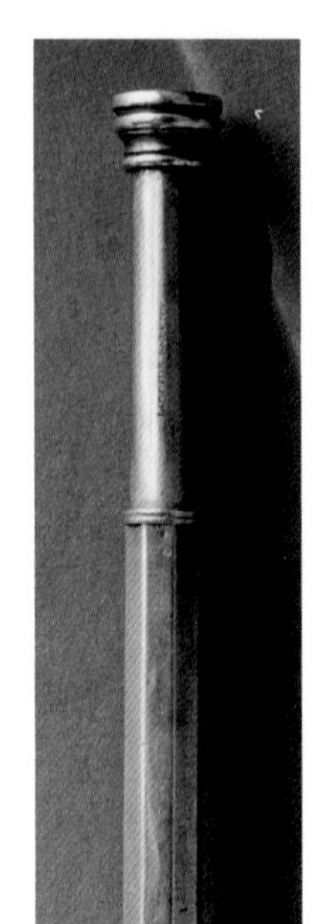

Lownd's patent pencil, shown open. 1836, $150-$200.

Lund, William: English, see illustration on page 189.

Mabie, Todd and Bard: The earliest pencils marked "Mabie's Patent" were based on a design for pen and pencil cases created by John Mabie, October 3, 1854.

The firm Mabie, Todd and Bard, also known as Mabie, Todd & Co., went through many changes. Edward Todd was involved with the company, as was another member of the Todd family (H. H.). Both John Mabie and George Whitfield Mabie (each holding patents for pen and pencil cases) were principals, as was Jonathan Bard. Bard Brothers were earlier established as gold pen (nib) manufacturers.

Mabie, Todd and Bard were known for their high quality products. In addition to their earlier merchandise, the company also made fountain pens (Swans) and pencils called "Fyne Point."

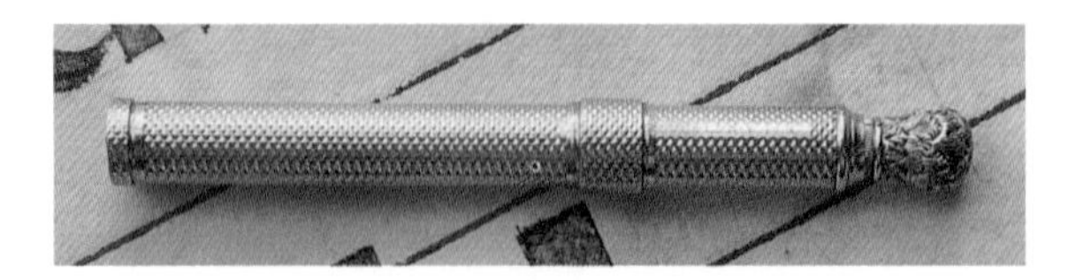

Solid gold pen-pencil in barleycorn, made by Edward Todd using Lownds' concept. $175-$225.

Same pen-pencil combination, with the pen extended.

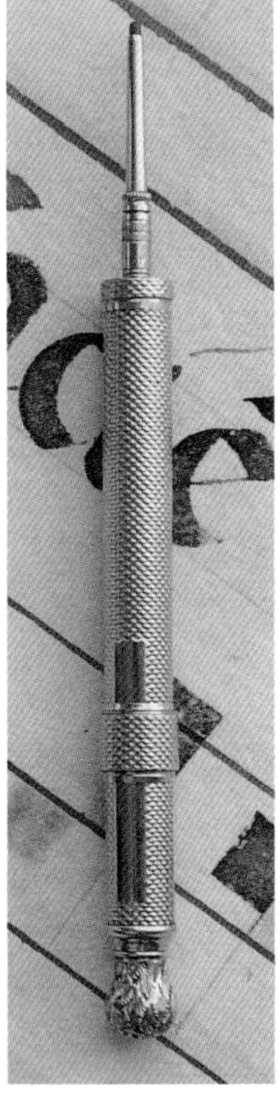

Same piece but with the pencil extended.

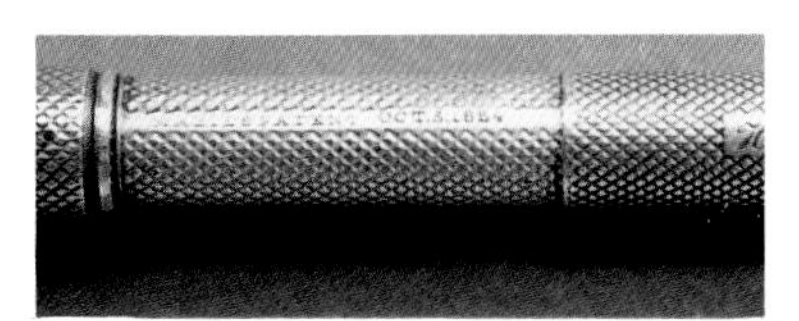

John Mabie's patent, October 3, 1854.

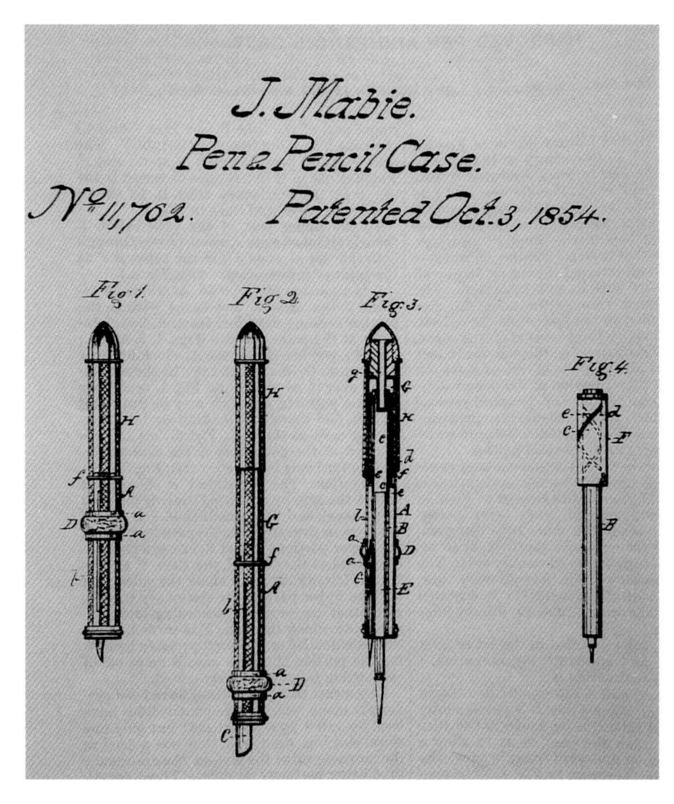

Patent drawing for the pen and pencil case designed by J. Mabie.

Ivory and dyed ivory pencil made by William Lund, circa 1848 (Eng.). The silver device twists up or down in the spiral groove of the barrel, to extend or withdraw the pencil. Invented and originally patented by Gabriel Riddle. $125-$175.

Silver barleycorn and rolled gold, fluted barleycorn pen-pencil combinations using Mabie's patent.

Full-length view of a pen-pencil combination using Mabie's patent, with both the pen and pencil extended. Rolled gold, $150-$175.

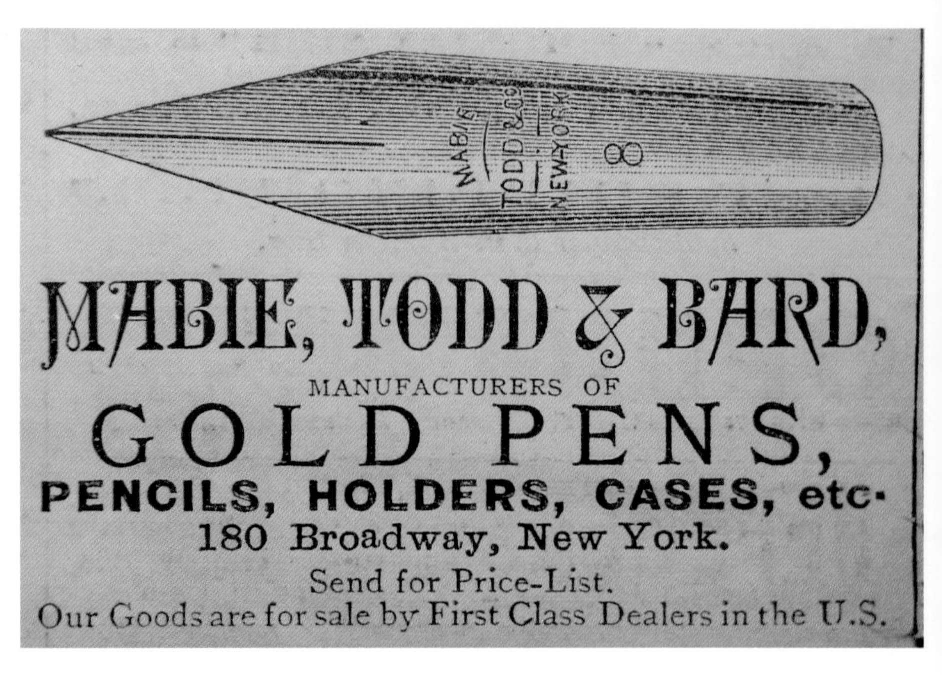

Ad for "Mabie, Todd & Bard."

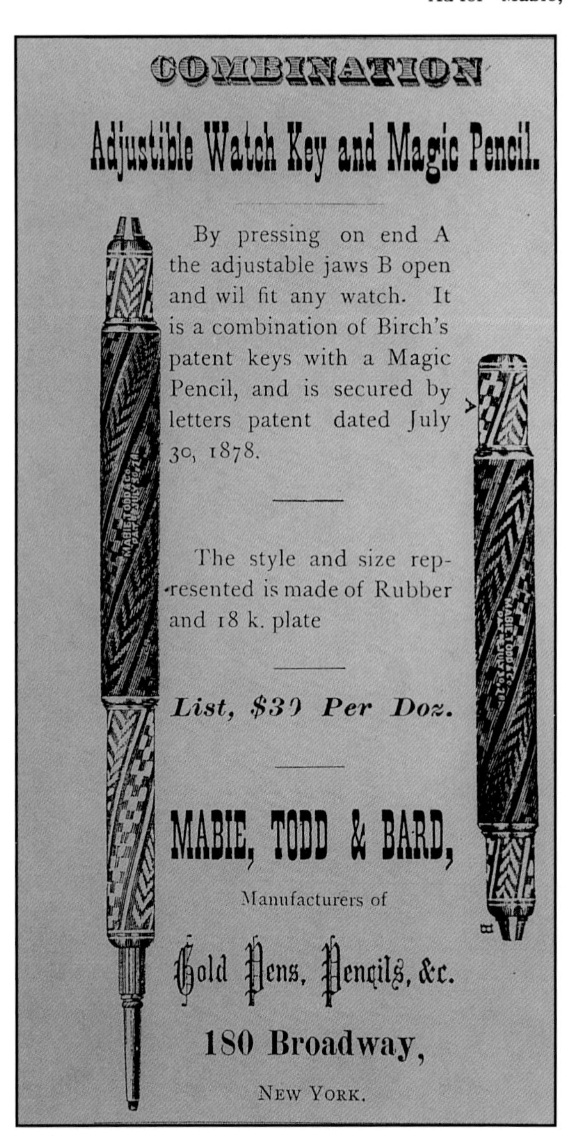

Ad for a combination "Adjustable Watch Key and Magic Pencil" made by Mabie, Todd & Bard, circa 1879.

Photograph of George Whitfield Mabie, circa 1898.

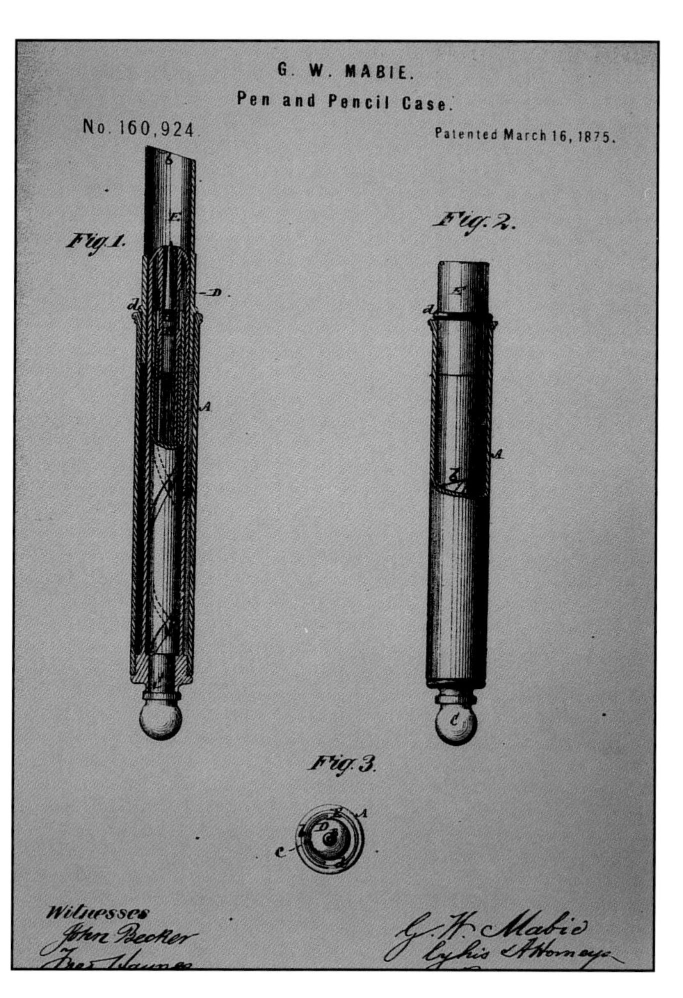

Pen and pencil case patent by George Mabie, 1875.

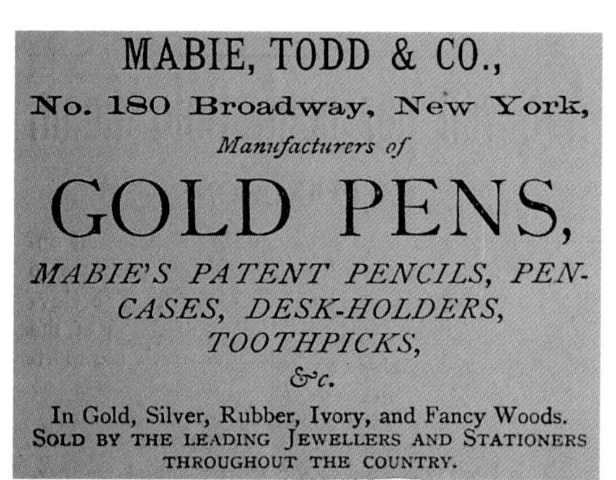

MABIE, TODD & CO.,
No. 180 Broadway, New York,
Manufacturers of
GOLD PENS,
MABIE'S PATENT PENCILS, PEN-CASES, DESK-HOLDERS, TOOTHPICKS, &c.
In Gold, Silver, Rubber, Ivory, and Fancy Woods.
SOLD BY THE LEADING JEWELLERS AND STATIONERS THROUGHOUT THE COUNTRY.

Ad from *Appleton's Journal*, November 27, 1869.

Joseph Monaghan: Along with Thomas Flynn, Monaghan patented an "Improvement in Pen and Pencil Cases" in 1871. "This invention relates to improvements in pen and pencil cases, in which both the pen and pencil are caused to project from one and the same end, and an extension holder is employed in the other end; and it consists in the arrangement of the extension tube with the pencil and its tube, and with the pen-holder, which is actuated by an exterior ring...this construction is very simple and cheap, and admits of making the cases much smaller than they are now made, which is highly desirable for such articles on account of the greater convenience of carrying it in the pocket. But the essential ad-

Telescopic pencil and chased cable twist sheath, made of 14kt gold, patent date March 10, 1891.

Patent for a pen and pencil case designed by J. Monaghan.

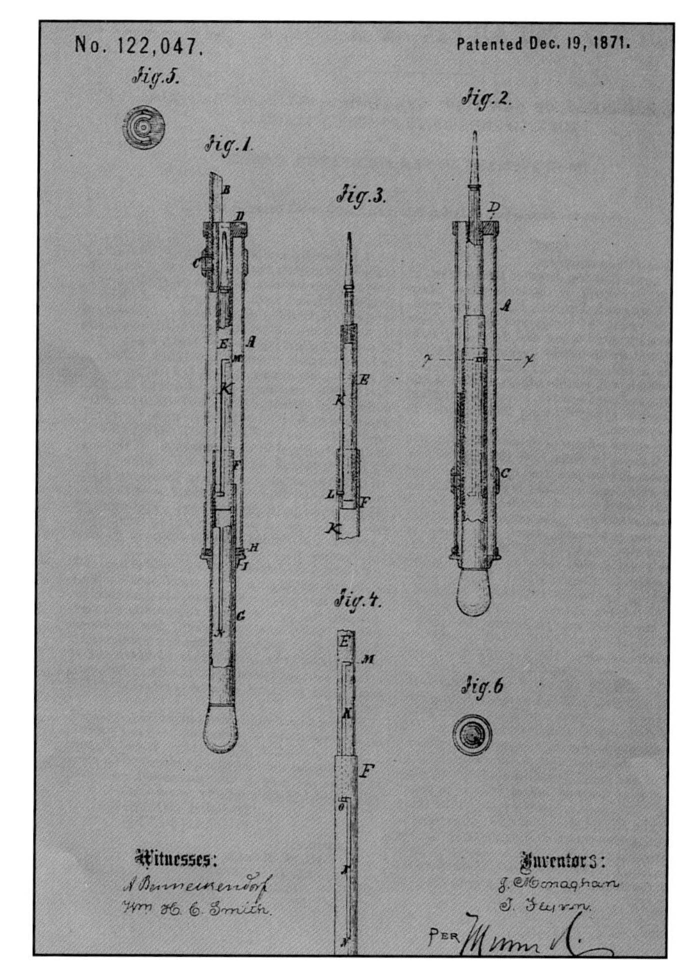

vantage is gained by its arrangement of a pen-holder and pencil to work in one end of a short case independently of each other, said case being provided with an extension handle for the pen." Monaghan was a partner of Edward Todd, and George W. Mabie (misspelled "Mabee") is listed as one of the witnesses to the 1871 patent, indicating that the three had business affiliations. (See Mabie, Todd & Bard.)

Sampson Mordan: produced some of the earliest mechanical pencils, as well as some of the most elegant. Pencils made by S. Mordan & Co. are finely crafted and today command high prices.

Sampson Mordan was born in 1790. He apprenticed with Joseph Bramah, who had a series of inventions to his credit (including a patent lock, a flushable toilet, a quill cutter, and a swivel type device for holding a precut quill in a pen staff). By 1815 he is said to have started his own business. In 1823, he bought out Hawkins' rights to the patent for the ever-pointed pencil and sold a share of it to a stationer named Gabriel Riddle (pencils produced during this partnership have the mark S. M. G. R.). By 1837 he registered the mark S. M. with the London Assay Office, indicating he was no longer in the partnership.

When Sampson Mordan died in 1843 at the age of fifty-four, his sons, Sampson II and Augustus, inherited his business. Another son, Francis, used the money he inherited from his father to establish a company that manufactured gold pens (nibs). It appears that S. Mordan & Co. was operated primarily by Augustus, although Sampson II was also listed with the firm until his death in 1881. (Sampson II had married a French woman named Victoire Bouchard, and spent most of his adult life living in Paris. It is probable that he was involved with the firm's manufactory there as well.) After Augustus retired, the firm was co-directed by George Edmund Johnson (who had become a partner with the two Mordan brothers in 1870, and who had married Augustus' daughter the same year), Horace Stewart (who had experience manufacturing jewelry), James Pulley, and Harry Lambert Symonds (who later became the director of S. Mordan Limited) until Pulley died in 1898.

S. Mordan & Co. produced a fascinating array of high quality items. In addition to a variety of pencils, they made pens, toothpicks, vinaigrettes, scent bottles, vesta cases, match safes, and cigar and cigarette cases. Mordan products are highly regarded, and are priced accordingly.

S. Mordan & Co. continued until its factories were bombed in 1941. Rights to Mordan patents were sold to Edward Baker, and are now owned by Yard O'Led.

Early Mordan pen-pencil combination, with a Bramah style quill (nib) holder on one end and a mechanical pencil on the other. Marked "Mordan & Co. Patent." Reeded barrel, hallmarked silver, 4.25" closed, 6.5" open (as shown, without a nib). Early piece, price not established, yet.

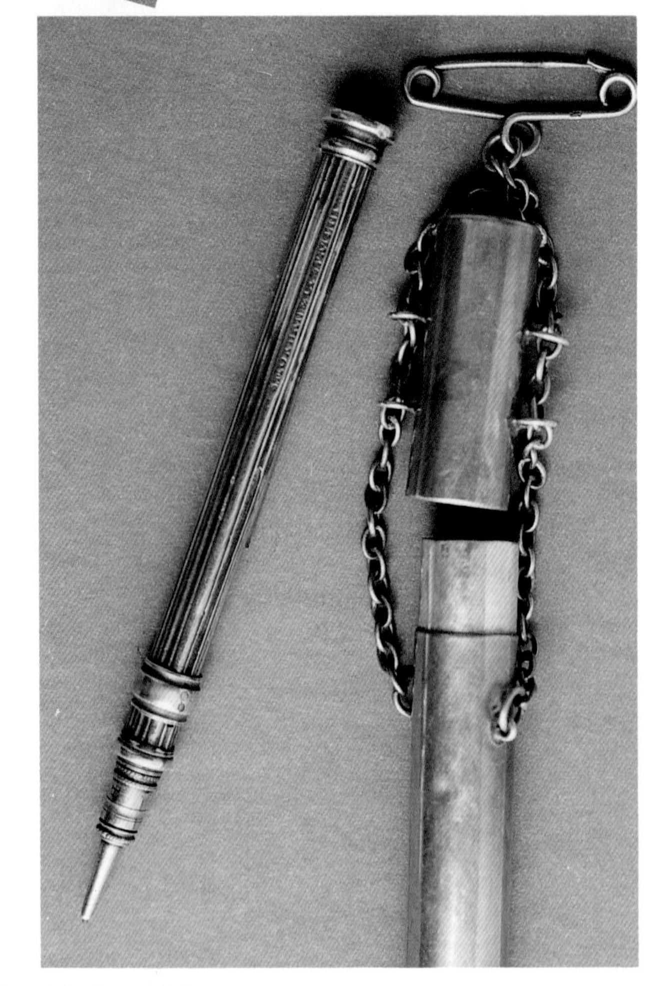

Reeded silver pencil, with wafer seal, marked "S. Mordan & Co.s Patent.," shown with a sterling silver chatelaine case hallmarked "D.L.R.L." $175-$225.

Early pen-pencil combination made by Mordan, marked "Mordan & Co. Patent."

Silver pencil marked "S. Mordan & Co.s Patent, S.M.G.R." with a cast finial and ring slide.

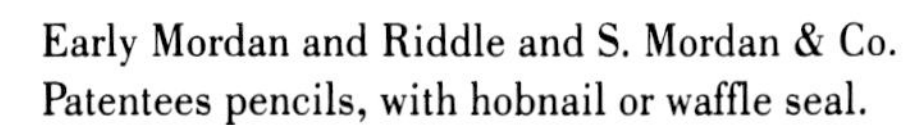

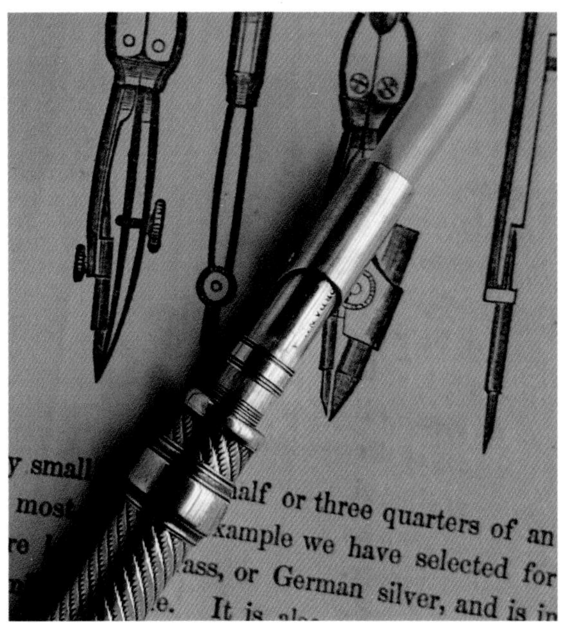

This is the other end of the pencil shown above, with the Bramah style quill holder marked "Mordan's Improv'd." 4" closed, 6.25" open (without nib). Shown with a quill..Very early piece, price not established..

Pencil marked "S. Mordan & Co.s Patent," with the "S.M.G.R" hallmark. Gabriel Riddle (G.R.) was Sampson Mordan's partner until 1837.

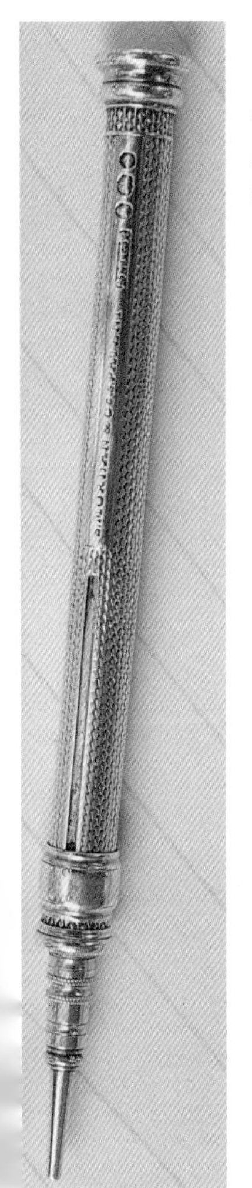

The same pencil as above, showing simple ring slide and terminal, $200-$225.

Early Mordan and Riddle and S. Mordan & Co. Patentees pencils, with hobnail or waffle seal.

Pencil made in the shape of a pistol. Silver, marked "S. Mordan," circa 1842. $400-$600.

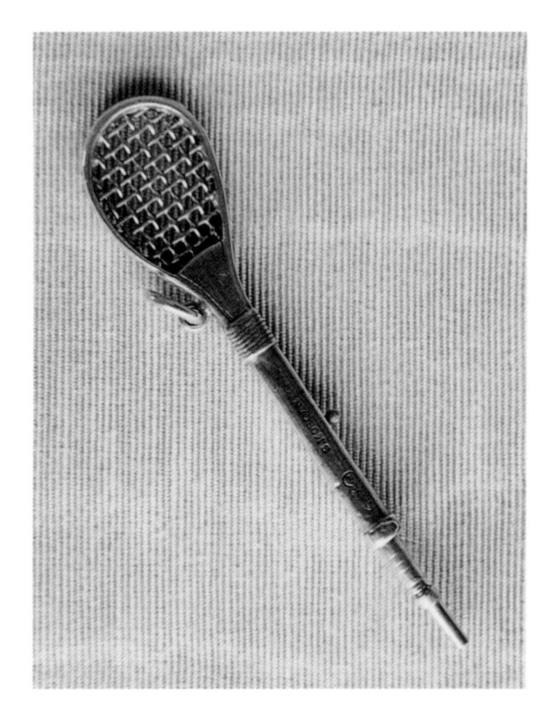

Pencil in the shape of a tennis racket. Slightly more than 2" long when closed, silver with registration mark, marked "S. Mordan & Co." Pencils like these were made between 1880 and 1900.

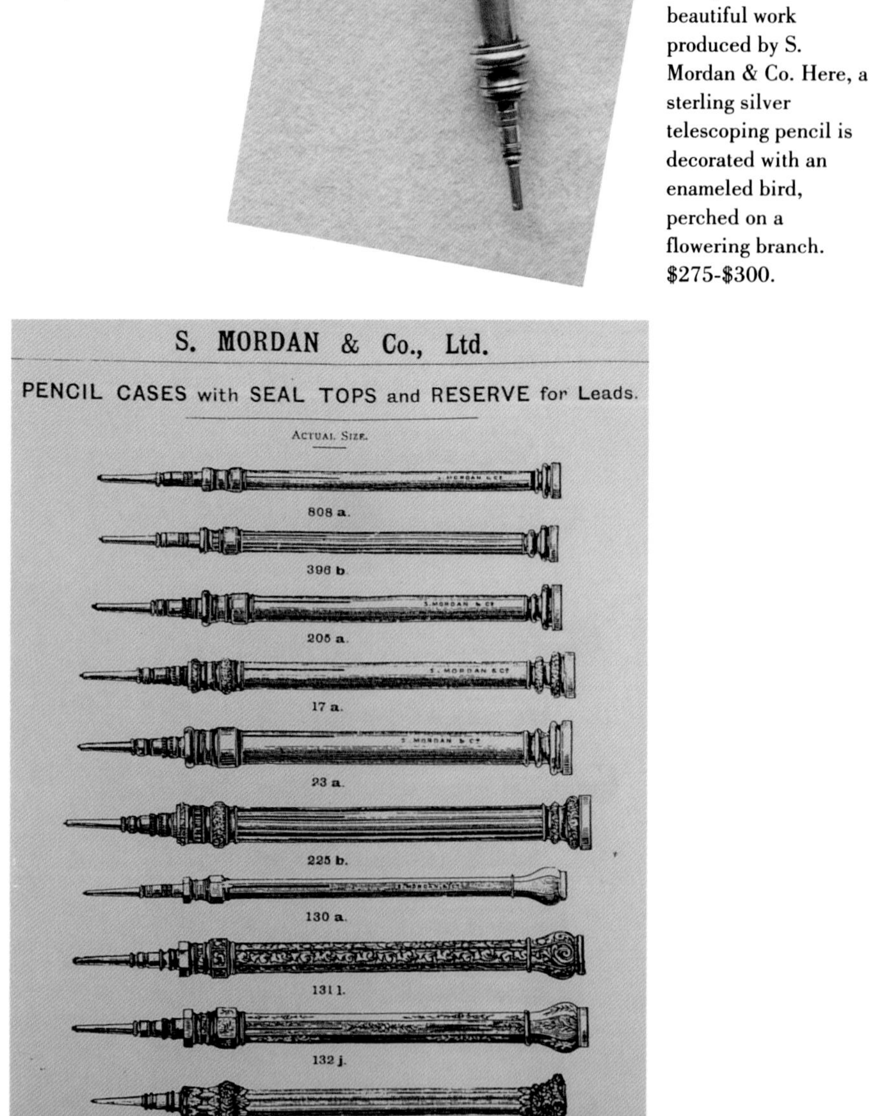

S. MORDAN & Co., Ltd.

PENCIL CASES with SEAL TOPS and RESERVE for Leads.

ACTUAL SIZE.

808 a.

396 b.

205 a.

17 a.

23 a.

225 b.

130 a.

131 l.

132 j.

184 b.

36 b.

Many of the above designs are also made in other sizes.

This pencil is an example of the beautiful work produced by S. Mordan & Co. Here, a sterling silver telescoping pencil is decorated with an enameled bird, perched on a flowering branch. $275-$300.

Page from a catalog describing Mordan pencils.

This pencil is similar to the one shown at the bottom of the catalog page, but with a line and dot pattern. Marked "S. Mordan & Co. Makers." $225-$300.

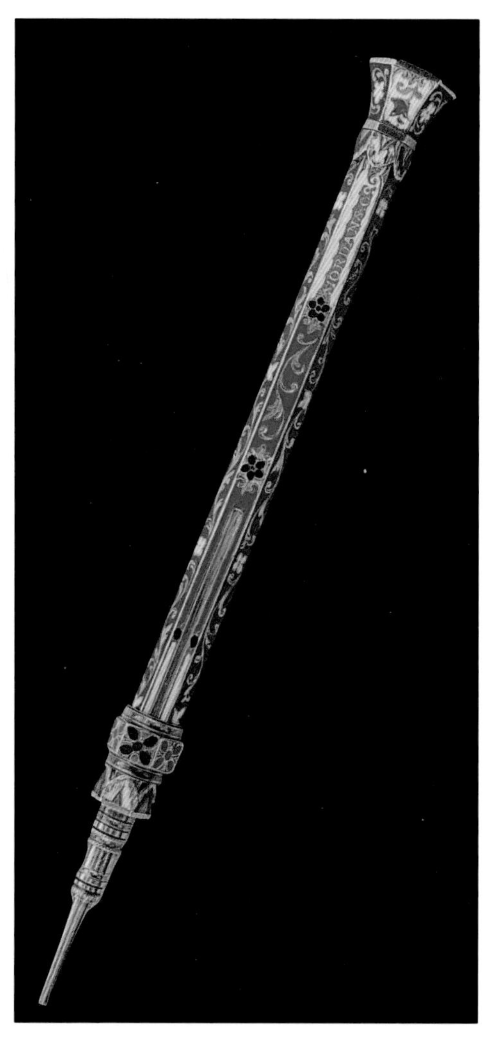

Magnificent gold and enamel pencil marked "Mordan's Co." Very rare, $1,200-$2,000. *Photo Courtesy of Chris Odgers.*

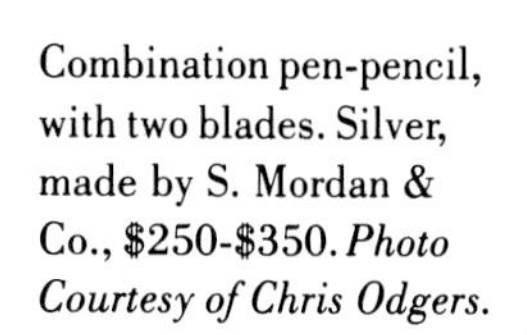

Combination pen-pencil, with two blades. Silver, made by S. Mordan & Co., $250-$350. *Photo Courtesy of Chris Odgers.*

Same pencil, shown in its original case. *Photo Courtesy of Chris Odgers.*

Two silver pencils with cast ring slides, both with seals cut into the stones mounted in the finials, and multi-chambered holders for spare leads. $200-$300.

Left: pencil marked "S. Mordan & Co. Makers and Patentees," with a lion carved in the onyx seal, 3.5" closed, 4.5" open. Right: also marked "S. Mordan & Co. Makers & Patentees," with the initials "R. H." carved in the amethyst seal.

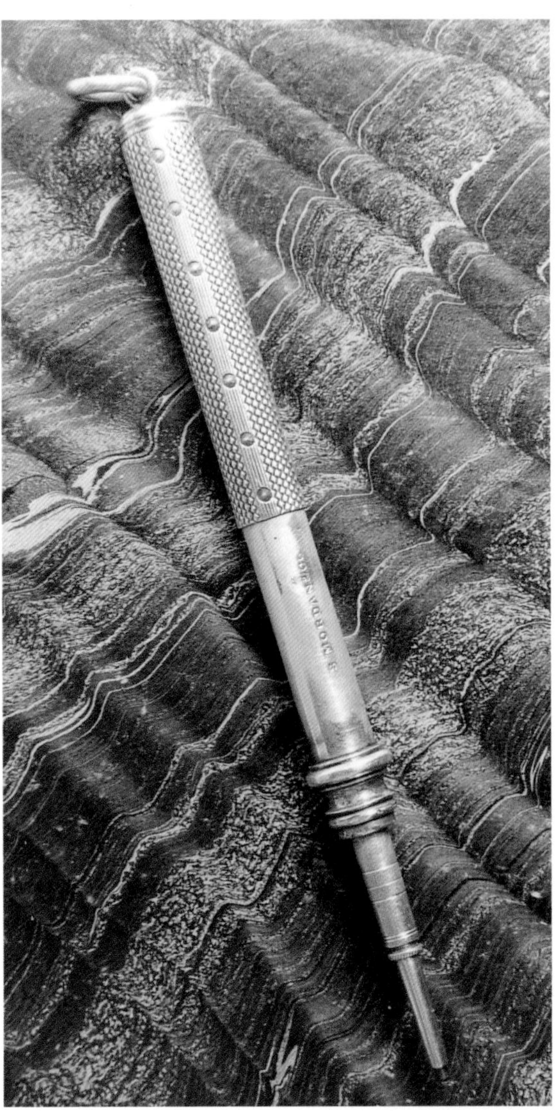

Silver telescopic pencil, made by S. Mordan & Co., with the line and dot pattern. $150-$175.

Barleycorn pencil with a ring slide, and floral engine turned pencil with a cast slider, both made by S. Mordan & Co. $150-$175.

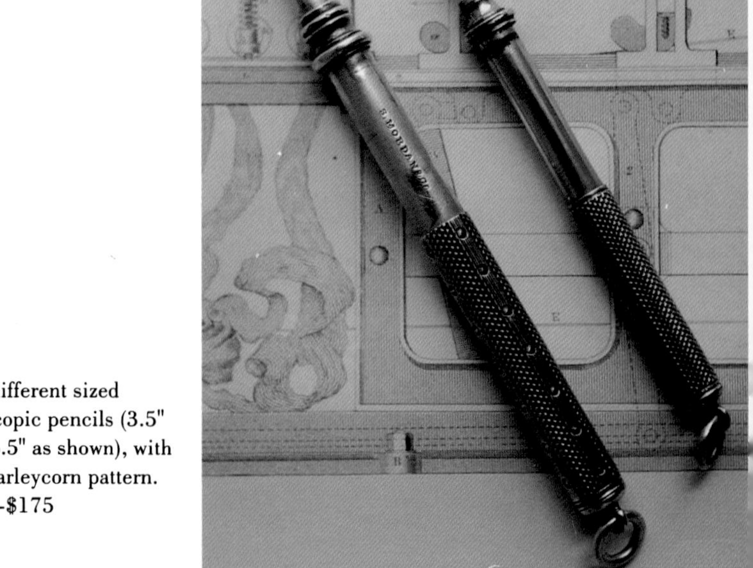

Two different sized telescopic pencils (3.5" and 4.5" as shown), with the barleycorn pattern. $125-$175

Hallmarked gold telescopic pencil, shown next to detachable, cable twist sheath. Made by S. Mordan & Co. The pencil is 2" closed, and 3.5" when open, $275-$325.

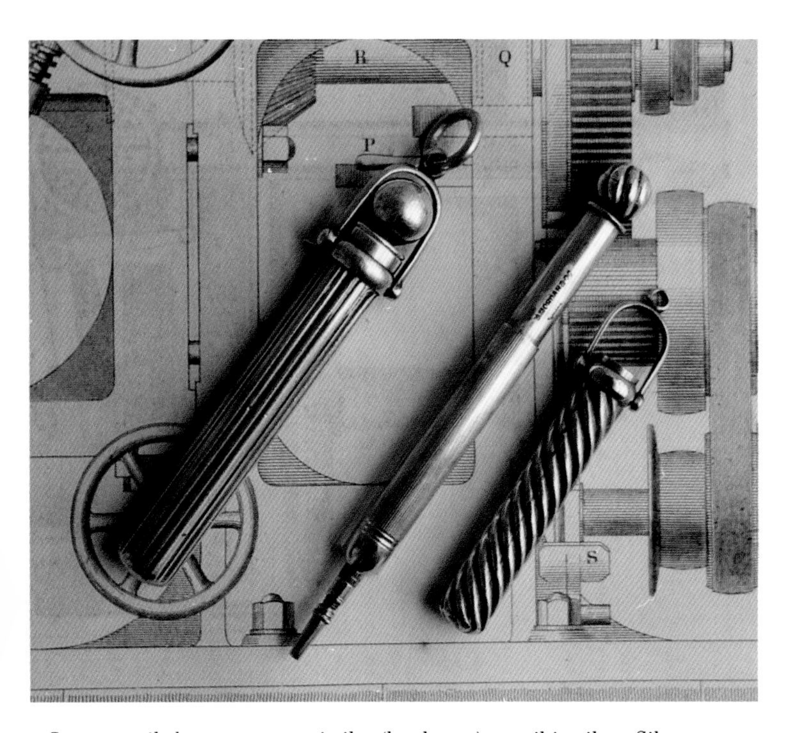

Same pencil shown next to a similar (but larger) pencil in silver. Silver pencil made by S. Mordan & Co., 2.75" closed, 4.25" open, $150-$200.

S. Mordan & Co. pencil with button slide, carved ivory, 3" closed, 4" open, $175-$200.

This small (just over 1" closed, 2.75" open) telescopic pencil has an unusual feature. The Mordan patent on this pencil is for the mechanism that locks the pencil onto a jump ring when the pencil is closed, and releases it when the pencil is opened, by pulling forth a pin located inside the slotted terminal. $125-$150.

Gold pencil made by S. Mordan & Co. $150-$225.

Twentieth century Mordan "Everpoint." $150-$225.

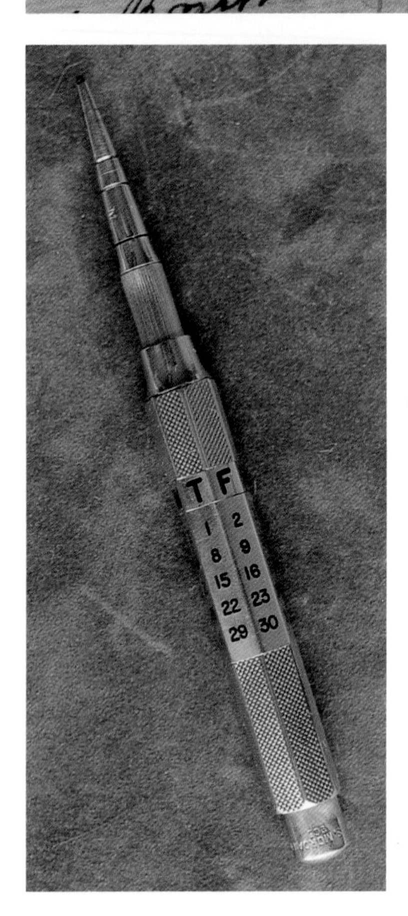

S. Mordan & Co. drop action pencil, in 9ct gold, with a perpetual calendar. $300-$350.

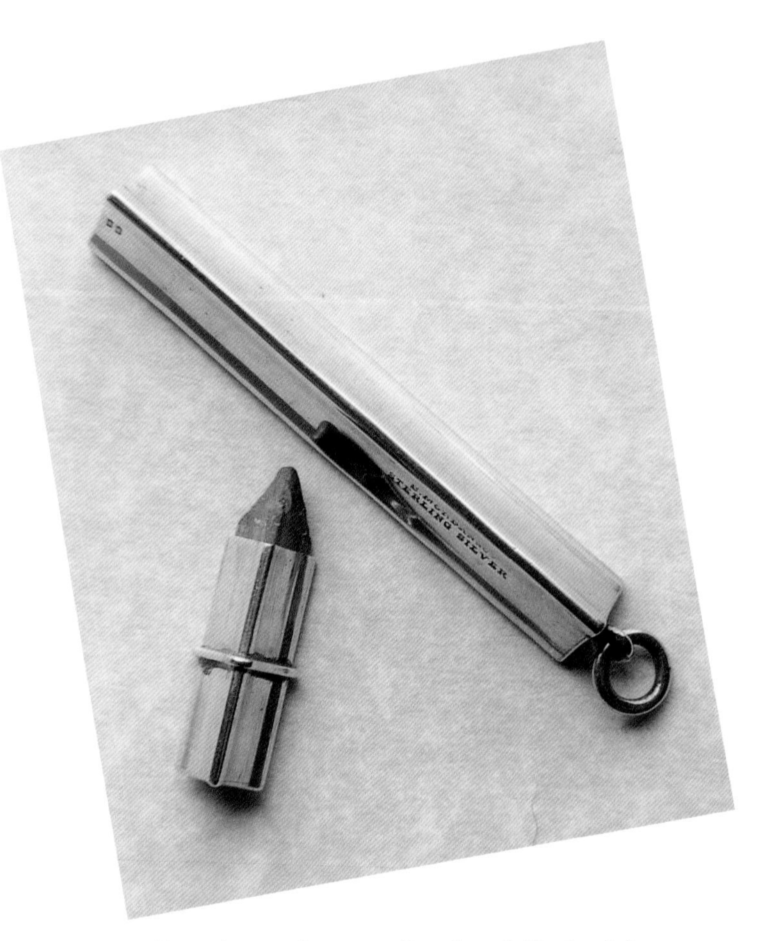

Triangular pencil case, sterling silver, S. Mordan & Co. $150-$175.

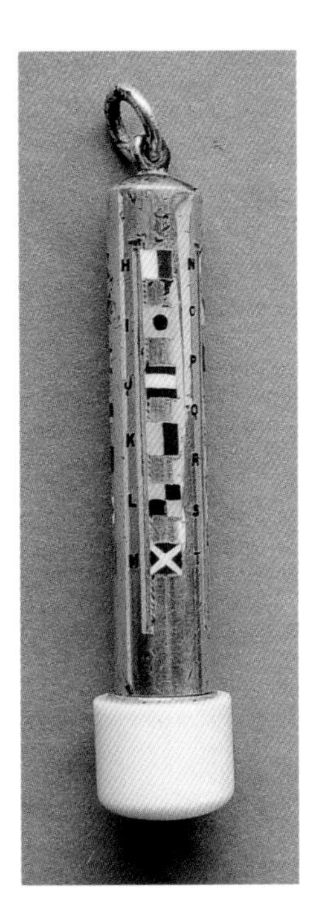

Heavy silver, ivory, and enamel sheath for a telescopic pencil, hallmarked, marked "SM & Co." 3.25" closed, 4.75" open. The enameled flags depict the international alphabet.

Mordan "Populist" pencil, circa 1898. $150-$175.

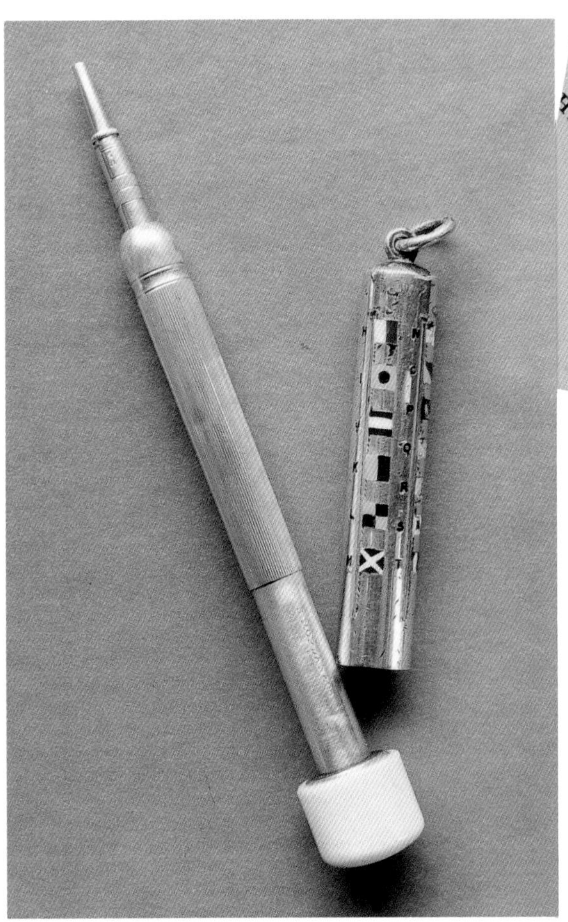

Same as above, shown open. $650-$900.

F. T. Pearce and Company: makers of gold pens, holders, pencils, toothpicks, cigar piercers, and fountain and stylographic pens. Located at 85 Spraque Street, Providence, Rhode Island, with an office at 3 Maiden Lane, New York (letterhead, date unclear). Pearce created beautiful pencils, including the one shown here with a clutch mechanism and a snake clip. He worked with A. T. Cross before forming his own company (see *Writing History*, Barbara Lambert).

Clutch point pencil, with a snake clip, in its original box, made by F. T. Pearce of Providence, Rhode Island. Before opening his own company, Pearce worked for A. T. Cross. $150-$175.

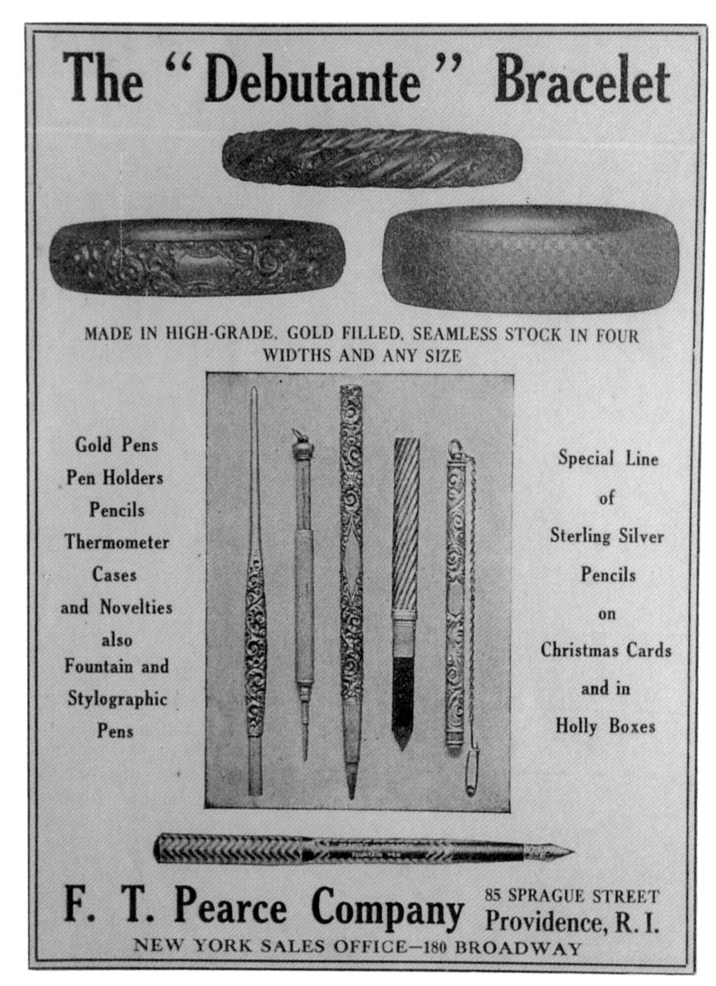

Ad for F. T. Pearce from the *Jewelers Circular*, circa 1890.

John Rauch: held many patents for combination pen-pencils. (American.)

Patent for a telescopic pen-pencil case by John Rauch.

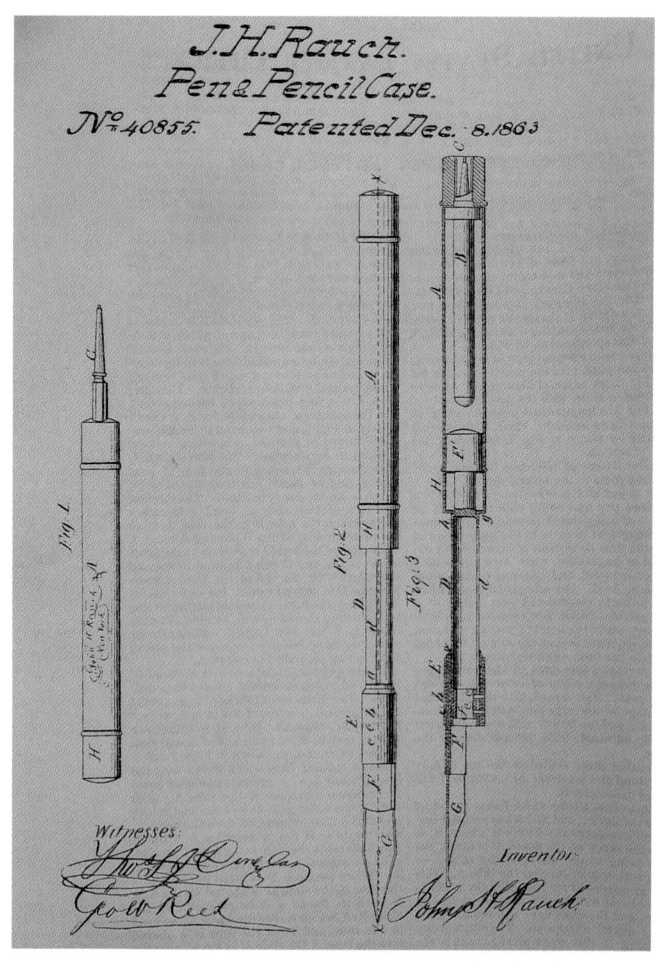

Pen-pencil combination with ring sliders and cast finial in the shape of an animal's head, with red stone eyes. Marked "Rauch," $225-$325.

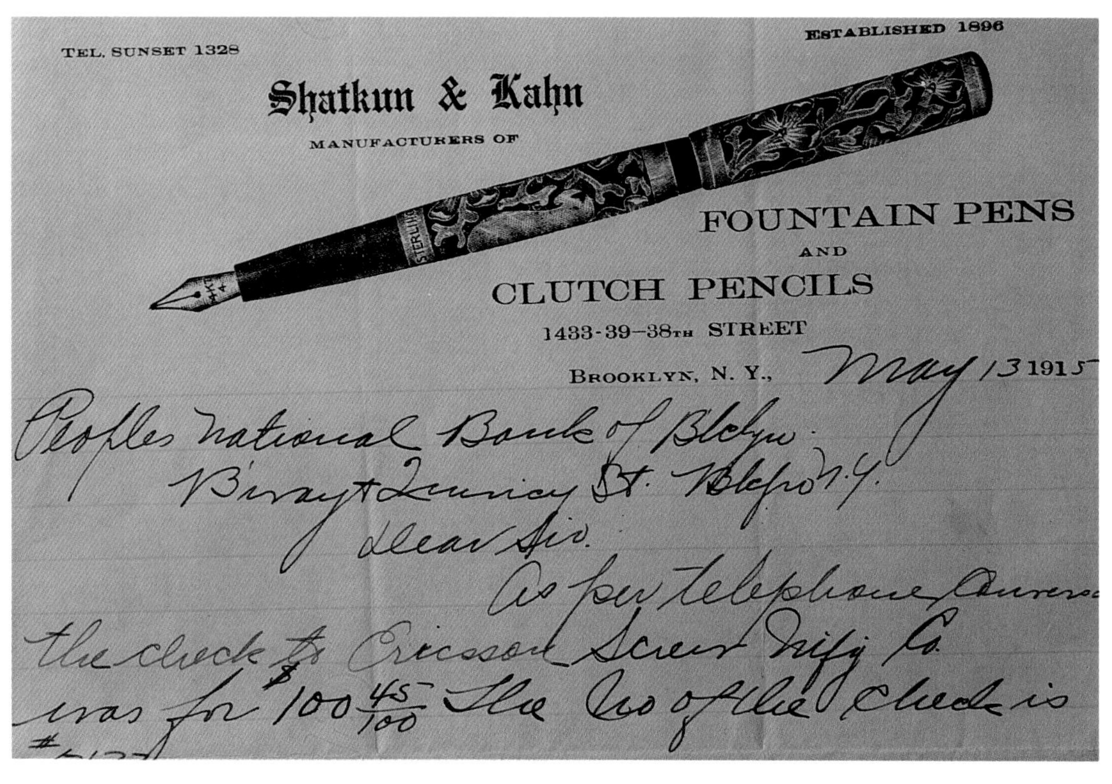

TEL. SUNSET 1328

ESTABLISHED 1896

Shatkun & Kahn

MANUFACTURERS OF

FOUNTAIN PENS

AND

CLUTCH PENCILS

1433-39—38TH STREET

BROOKLYN, N. Y., May 13 1915

Peoples National Bank of Bklyn

Broadway & Quincy St. Bklyn N.Y.

Dear Sir.

As per telephone conversation the check to Ericsson Screw Mfg Co. was for $100 45/100 The No of the check is #...

Letterhead for Shatkum and Kahn, manufacturers of clutch pencils.

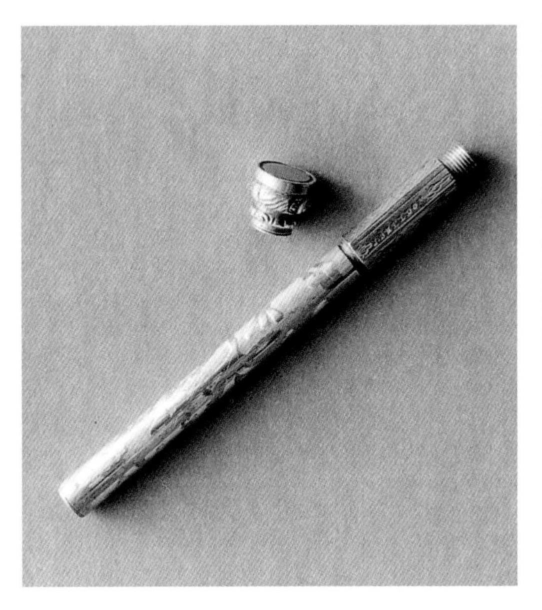

Unusual, early stamped and engraved silver pencil made by J. Sheldon. This pencil has a screw mechanism. $225-$350.

Same as above, open and with finial screwed in place.

Detail showing imprint "John Sheldon 1086 Feb. 8, 1842" from a combination pencil. Sheldon made combination pencils and postal scales, as well as combination pencils with postal scales inside (see Chapter 6).

H. Ropes and Company: wholesale stationer and manufacturer of lead pencils and crayons, 82 Nassau Street, New York (letterhead, 1859).

John Sheldon: made combination pencils and postal scales, as well as combination pencils with postal scales inside.

Shreve and Company: sold flat pencils made of sterling or 14 kt gold, made to commemorate "Panama Pacific International Exposition 1915." (American.)

Sims and Company: black lead pencils, "Makers to His Majesty's Honorable Board of Ordnance" (advertisement, undated).

H. M. Smith and Company: manufacturers of a variety of gold pens, holders, pencils, etc. Dealer of Paul E. Wirt fountain pens. Located at 83 Nassau Street, New York (letterhead, circa 1890).

702

H. M. SMITH & CO.

83 Nassau St. New York City

Manufacturers of Every Description of

GOLD PENS, HOLDERS, PICKS, PENCILS, ETC.

The Paul E. Wirt and Black Diamond Fountain Pens.

Ad for H. M. Smith & Co., maker of pencils and authorized dealer of Paul Wirt fountain pens, from *The Keystone*, September 1894.

A. J. Strachan: London, made holders for lead pencils that slid out through the opening for a gold nib as early as 1801. (See *A Loan Collection of Writing Implements & Accessories in Silver & Gold c. 1680-1880*, Gerald Sattin Ltd.)

Tiffany: Beginning with their first catalog, "Blue Book," published in 1845, Tiffany's advertised "Gold and Silver Sand, Quills, Steel Pens, Pen Holders, Seals, Pencils...Pen and Pencil Case united, Knife and Pencil Case united..." In 1876, under "Presents for Ladies," they offered a "shopping bracelet, with pencil attached, gold and silver, some richly jeweled." In their 1878 catalog, they offered "Double vest chains, extending from the button hole to both pockets, one end for watch and the other for pencil, safe key, night key, or other article," and "Magic Pencils. Plain gold, and some studded with diamonds pearls, turquoise and other stones." In 1881-82, "Pencils. The variety of magic pencils that project the points and elongate the pencil with one movement is larger than ever before, and includes many new charm pencils in the form of jockey devices, such as caps, whips and horseshoes; others with frogs, lizards, or fish, of various colors of gold, besides the usual variety of gold, silver, enameled, ivory, rubber, and leather, from $2.50 upwards. A variety of handsome gold pencils to attach to watch chains, $8 to $18. Some richly jeweled, up to $300. Silver, inlaid with niello enamel, $4 and upward." In 1892, under the category of pencils, they describe "all gold or part platinum, in a variety of designs, with emblems of riding, boating, sporting, ball playing, etc...with precious stones...$70 to $150."

The mechanisms for pencils made by Tiffany were usually made by other makers, including W. S. Hicks and Aikin, Lambert & Co. One such pencil is a wonderful three-dimensional image of the Metropolitan Life Building, in New York.

Magic pencil sold by Tiffany's, made of 14kt gold in its original case. $150-$200. W. S. Hicks made the mechanisms for Tiffany pencils.

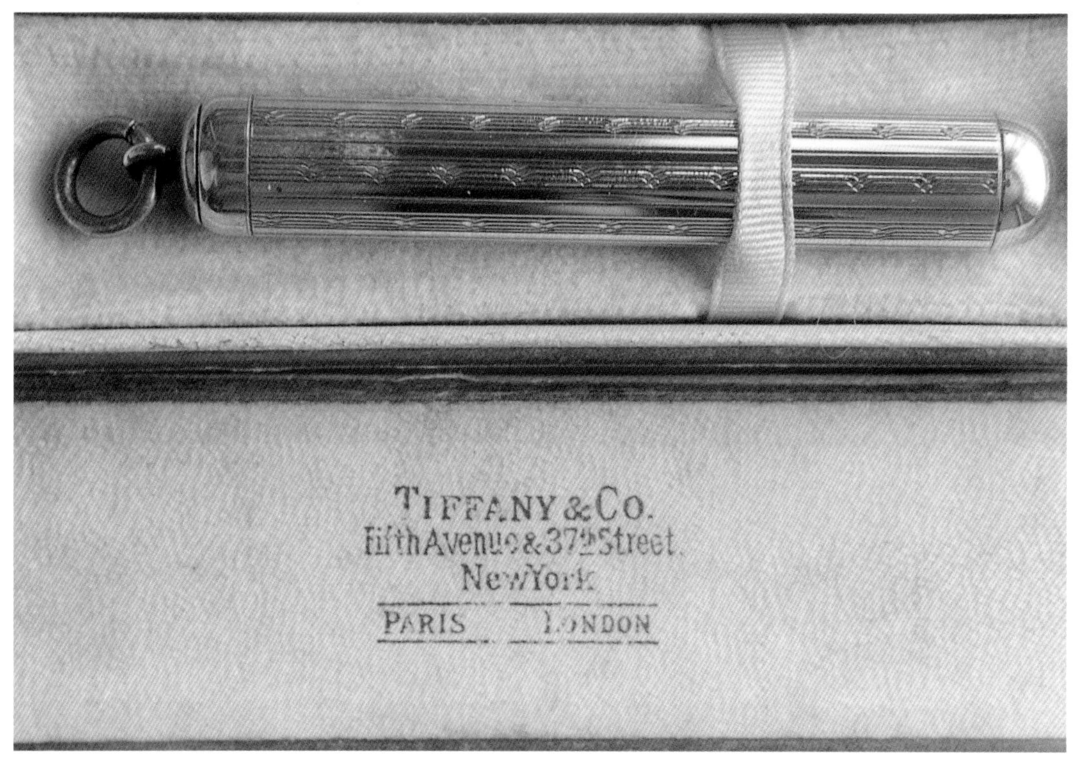

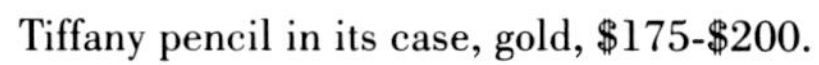
Tiffany pencil in its case, gold, $175-$200.

Sterling silver pencil, chased floral design. Marked "Tiffany & Co." $175-$225. *Photo Courtesy of Chris Odgers.*

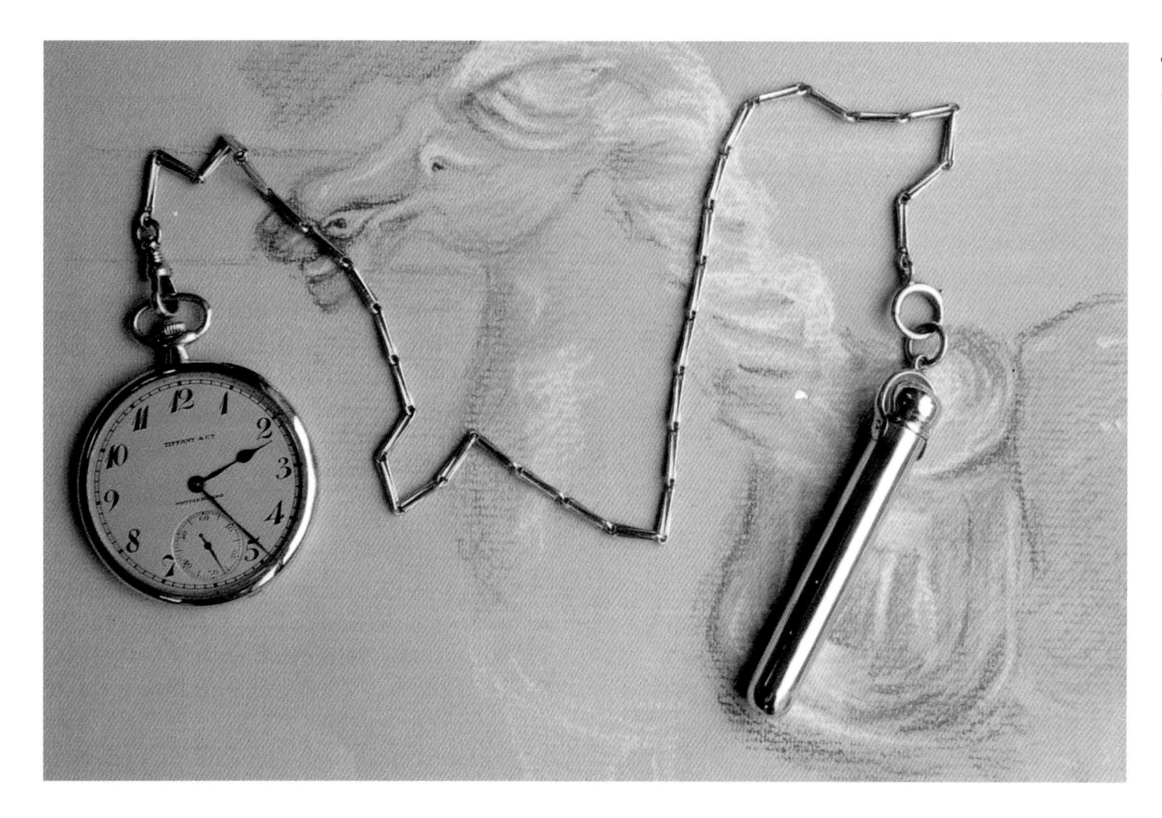
Tiffany: gold pocket watch, watch chain and 14 kt gold pencil in removable sheath. Pencil, $175-$225.

Edward Todd: There is no question that Edward Todd & Co. produced writing instruments par excellence. The Edward Todd mark on a pen or pencil distinguishes the piece as well made. What has seemed a bit unclear, at least to this point, is who exactly was Edward Todd? Some recently discovered information may help to shed some light on that subject.

Firstly, it should be noted that there was not one Edward Todd, but two.

Edward Todd (senior) was born in Plymouth, Connecticut, on either September 26, 1826 (according to one source), or on August 21, 1822, and died in 1898 in New York City. Edward Todd (junior) was born in New Haven, Connecticut, in July 1857 and died on December 24, 1937. Edward Todd Jr. joined the firm founded by his father at the age of twenty-one in 1878.

Edward Todd Sr., who was relatively well educated for his time, entered business as a dry-goods merchant in Hartford, Connecticut. He also was engaged in the mercantile business in Pittsburgh, Pennsylvania, for a brief period. In 1848 he joined Bard Brothers of New York City, and there learned about gold pen manufacturing. By 1851 he had established his own firm, called Smith and Todd, which eventually became Mabie, Todd and Company. Another member of the Todd family, George, was also involved in Mabie, Todd & Co.

Edward Todd display in plush and silk. $75-$135.

In 1868, Edward Todd Sr. traveled through Europe with his family. Although what he did during this period isn't documented, it's possible that while in Europe he investigated the pen and pencil-case industry there. When he returned to the United States (in 1869, according to one source) he worked with another manufacturer of gold pens, Newton, Kurtz & Co., where he became a partner. This company became known as Edward Todd & Co., with Joseph Monaghan as his partner. Edward Todd Sr. "acquired a wide reputation as a manufacturer of gold pens and pencil-cases, being largely instrumental in devising improvements in the goods, and developing new methods of manufacture."[2] Although the only patent listed in his name is for an "Improvement in Stylographic Fountain Pens" (applied for on July 22, 1879, and granted on August 25, 1879)[3], his partner Jonathan Monaghan patented a pen and pencil case with a T. Flynn in 1871.

In a trade catalogue, circa 1876, the company advertised "Gold Pens, Gold, Silver and Rubber Pencils, Pen and Pencil Cases, Toothpicks, Etc." and listed their address as 652 Broadway, New York City.[4] Interestingly, one of their gold pens (nibs) is described thusly: "Letter patent [sic] have just been issued to us on the name and shield shape of this pen. Beautifully engraved with the stars and stripes, which, with its very fine finish, makes it 'the pen' of the year 1876." There is an entire category for "Centennial Goods," including "18-karat plated "Centennial Charm Magic Pencils" and "Rubber (18-karat mountings), 'Centennial' Charm Magic Pencils." The catalogue lists a wealth of pens, some stamped "Edward Todd & Co.," some marked "Joseph Monaghan," in a wide variety of shapes and sizes. Desk holders and pen cases (with 16-karat gold pens) are made of ebony, snakewood, coin silver, gold plate, ivory, pearl, India Rubber, and 18-karat gold. Screw and slide, 18-karat gold-mounted rubber pen and pencil cases are offered, as well as

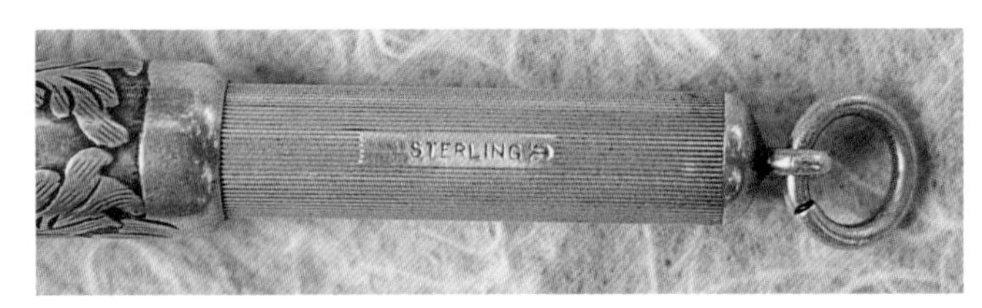

Detail showing Edward Todd mark.

Chased silver pencil-letter opener, marked Edward Todd. $150-$200.

Edward Todd mark on a silver aide mémoire and matching pencil (see Chapter 7).

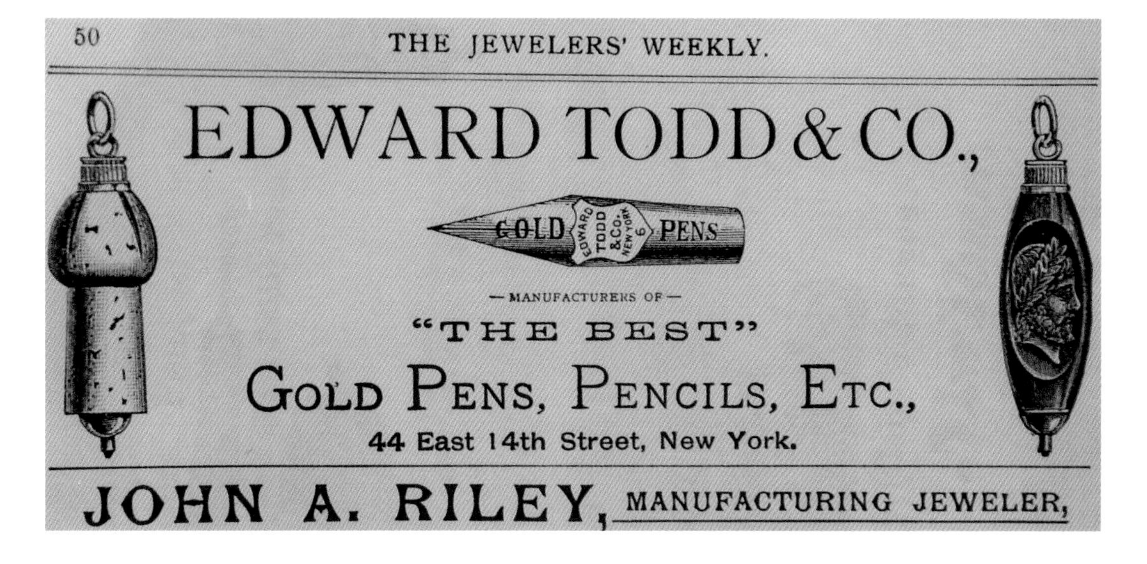

Ad showing figural pencils made by Edward Todd & Co., *The Jewelers' Weekly,* circa 1880.

"Patent Paragon" pencils, with 18-karat gold mountings. Listed are telescopic, reverse, and slide holders in black or red rubber (at the same price), screw charm and "fancy" pencils; round, fluted, enamel engine-turned, magic extension; pearl, plated, solid gold pen and pencil cases with "duplex screw movements," coral and stone charm pencils in addition to solid gold charm pencils in red or yellow gold. Unfortunately, the catalogue is not illustrated, as a drawing of the solid gold "Hexagon Patent Paragon Stone Head, Extra Engraved...$10.00 each" would be most useful in determining what such a pencil would look like.

One delightful figural pencil made by Edward Todd & Co. is a double-sided watch chain charm, made in the form of two people. On one side, a plump but demure woman wearing a patterned kimono (with her hair tucked behind one ear) smiles above her open fan. On the other side, an also plump male figure wearing a kimono of a different pattern uses his right hand to clean his teeth with a tiny toothpick. The pencil appears to be of bronze plated with silver. The bronze color shows through the silver on the hands and faces, differentiating which areas were supposed to portray skin as opposed to fabric or hair. The watch chain still attached to the pencil is marked "coin." Because the company used the word "coin" to describe pieces offered in their 1876 catalogue, and because of the popularity of figural charm pencils during the Victorian era, it can be surmised that this pencil dates from circa 1875-1900.

Edward Todd came from a distinguished American family, which was instrumental in the early development of several communities in the northeast. Documented in the Todd family genealogy, entitled *Todd Family in America or the Descendants of Christopher Todd 1637-1919*, his ancestors emigrated from Yorkshire, England, and were instrumental in the establishment of New Haven, Connecticut. Edward Sr. married Lydia Alden on January 29, 1853, thus joining the Todd

Favorites as Souvenirs

for banquets and other occasions

NEW and useful articles in gold and silver, New and attractive patterns in pencils, penholders, novelties, etc.

We cordially invite the trade to inspect our line at our office or when our traveler calls. We have many new and attractive goods this season. Catalogues sent on application.

Edward Todd & Co.,

CENTURY BUILDING,

1 West 34th St., NEW YORK.

Ad for Edward Todd & co., circa 1905.

family with descendants of John Alden, one of the original Plymouth pilgrims. When Edward Sr. died in January 1898, four children survived him: Charlotte Lydia, Edward Jr., Ambrose Giddings, and Jennie Alden. The firm he founded was then incorporated as Edward Todd & Co., with Edward Jr. eventually becoming president.

As a successful businessman, Edward Todd Sr. helped establish the Brooklyn Life Insurance Co., and became a member of the board of directors of both that company and of the Bank of North America. Additionally, he used his talents and substantial wealth for philanthropic purposes, serving as vestryman and later superintendent of Brooklyn's Holy Trinity Church and Seminary, trustee of the Church Charity Foundation and Homeopathic Hospital, as well as being an active member of the Metropolitan Museum of Art and multiple charitable organizations.

Edward Todd, Jr. married Bessie M. Tarbell on June 8, 1887, and had two children, Fannie L. and Edward T. Not only did he continue in his father's business (which he apparently expanded), but also in his footsteps as trustee or director of many organizations, in addition to continuing the family tradition of involvement with the Holy Trinity Church of Brooklyn. According to his obituary in the *New York Times* on December 25, 1937, he had been with Edward Todd & Co. for approximately sixty years. The company made "gold pens and other articles" until the firm (located at 1 West 34th Street until at least 1928) went out of business in 1932.

Although it is not yet clear exactly who was responsible for the especially beautiful figural pieces made by the company, it is apparent that the quality of the writing implements they made was exceptionally high. While only one trade catalogue has so far surfaced, there are undoubtedly other documents that will allow us better understanding of this intriguing father-son business.

Walter Thornhill & Company: (English) was first established as Morley & Thornhill in 1805. From 1820 until about 1912 (guided by at least three generations of Thornhills), the firm was essentially called Thornhill & Co. or Thornhill Ltd. Listed as cutlers to the Queen, the firm employed as many as ten men by 1851. In the Paris Exposition of 1878, they were described as both cutlers and silversmiths, and exhibited (among other things) patent gold and silver pencil-cases. In a Christmas catalogue dated 1876, they described a long list of figural pencils, including "the silver racket pencil...the silver champagne bottle pencil...the silver post-horn pencil...silver screw pencil-case...silver nail with gilt head...silver peg top pencil-case...Thornhill's registered Silver Fish Knife, Pencil

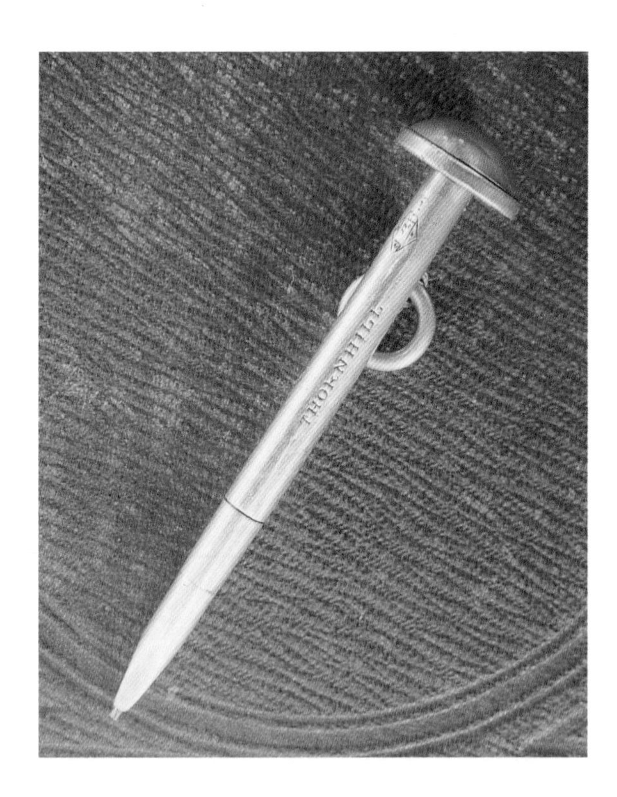

Silver pencil in the shape of a round-headed nail, made by Thornhill, with registration mark. $200-$250.

and Whistle, comprises catcall, pencil and two blades in a flat and convenient form..." (see John Culme, *The Directory of Gold and Silversmiths, Jeweller's and Allied Trades*). Most of their figural pencils are believed to have been made by Sampson Mordan & Co.

Vale & Co.: held a patent for a button or gravity pencil, in the shape of a bottle.

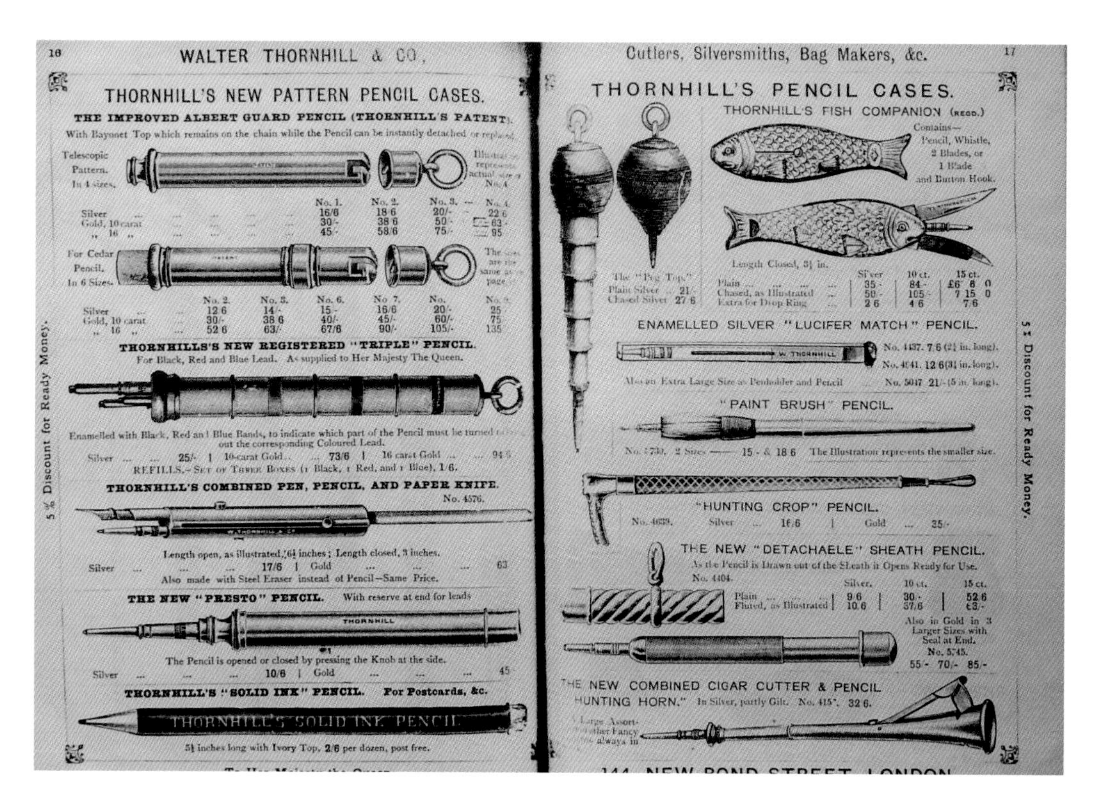

Page from Walter Thornhill (London) catalog, circa 1890. Notice the similarity between what Thornhill advertised and what S. Mordan & Co. made.

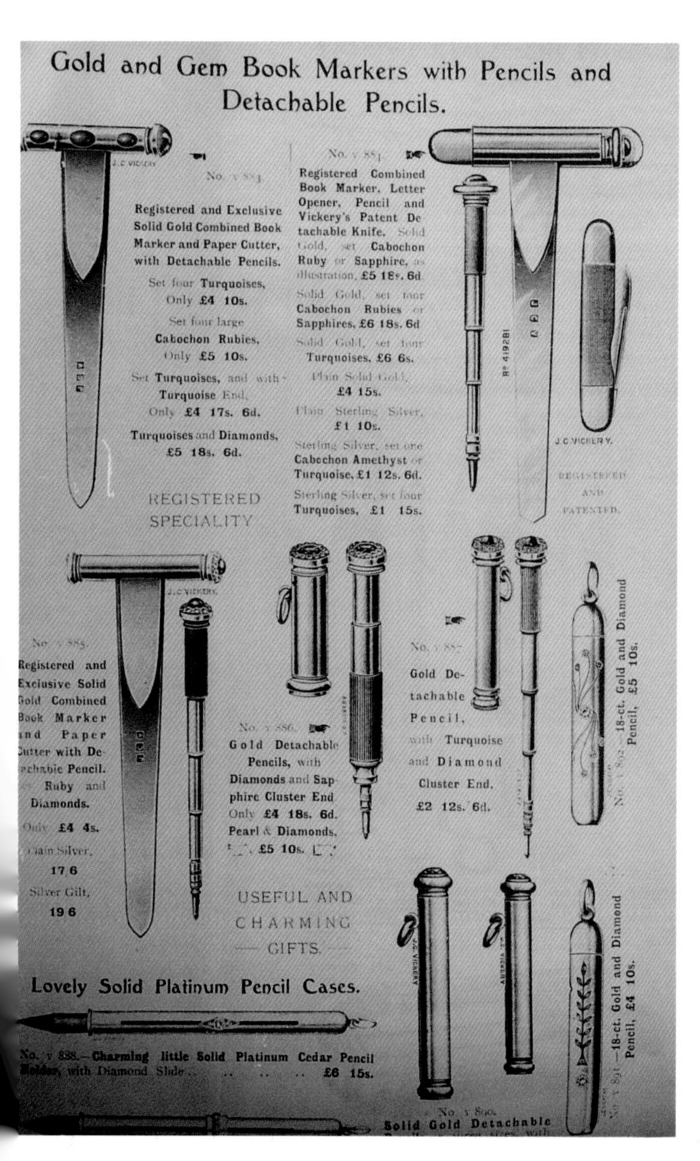

J. C. Vickery: "Their Majesty's Jeweller, Silversmith, and Dressing Case Manufacturer... American visitors to England are most cordially invited to call and see Vickery's collection of charming novelties for gifts." Maker's of Vickery's Calendar Pencils (*Town and Country*, April 1914).

Page from J. C. Vickery (Regent Street, London) catalog, circa 1915.

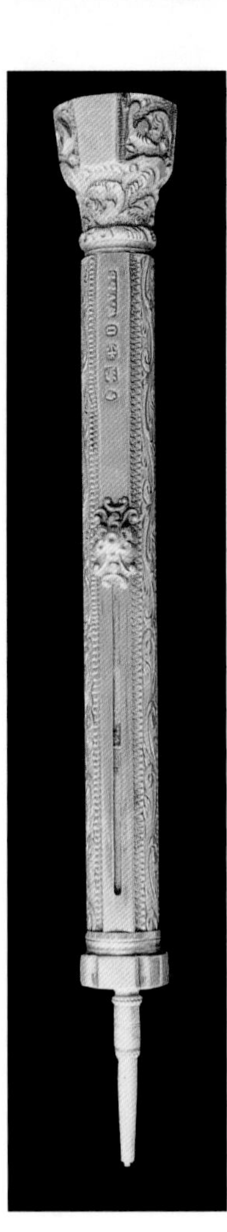

Silver pen-pencil hallmarked W. V. & S. $200-$225.

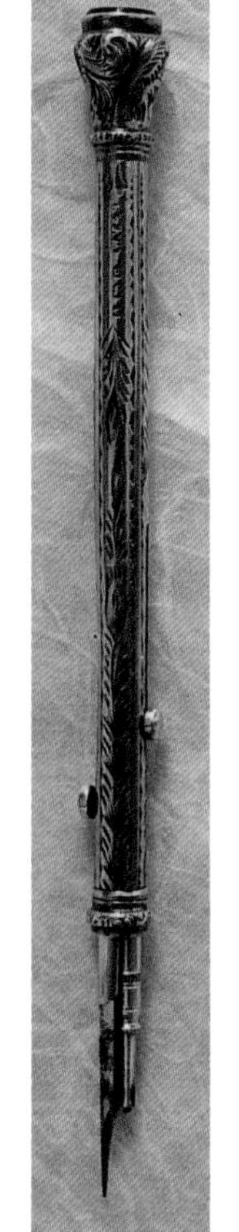

Sterling silver pen-pencil combination, with button slides, hand engraved, and mounted with an amethyst. Hallmarked, W. V. & S.,3.5" closed, $175-$225.

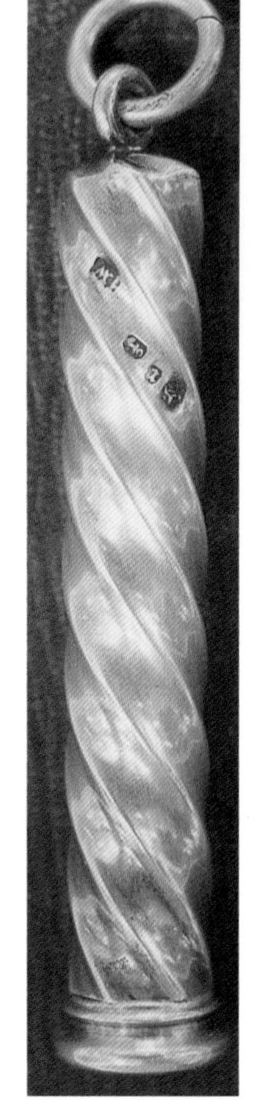

Silver telescopic pen-pencil combination, hallmarked, "H. W." $175-$225.

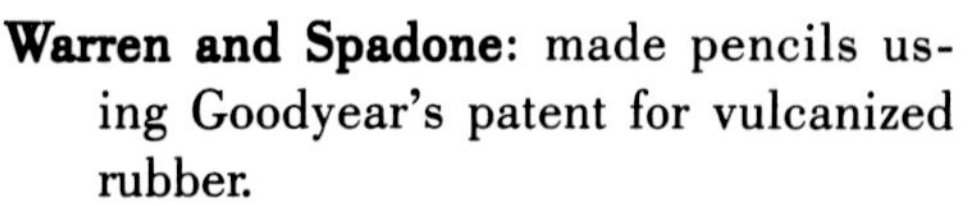

Warren and Spadone: made pencils using Goodyear's patent for vulcanized rubber.

Unger Brothers: This firm was established in Newark, New Jersey in 1881. They were known primarily for the graceful Art Nouveau designs they used (including pattern's entitled "Stolen Kiss," "Dawn," and "Man in the Moon"). They made exquisite desk items, including pens, pen holders, inkwells and pencil extenders, some decorated with a Native American motif.

Webster Company: Silver company established in North Attleboro, Massachusetts, in 1869. They made sheath pencils and aide mémoires, as well as silver dresser ware and frames. It is now a subsidiary of Reed & Barton.

Silver pencil, hallmarked, "T. H. V." Made in the shape of a thistle, $125-$150.

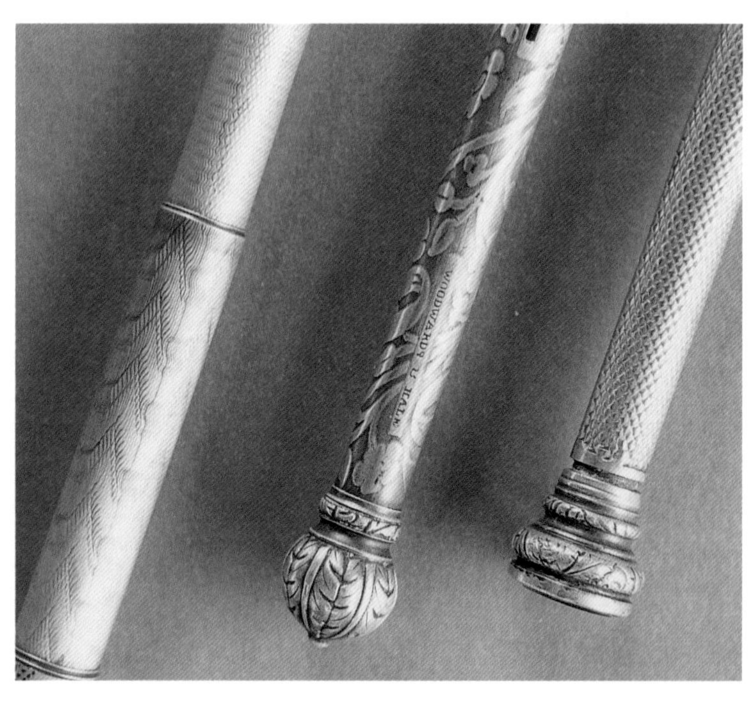

Center: Silver pencil with engine turning, marked "Woodwards Hale." $100-$130.(Nothing is known about this company.)

Chapter Nine

Display and Care

One of the nicest ways of displaying Victorian pencils is to wear them. The pencil can be hung from a ribbon or chain, or attached to a chatelaine pin and worn on a lapel. Many different types of chains work well with the pencils. They can be worn on a long antique watch chain (which may be lengthy enough to double around the neck) or on a shorter antique watch chain, one originally made to hold a pocket watch. Modern "snake" chains are strong enough to hold, yet simple enough to not detract from the pencil, and come in various lengths. Many of the pencils were sold with black grosgrain ribbon already attached, and some are still found that way (make sure the ribbon is still strong enough to hold the pencil before it's worn).

Before one wears a pencil, be sure to ascertain that no parts are loose and that it can stand up to being worn. Check the jump link to make sure it's secure and that the pencil hangs from it correctly (the pencil should face out, not sideways.) As with anything worn as a pendant, the wearer should be careful not to catch or knock the piece. Some pencils can be worn together, and some can be combined with other examples of Victorian jewelry (see illustrations).

Pencils can also be displayed in cabinets. Those shown here were designed specifically for John Holland & Co., a company that manufactured pens and pencils. Although it's best to keep the pencils protected and in a dust-free environment, uncovered manufacturers' display trays occasionally become available. These work especially well when they're kept in glass cabinets.

Antique pencils should be kept clean and may require a careful polishing with a non-abrasive silver polish. (I've found that Cape Cod Polish works.) Do not ever force a pencil to repel or retract. Sometimes, gentle pressure on both the nozzle and the barrel end is all that's needed to guide a pencil back into its case. Also, don't try to pull the nozzle out of a barrel with a pair of pliers. Even if one can find small enough pliers, the tube may be crushed in the process and it will be difficult to repair. Although some people use various lubricants, it's probably best to do as little as possible to the pencil unless you've done considerable experimenting with broken or less valuable pieces. Some lubricants can actually gum up the mechanism and create more problems in the long run.

Many of the pencils have parts held together with soft (or lead) solder. This solder melts at very low temperatures, and pieces can either fall apart or become joined where they shouldn't be if too much heat is applied. (One dealer told me that she put a working pencil in her toaster oven to dry on low heat after it had gotten wet. The unfortunate result was that solder in the mechanism melted, rendering the pencil unworkable.)

Some people display their pencils in showcases, like this one originally made for John Holland & Co.

Testing the pencil for gold content can be destructive. Notice the large gouges on the two pencils shown, where someone shaved off metal to be tested. It's probably better to understand that many of the gold colored pencils have either a thin layer of gold over a stronger base metal, or the pencil-case (certainly not the mechanism) may be made of gold and not marked. Sometimes dealers weigh an entire pencil and try to sell it for its value based on weight. Because the mechanism is the heaviest part of the pencil, and because mechanisms were never made of precious metal, it is unwise to buy a pencil priced by this system.

Finding lead that fits the antique pencils can be a challenge. Trial and error is perhaps an inefficient system, but be prepared to try many pieces of graphite before you find one that fits perfectly. Don't be surprised if the pieces are very short (perhaps less than 1/2" long). This is sometimes the appropriate size, based on how the pencils were originally designed. (See Issue No. 3 of *The Pen & Pencil Gallery- Marestin Magazine* for more information on sizing graphite.)

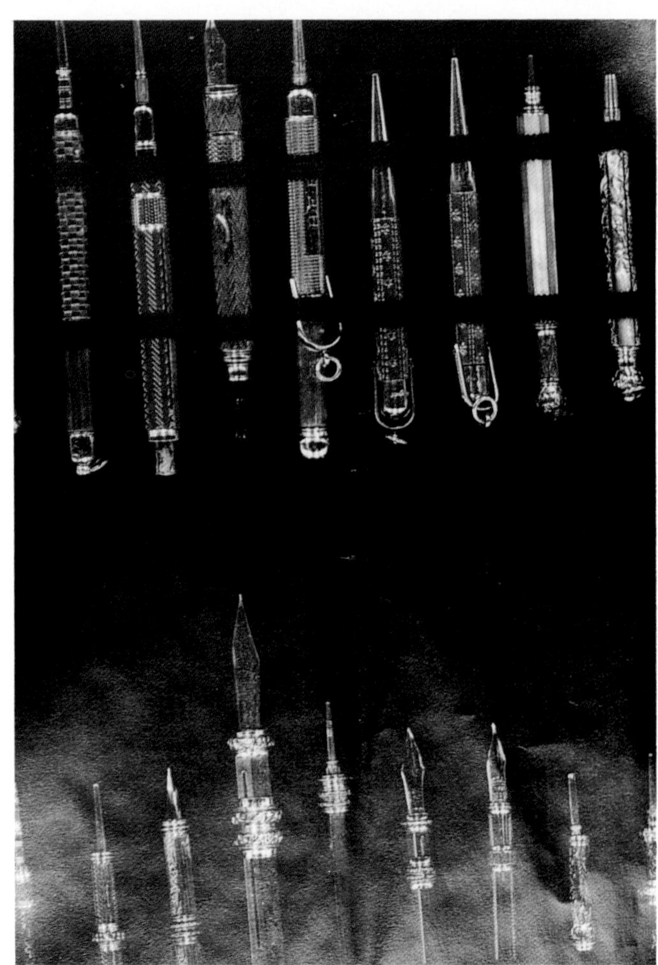

More pencils carried in a case with elastic straps.

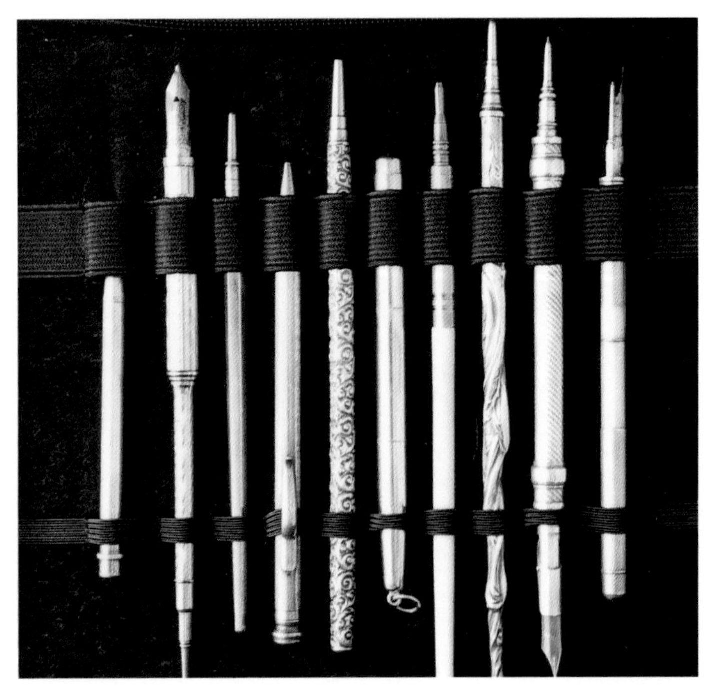

There are pen cases available commercially, made to hold pens and pencils.

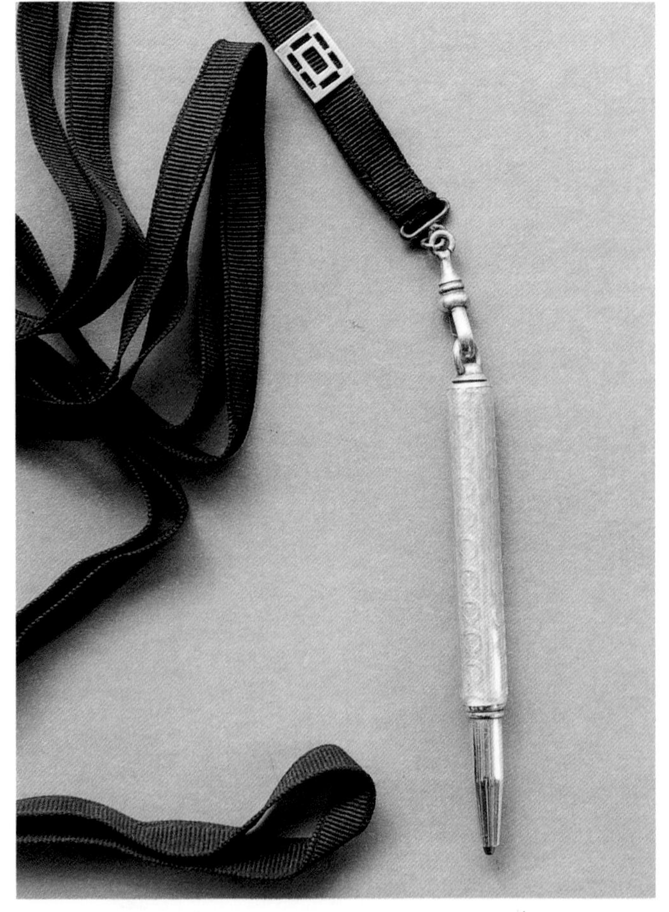

Enamel and silver pencil made by A. T. Cross, with the original grosgrain ribbon and silver findings. $75-$125.

This unusual piece consists of several figural components, all representing mining or panning for gold. It was intended to be worn from the waist as a dance card, with both the pencil and the pad of paper behind the cloth and metal cover. Circa 1880, $225-$250.

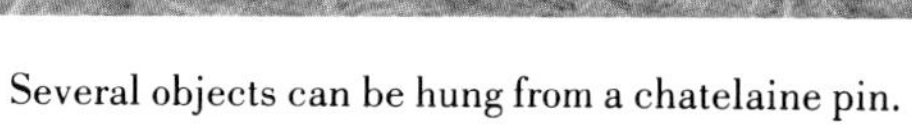

Several objects can be hung from a chatelaine pin.

Back view of above.

Solid 14kt Gold Chatelaine Pins.

Illustrations are actual size. The pins are strong and well made. The stones are genuine.

7885	Rose gold, with diamond, $23.00; with pearl	$11 00
7886	Rose gold, with diamond, $13.50; with pearl	8 00
7887	Rose gold, with diamond, $11.50; with pearl	6 00
7888	Rose gold, with diamond, $13.50; with pearl	7 50
7889	Rose gold, with diamond, $13.50; with pearl	[illegible] 75
7890	Roman or polished gold	4 25
7891	Rose gold, ruby eyes	[illegible] 50
7892	Rose gold, with diamond, $13.50; with pearl	7 50
7893	Rose gold, with diamond, $19.00; with pearl	11 50
7894	Roman or polished gold	3 00
7895	Rose gold, ruby eyes	6 00
7896	Roman or polished gold	2 75
7897	Roman or polished gold, $3.00; with 3 diamonds	11 50
7898	Roman or polished gold	3 75
7899	Roman or polished gold, $4.25; with 6 diamonds	53 00
7900	Roman or polished gold	4 25
7901	Roman or polished gold, $3.75; with 6 diamonds	46 00
7902	Roman or polished gold	5 50
7903	Roman or polished gold	3 50
7904	Roman or polished gold, $3.75; with 3 diamonds	12 00
7905	Polished gold	5 25
7906	Polished gold, hand-engraved	5 25
7907	Polished gold	4 50
7908	Polished gold	5 50
7909	Polished gold	$5 00
7910	Roman or polished gold, $4.25; with 3 diamonds	13 50
7911	Rose gold	6 00
7912	Rose gold, 1 sapphire	8 75
7913	Rose gold, 1 diamond	12 00
7914	Rose gold, 4 pearls and 1 topaz	11 00
7915	Rose gold, 3 pearls	7 75
7916	Polished or Roman gold	5 50
7917	Polished or Roman gold	4 75
7918	Rose gold	5 50
7919	Rose gold, 1 diamond and 1 amethyst	18 00
7920	Rose gold	5 50
7921	Polished gold, engraved	6 50
7922	Polished gold	4 50
7923	Polished gold	4 50
7924	Polished gold, $5.00; hand-engraved	5 75
7925	Polished gold	4 50
7926	Polished gold	4 50
7927	Polished gold, $6.75; smaller size	5 50
7928	Polished gold, $4.00; hand-engraved	4 50
7929	Polished gold, $4.75; larger size	5 75
7930	Polished gold, $5.50; hand-engraved	6 50
7931	Polished gold, $4.75; hand-engraved	5 50

Page from S. Kind & Sons (Chestnut St., Philadelphia) catalog, dated 1911, showing gold chatelaine pins.

Sterling silver, cable twist magic pencil on chatelaine pin with a retractable chain. $125-$135.

Gold-filled, chased mechanical pencil made by W. S. Hicks, hung from a chatelaine pin with a retractable chain. $125-$150.

Fleur de lis chatelaine pins: gold-filled and green enamel ($35), sterling silver ($35), and gold ($75).

Magic pencil on a gold-filled watch chain that's long enough to wear as a necklace ($175-$225 together).

Gold-filled and enamel fleur de lis chatelaine pin, gold-filled pencil ($125).

Gold pencil worn on a long watch chain.

A sterling silver pencil on a long chain.

Tri-colored pencil in sterling silver on a contemporary silver snake chain ($200-$250 together).

Pharoah pencil on a contemporary silver chain ($200-$250).

Gold collapsible pencil with seed pearls, in sheath, hung from a nineteenth-century ladies watch chain. Pencil, $325-$400.

Gunmetal match safe, heart-shaped coin holder, telescoping pencil on a nineteenth-century chain ($300-$350).

Ring slide pen-pencil combination on a contemporary chain. Pencil, $125-$175.

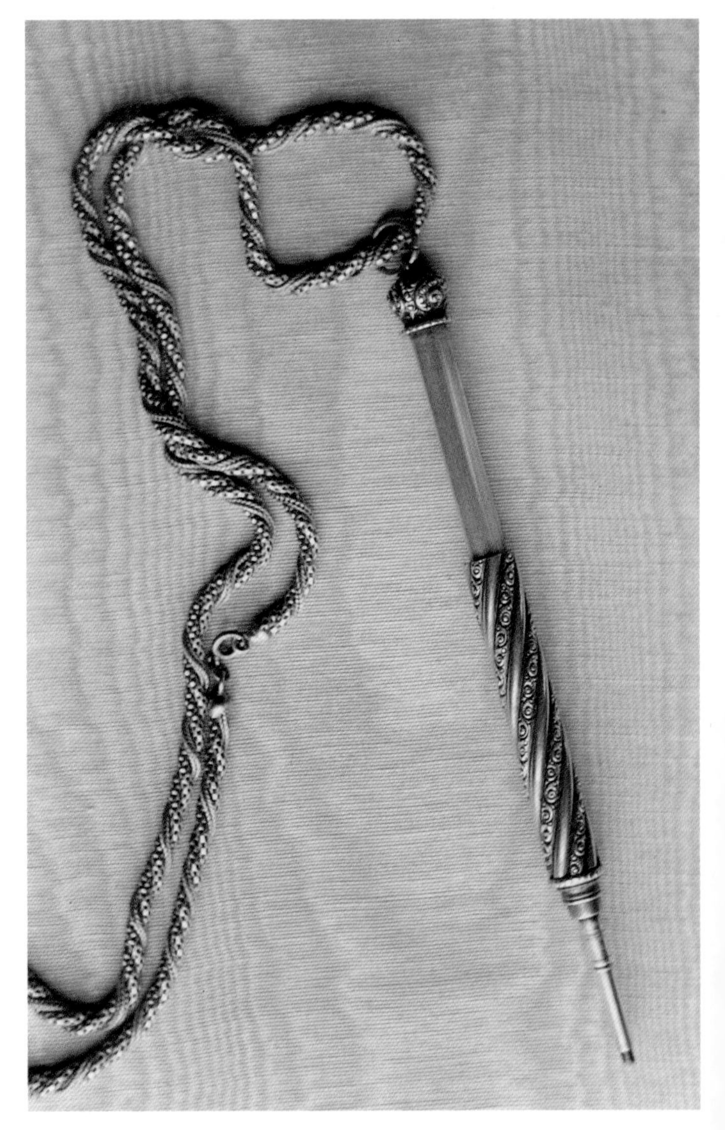

Sterling silver magic pencil on a contemporary silver chain. Pencil, $125-$150.

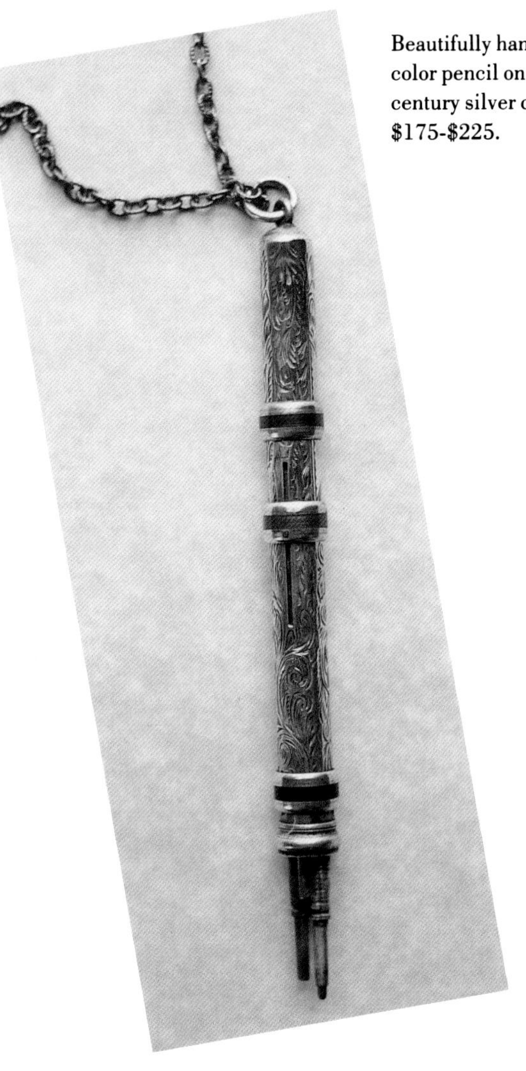

Beautifully hand-engraved tri-color pencil on a nineteenth century silver chain. Pencil, $175-$225.

Cherub holding a dove atop a pencil holder on a filigree chain. Pencil, $75-$85.

Gold-filled pencil on chatelaine pin. Pencil, $75-$125.

Gold-filled pencil by A. T. Cross on a chatelaine pin with a retractable chain. Pencil, $75-$85.

Someone made a serious gouge in the tip of this pencil to test it for gold content. Testing of this nature is invasive and irreparable.

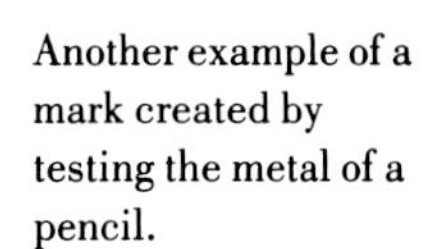

Another example of a mark created by testing the metal of a pencil.

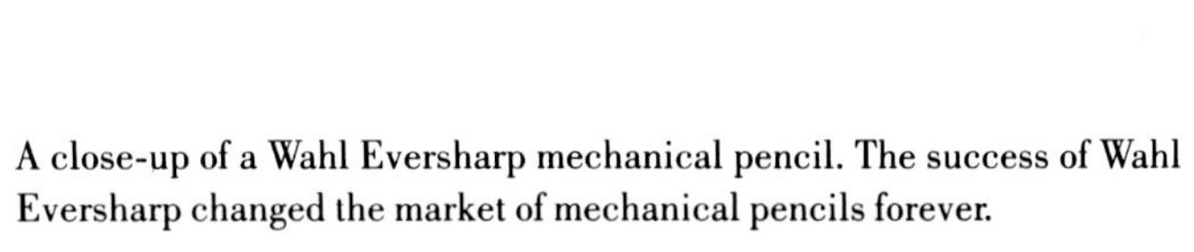

A close-up of a Wahl Eversharp mechanical pencil. The success of Wahl Eversharp changed the market of mechanical pencils forever.

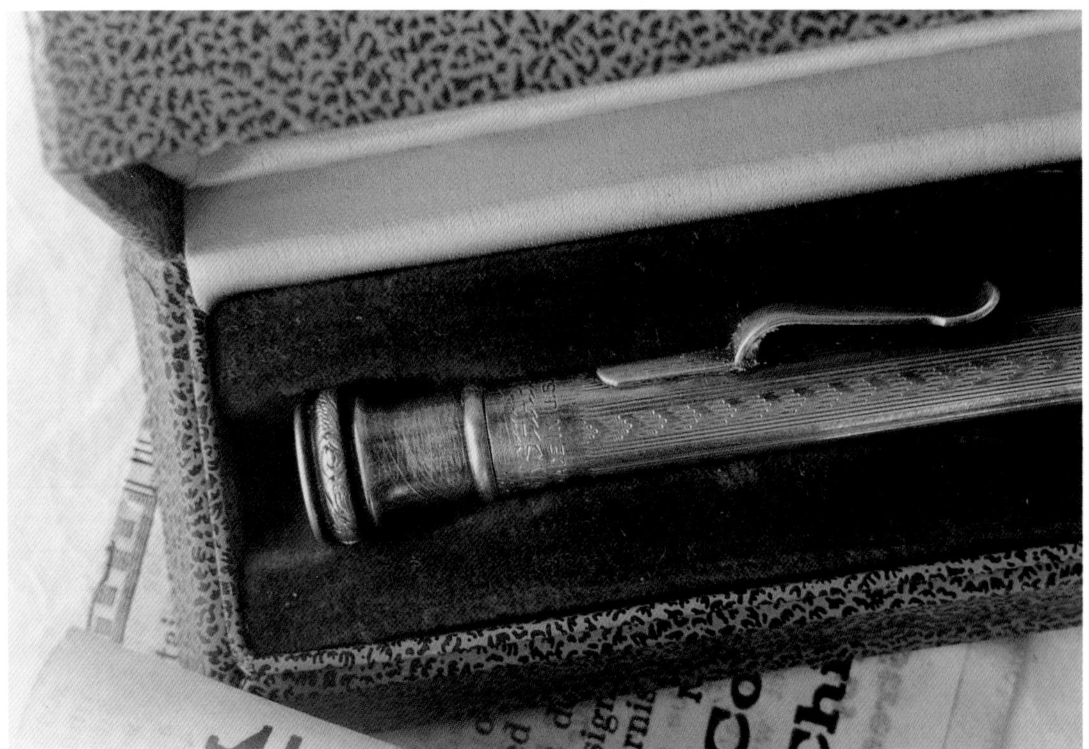

Conclusion

There is still much to be learned about mechanical pencils. It has been gratifying to note the increased interest in these pencils during the past year, and it is hoped that this book will lead to the discovery of more of these beautiful writing instruments and additional information about them. It is also hoped that, as more people take an interest, there will be an increased number of craftsmen who specialize in their repair.

With the immense popularity of the Wahl-Everharp pencil and the remarkable success of that company in the twentieth century, a transformation took place in the pencil industry; a transformation which could serve as the starting point of another volume.

Bibliography

Bicknell, Thomas Williams. *The History of the State of Rhode Island and Providence Plantations, Vol. 2.* New York: American Historical Society, Inc., 1920.

Boyd, Andrew. *Boyd's Business Directory, 1869-1870.* Albany, New York: Andrew Boyd, 1869.

_____. *Boyd's New York State Directory, 1872, 1873, 1874.* Albany, New York: Truair, Smith & Co., 1873.

Bruce, Edward C. *The Century; Its Fruits and Its Festival: The Centennial Exhibition.* Philadelphia: J. B. Lippincott & Co., 1877.

Carpenter, Charles. *Gorham Silver, 1831-1981.* New York: Dodd, Mead & Co., 1982.

Caunt, Pamela. *Victorian Commemorative Jubilee Jewellery, A Guide For Collectors.* London: ARBRAS, 1997.

Cirlot, J. E. *A Dictionary of Symbols*, trans. Jack Sage. 2nd ed. New York: Philosophical Library, Inc., 1971.

Clayton, Michael. *Christie's Pictorial History of English and American Silver.* Oxford: Phaidon-Christie's, 1985.

Culme, John. *The Directory of Gold and Silversmiths, Jeweller's and Allied Trades, 1838-1914.* Volumes I and II. New York: Antique Collectors' Club, 1987.

Cummins, Genevieve E., and Nerylla D. Taunton. *Chatelaines: Utility to Extravagance.* Woodbridge, Suffolk: Antique Collectors' Club, 1996.

Dietz, Ulysses Grant, ed. *The Glitter and the Gold.* Newark: The Newark Museum, 1997.

Elliott, Brenda J. *The Best of Its Kind: The Incredible American Heritage of the Dixon Ticonderoga Company.* Heathrow, Florida: The Dixon Ticonderoga Company, 1996.

Ensko, Stephen G. C. *American Silversmiths and Their Marks, III.* New York: Robert Ensko, Inc., 1948.

Field, Edward. *State of Rhode Island and the Providence Plantations at the End of the Century: A History, Vol. 3.* Boston: The Mason Publishing Company, 1902.

Findlay, Michael. *Western Writing Implements in the Age of the Quill Pen.* Wetheral, Carlisle, Cumbria: Plains Books, 1990.

Fischler, George, and Stuart Schneider. *Fountain Pens: The Golden Age of Writing Implements.* West Chester, Pennsylvania: Schiffer Publishing Ltd., 1990.

Geyer, Dietmar. *Collecting Writing Instruments.* Translated by Dr. Edward Force. West Chester, Pennsylvania: Schiffer Publishing Ltd., 1990.

Gobright, John C. *The New York Sketch Book and Merchants' Guide.* New York: J. C. Gobright & Company, 1859.

Green, Robert Alan. *Marks of American Silversmiths, Revised, 1650-1900.* Key West, Florida: Robert Alan Green, Publisher, 1977.

Hall, W. W. *Health at Home, Or Hall's Family Doctor.* Albany: Russell Publishing Company, 1872.

Hambly, Maya. *Drawing Instruments, 1580-1980.* London: Sotheby's Publications, 1988.

Harding, J. D. *Elementary Art; or, the Use of the Lead Pencil Advocated and Explained.* London: Charles Tilt, 1834.

Helliwell, Stephen. *Collecting Small Silverware.* Oxford: Phaidon-Christie's Limited, 1988.

Hughes, Bernard and Therle. *Three Centuries of English Domestic Silver, 1500-1820.* New York: Wilfred Funk, Inc., 1952.

Jackson, Donald. *The Story of Writing.* New York: Taplinger Publishing Co., Inc., for Parker Pen International, 1981.

Jull, Douglas. *Collecting Stanhopes.* West Sussex: D. S. Publications, 1997.

King, Moses. *Notable New Yorkers, 1896-1899.* New York: Bartlett & Company, The Orr Press, 1899.

Knight, Edward H. *Knight's American Mechanical Dictionary, Volumes 1 and 2.* Boston: Houghton, Osgood and Company, 1880.

Lambert, Barbara. *Writing History: 150 Years of the A. T. Cross Company.* Lincoln, Rhode Island: A. T. Cross Company, 1996.

Leonard, John W., ed. *Who's Who in New York City and State.* 3rd ed. New York: L. R. Hamersly & Co., 1907.

McKinstry, E. Richard. *Trade Catalogues at Winterthur: A Guide to the Literature of Merchandising, 1750-1980.* A Winterthur Book, New York: Garland Publishing, Inc., 1984.

Montgomery Ward and Co. Catalogue and Buyer's Guide No. 57, Spring and Summer 1895. Reprint, with an introduction by Boris Emmet, New York: Dover Publications, Inc. 1969.

New York State Business Directory, 1864. Boston: Adams, Sampson & Co., 1863

Nickell, Joe. *Pen, Ink and Evidence.* Lexington, Kentucky: the University Press of Kentucky, 1990.

Nissenson, Marilyn, and Susan Jonas. *Snake Charm.* New

York: Harry N. Abrams, Inc. 1995.

Pearsall, Ronald. *Collecting and Restoring Scientific Instruments.* New York: Arco Publishing Co., Inc., 1974.

Petroski, Henry. *The Pencil: A History of Design and Circumstance.* New York: Alfred A. Knopf, 1990.

Pickering, Thomas R. *Official Catalogue of the United States Exhibitors.* London: Chiswick Press, 1858.

Priestly, J. B. *Victoria's Heyday.* New York: Harper & Row, Publishers, 1972.

Rainwater, Dorothy T. *American Jewelry Manufacturers.* West Chester, Pennsylvania: Schiffer Publishing Ltd., 1988.

_____. *Encyclopedia of American Silver Manufacturers, Third Edition, Revised.* Atglen, Pennsylvania: Schiffer Publishing Ltd., 1986.

Romaine, Lawrence B. *A Guide to American Trade Catalogs, 1744-1900.* New York: R. R. Bowker Company, 1960.

Rulau, Russell. *Standard Catalog of United States Tokens, 1700-1900.* 2nd ed. Iola, Wisconsin: Krause Publications, 1997.

Sattin, Gerald. *A Loan Collection of Writing Implements in Silver and Gold, c. 1680-1880.* London: Gerald Sattin Ltd., 1993.

Schneider, Stuart, and George Fischler. *The Book of Fountain Pens and Pencils.* West Chester, Pennsylvania: Schiffer Publishing Ltd., 1992.

Swenson, Evelyn. *Victoriana America: The Customs and Costumes of Victorian America: Fashion Dolls and Fashion Plates, Passementerie, Jet Jewelry, Mourning Fashions.* Matteson, Illinois: Greatlakes Living Press, 1976.

The Boston Almanac and Business Directory, 1888, Vol. 54. Boston: Sampson, Murdock, & Co., 1888.

The New York State Business Directory, 1867. Boston: Sampson, Davenport & Co., 1866.

Thompson, Eleanor McD. *Trade Catalogues at the Winterthur Museum (a Guide to the Microfische Edition of, Part 2).* Bethesda, Maryland: University Publications of America, 1991.

Todd, John Edwards. *Todd Family in America; or: The Descendants of Christopher Todd, 1637-1919.* Northampton, Massachusetts: Press of Gazette Printing Co., 1920.

Trademarks of the Jewelry and Kindred Trades, 3rd ed. Reprinted from the 1915 edition. New York: The Jeweler's Circular Publishing Co., 19

Trow, John F. *New York State Business Directory, 1859.* New York: Adams, Sampson & Co., 1858.

Turner, Noel D. *American Silver Flatware, 1837-1910.* New York: A. S. Barnes and Co., 1972.

Venable, Charles L. *Silver in America: A Century of Splendor, 1840-1940.* Dallas Museum of Art; New York: distributed by Harry N. Abrams, Incorporated, 1995.

Voice, Eric H. "The History of the Manufacture of Pencils," *Newcomen Society for the Study of History and Engineering and Technology, 1949-50 and 1950-51, Volume 27.* London: The Courier Press, 1956.

Wardle, Patricia. *Victorian Silver and Silver-Plate.* New York: Thomas Nelson & Sons, 1963.

Whalley, Joyce Irene. *Writing Implements and Accessories: From the Stylus to the Typewriter.* Detroit: Gale Research Company, 1975.

Willson, Marcus. *The Drawing Guide: A Manual of Instruction in Industrial Drawing.* New York: Harper & Brothers, Publishers, 1873.

Edward Todd

Catalogs

Babcock & Co., c.1890
Bailey, Banks and Biddle, 1894
Baird-North Company, c. 1880
Edward Baker, 1915 (facsimile)
Dixon, Joseph Crucible Co., 1891
W. F. Doll Mfg. Co., c. 1890
John Foley, c. 1875
Gorham Photo Book of Silver Toiletries, Novelties, c. 1930
Johnson, E. S., 1895
Koch & Sons, 1876
Lamos & Co., 1878
Low, Daniel, & Co., 1913
Lynn, J., & Co., 1903
McChesney Co., c.1890
Miller, Bradley & Hall, 1857
Sampson Mordan & Co., 1898 (facsimile)
Morton's Celebrated Gold Pens
Royal Manuf. Co., c. 1911
Shreve & Co., 1915
Tiffany & Co. Blue Books, 1845 on
Tiffany Studios, 1910
Todd, Edward & Co., c. 1870
Unger Bros., c. 1915 (reproduction)
Young, Otto, & Co. 1900

Periodicals

The Cosmopolitan
Eagle News
Jeweler's Circular and Horological Review
Jeweler's Review
Jeweler's Weekly
Journal of the Writing Equipment Society
McClures
The Pen & Pencil Gallery (Marestin) Magazine
Pen Fancier's Magazine
Pen World International
Scientific American
Silver Magazine

Index

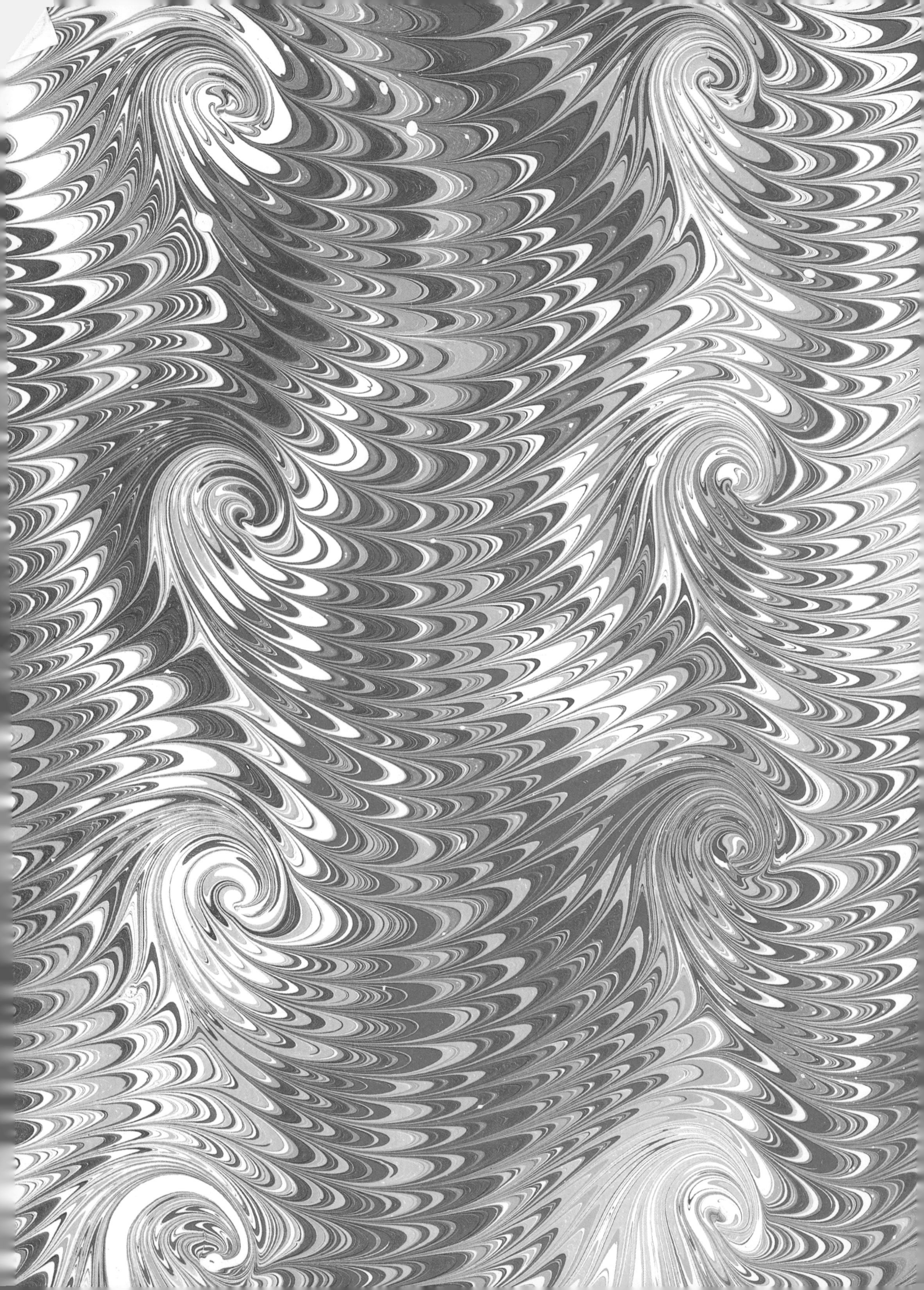